Partial Differential Equations: Spectral and High Order Methods

Partial Differential Equations: Spectral and High Order Methods

Edited by
Patrick McCann

www.murphy-moorepublishing.com

Published by Murphy & Moore Publishing,
1 Rockefeller Plaza,
New York City, NY 10020, USA

Copyright © 2022 Murphy & Moore Publishing

This book contains information obtained from authentic and highly regarded sources. Copyright for all individual chapters remain with the respective authors as indicated. All chapters are published with permission under the Creative Commons Attribution License or equivalent. A wide variety of references are listed. Permission and sources are indicated; for detailed attributions, please refer to the permissions page and list of contributors. Reasonable efforts have been made to publish reliable data and information, but the authors, editors and publisher cannot assume any responsibility for the validity of all materials or the consequences of their use.

Trademark Notice: Registered trademark of products or corporate names are used only for explanation and identification without intent to infringe.

ISBN: 978-1-63987-421-7

Cataloging-in-Publication Data

    Partial differential equations : spectral and high order methods / edited by Patrick McCann.
       p. cm.
    Includes bibliographical references and index.
    ISBN 978-1-63987-421-7
    1. Differential equations, Partial. 2. Differential equations, Partial--Numerical solutions.
    3. Spectral theory (Mathematics). I. McCann, Patrick.
QA377 .P37 2022
515.353--dc23

For information on all Murphy & Moore Publications
visit our website at www.murphy-moorepublishing.com

Contents

Preface VII

Chapter 1 **Direct estimation of the parameters of a delayed, intermittent activation feedback model of postural sway during quiet standing** 1
Kevin L. McKee, Michael C. Neale

Chapter 2 **A novel power-driven fractional accumulated grey model and its application in forecasting wind energy consumption of China** 33
Peng Zhang, Xin Ma, Kun She1*

Chapter 3 **Risk factors in the illness-death model: Simulation study and the partial differential equation about incidence and prevalence** 70
Annika Hoyer, Sophie Kaufmann, Ralph Brinks

Chapter 4 **Parameter identification for gompertz and logistic dynamic equations** 82
Elvan Akın, Neslihan Nesliye Pelen, Ismail Uğur Tiryaki, Fusun Yalcin

Chapter 5 **Mean almost periodicity and moment exponential stability of semi-discrete random cellular neural networks with fuzzy operations** 107
Sufang Han, Guoxin Liu, Tianwei Zhang

Chapter 6 **Local Riemannian geometry of model manifolds and its implications for practical parameter identifiability** 137
Daniel Lill, Jens Timmer, Daniel Kaschek

Chapter 7 **Investigation of singular ordinary differential equations by a neuroevolutionary approach** 151
Waseem Waseem, Muhammad Sulaiman, Poom Kumam, Muhamad Shoaib, Muhammad Asif Zahoor Raja, Saeed Islam

Chapter 8 **Analytical solution to swing equations in power grids** 178
HyungSeon Oh

Chapter 9 **Neural minimization methods (NMM) for solving variable order fractional delay differential equations (FDDEs) with simulated annealing (SA)** 211
Amber Shaikh, M. Asif Jamal, Fozia Hanif, M. Sadiq Ali Khan, Syed Inayatullah

Permissions

List of Contributors

Index

Preface

Partial differential equations (PDE) are equations that establish relationships between different partial derivatives of multivariable functions. These functions are considered to be unknown, and need to be solved. There is a large amount of modern mathematical and scientific research on methods for numerically approximating solutions of selected partial differential equations using computers. Scientific fields concerned with mathematics, such as physics and engineering, depend heavily on PDEs. They also play an important role in understanding sound, heat, diffusion, general relativity and fluid dynamics. This book includes a detailed explanation of the various concepts and applications of partial differential equations. Different approaches, evaluations, methodologies and advanced studies on this topic have been included herein. This book will serve as a reference to a broad spectrum of readers.

This book is a comprehensive compilation of works of different researchers from varied parts of the world. It includes valuable experiences of the researchers with the sole objective of providing the readers (learners) with a proper knowledge of the concerned field. This book will be beneficial in evoking inspiration and enhancing the knowledge of the interested readers.

In the end, I would like to extend my heartiest thanks to the authors who worked with great determination on their chapters. I also appreciate the publisher's support in the course of the book. I would also like to deeply acknowledge my family who stood by me as a source of inspiration during the project.

Editor

Direct estimation of the parameters of a delayed, intermittent activation feedback model of postural sway during quiet standing

Kevin L. McKee *, Michael C. Neale

Virginia Commonwealth University, Virginia Institute of Psychiatric and Behavioral Genetics, Richmond, Virginia, United States of America

* mckeek@vcu.edu

Editor: Jonathan David Touboul, Brandeis University, UNITED STATES

Funding: MCN was funded under NIDA 5 R25 DA026119 10 (drugabuse.gov). The funders had no role in study design, data collection and analysis, decision to publish, or preparation of the manuscript.

Competing interests: The authors have declared that no competing interests exist.

Abstract

Human postural sway during quiet standing has been characterized as a proportional-integral-derivative controller with intermittent activation. In the model, patterns of sway result from both instantaneous, passive, mechanical resistance and delayed, intermittent resis- tance signaled by the central nervous system. A Kalman-Filter framework was designed to directly estimate from experimental data the parameters of the model's stochastic delay dif- ferential equations with discrete dynamic switching conditions. Simulations showed that all parameters could be estimated over a variety of possible data-generating configurations with varying degrees of bias and variance depending on their empirical identification. Appli- cations to experimental data reveal distributions of each parameter that correspond well to previous findings, suggesting that many useful, physiological measures may be extracted from sway data. Individuals varied in degree and type of deviation from theoretical expecta- tions, ranging from harmonic oscillation to non-equilibrium Langevin dynamics.

Introduction

Several previous studies have analyzed bodily sway patterns in quiet standing, and a variety of models have been proposed. In this study, we designed and tested a method of directly estimating the parameters of the Asai et al. [1] intermittent feedback control model of posture from experimental data. We begin with a brief review of prior models and the rationale for choosing a model of intermittent postural control (IPC). In the second section, we describe the current model in more detail and explain our framework for the estimation of its parameters. The third section describes simulation studies that tested the estimation capabilities of our frame- work when the data-generating parameters are known and the model is specified accurately. In the fourth section, we applied the model to experimental data and estimated sampling dis- tributions for each parameter.

Observed trajectories of postural sway have largely been studied as a problem of stochastic be- havior, though some studies have focused on its chaotic properties [2]. In this study, we too re- garded postural sway as a random process subject to statistical analysis. The center of pres- sure (COP) on a force plate during quiet standing has been shown to exhibit the features of a bound- ed, random walk, or correlated noise [3]. Center of mass (COM) is one of the most com- mon metrics of body sway but has to be inferred from other position and force metrics such as the COP [4]. For the small radius in which postural sway occurs, body tilt angle is nearly equiv- alent to COM and has likewise been used to develop models of posture [1].

Many authors have observed that sway follows a two-frequency oscillation scheme, with fast oscillations of the COP around a drifting center point [3, 5–7]. Collins and De Luca [3] regarded these patterns as a combination of short-term, open-loop system with long-term closed-loop control. Alternatively, the "rambling and trembling" hypothesis sug- gests that short-term trem- ors result from corrective, closed-loop feedback that is activated with deviation of the COP from the ground projection of the COM, which is itself allowed to drift [7].

Broadly, more recent debate over the control scheme of human balance has focused on two kinds of models: continuous and intermittent feedback controllers. Continuous control is ex- erted through a proportional-integral-derivative (PID) or closed-loop system often charac- ter- ized by a second order linear differential equation, sometimes including delayed propor- tional and derivative feedback. For instance, Maurer and Peterka [8] tested a PID inverted pendulum model that distinguishes passive, instantaneous feedback from sources such as ankle joint stiff- ness, from delayed, active feedback from the central nervous system and subsequent muscular response. Others have argued that human postural movement is better described by intermit- tent feedback mechanisms due to a smaller reliance on process noise, reproduction of cyclical behavior over multiple timescales, more efficient energy expenditure, and greater robustness to disturbances and instability caused by delays in neural signaling [9]. Simulations [10] and rein- forcement learning [11] have been used to show that an upright pen- dulum, taken as a simple model of the standing human body, can exhibit stability and the observed slow oscillation pat- terns as a result of learned, time-delayed, intermittent feedback.

Intermittent activation models have taken multiple forms. Gawthrop and Wang [12] ini- tially proposed clock-driven muscular feedback, but later considered event-driven models [13]. Event-driven models are generally defined by a combination of stable and unstable mani- folds in the phase space of the body's position or angle. Gawthrop et al. [13] and Eurich and Milton [14] describe the behavior of systems with position-based thresholds that result in two stable equilibria. A model by Bottaro et al. [15] proposes boundary functions of both position and ve- locity that jointly determine probabilistic bursts of negative feedback. Asai et al. [1] reproduced a commonly observed double power-law structure in the PSD of sway [6] using similar control manifolds but with deterministic rules for sustained feedback activation. Their model requires only a simpler, Gaussian distribution of process noise with a smaller variance as compared to continuous PID models. Nomura et al. [16] showed that the same intermittent activation feed- back model is capable of reproducing both chaotic and stochastic patterns that resemble human postural sway as a function of small hemodynamic perturbations, while con- tinuous feedback models cannot.

A common method of estimating the parameters of each model is to simulate data that opti- mal- ly resemble the experimental data. This is accomplished by varying parameters over itera- tions of simulation until resulting disparities on a set of key summary statistics have been minimized. Bottaro et al. [15] used the the Root Mean Square (RMS) of both the COP and COM series and each of its derivatives, unimodality of the series histogram, the length of larg- est oscillations calculated from zero crossings, and the PSD of the COP series. Maurer and Peterka [8] estimat- ed parameters from observed data in a similar manner using mean velocity, RMS distance and velocity, spectral properties such as mean frequency, frequency dispersion, and total power. Asai et al. [1] used the double power law structure of the frequency spectrum as a criterion for the success of their model but did not demonstrate a direct empirical application. To obtain statis- tical information about estimated parameters, summary statistic methods have been combined with approximate Bayesian inference [17]. This method was used to acquire empirical posterior distributions of five out of the eight parameters of interest [18].

While the simulation approach is flexible for a wide range of model specifications and lev- els of complexity, it risks overlooking attributes of the data that do not have specific effects on the chosen summary statistics. Bottaro et al. [15] notes, for instance, that "The intermit- tent nature of the control process cannot be detected by global descriptors of the sway pat- terns, like the PSD of the COP, because they cannot distinguish between asymptotic and bounded stability". Furthermore, the amplitude spectrum is invariant to reversal of the sig- nal, giving identical results for potentially different mechanisms of variation. This is prob- lematic when the system includes discontinuous dynamics, such as a sharp impulse followed by a more gradual decay. An alternative approach that better accounts for fine-grained sequential dependence is to estimate the structural parameters directly from the data using Kalman filtering or other iterative tech- niques. No simulation or descriptive statistics are necessarily used, rather the structural param- eters are estimated by minimizing an objective function such as the squared prediction error, or by maximizing the likelihood of the data according to an expected noise distribution. The results obtained by this approach can be sensitive to the exact predictive mechanisms specified in the model, and post-hoc analyses of the estimates can be highly informative about the types and de- grees of misspecification. Direct estimation (sometimes called *exact* estimation by comparison

[19]) may be particu- larly useful when the dynamic structure cannot represented by any descriptive statistics with sufficient specificity. The Asai et al. [1] model of posture may present one such case in that it postulates dependence of the spectral power-law property upon nonlinear, physiological mechanisms of feedback control and their properties. Such properties include the delay in neural signaling, the sensitivity of feedback activation, and the strength of passive versus active corrective forces. Furthermore, the process noise distribution of the Asai et al. [1] model is a Gaussian process and thus accords with the statistical assumptions of the Kalman filter. A drawback of direct estimation is that a misspecified model is not guaranteed to result in any interpretable or accurate parameter estimates if the parameters are highly dependent. If the parameter estimates deviate significantly from their theorized values, we may nonethe- less analyze the behaviors they imply and draw general inferences.

The aim of this study was to validate and apply a method of directly estimating parameters for event-driven control with specific focus on the popular intermittent control model by Asai et al. [1]. Validation of this analytic strategy will set a foundation for estimating the parameters of alternative models and more comprehensive comparisons. Following the validation study, we estimated empirical values of each parameter from publicly available COM data [20] and compared our results with theoretically expected values from the literature. We included two previously demonstrated covariates in our analysis, visual feedback and age, to attempt to rep- licate previous findings as further evidence for the validity of the model.

Model

The intermittent postural control (IPC) model by Asai et al. [1] describes a tension between toppling torque due to gravity and a combination of active and passive resistance mechanisms. Passive resistance is proposed to come from leg stiffness and joint friction and is modeled with instantaneous relations between position, velocity, and acceleration. Active feedback control is proposed to arise from motor responses signaled by the central nervous system and is consequently delayed by about 190-210 ms [21].

The model is provided in terms of body tilt angle (θ) as follows:

$$I\ddot{\theta}_t = mgh\theta_t - T, \tag{1}$$

$$T = mghK\theta_t + B\dot{\theta}_t + mghf_P(\theta_{t-\tau}) + f_D(\dot{\theta}_{t-\tau}) + \sigma w_t, \quad w_t \sim N(0,1), \tag{2}$$

where I is the rotational inertia, m is the body mass (kg), g is gravity ($\approx 9.81 m/s^2$), and h is the height of the COM. T includes all the terms representing mechanisms of resistance to the angular toppling force. w_t is a Gaussian, independent and identically distributed random variable accounting for stochastic variation in acceleration, with standard deviation σ. The total passive forces may be written as $mgh(1 - K)\theta_t$, as K is the percentage of the gravitational acceleration counteracted by passive resistance. While a certain definition of B is not given, its effects are non-trivial and an interpretation may be taken from the common use of the second- order damped oscillator equation, in which the velocity coefficient represents negative feed- back due to friction. In this case, it may be regarded as a measure of ankle and knee joint friction.

The active control terms, f_P and f_D, intermittently respond to θ on a time lag of $\tau \approx 200$ ms according to the conditions:

$$\text{if } \theta_{t-\tau}(s\dot{\theta}_{t-\tau} - \alpha\theta_{t-\tau}) > 0, \text{ and } \theta_{t-\tau}^2 + (s\dot{\theta}_{t-\tau})^2 > r^2 \quad \begin{cases} f_P(\theta_{t-\tau}) = P\theta_{t-\tau} \\ f_D(\dot{\theta}_{t-\tau}) = D\dot{\theta}_{t-\tau} \end{cases} \text{(Active)}, \quad (3)$$

$$\text{otherwise} \quad \begin{cases} f_P(\theta_{t-\tau}) = 0 \\ f_D(\dot{\theta}_{t-\tau}) = 0 \end{cases} \text{(Inactive)} \quad (4)$$

The first condition represents a threshold dividing the saddle-type attractor of the toppling acceleration into stable and unstable manifolds. The stable manifold briefly occurs when the tilt angle is moving toward zero, while the unstable manifold is characterized by falling away from zero. The angle of the dividing line is given by the slope parameter α. The second condition describes a radius (r) about the origin within which the tilt angle is too small to be detected or too stable for immediate correction (note that r has conventionally been used to denote the delay time interval in the delay differential equation literature. Here we have pre- ferred τ for that purpose.). By converting the switching threshold slope α into the angle a as $\alpha = \frac{\sin(a\pi)}{\cos(a\pi)}$, we change the upper and lower estimation bounds from $[-\infty, \infty]$ to $[0, 1]$. This way, the parameter a represents the percentage of the phase space, not including the insensitiv- ity radius, for which the active control parameters are non-zero.

The estimable parameters of the SDDE are summarized in Table 1. Many of the parameters have previously been estimated in a variety of ways, sometimes with highly varied results. Tietävä̈inen et al. [18] used the approximate Bayesian inference [17] with data simulation to estimate P, D, a, τ, and σ. Among these, the method failed to obtain precise distributions for D in both simulations and empirical application. It is also not clear whether fixing the other parameters to uncertain theoretical priors ($K = .8$, $B = 4$, and $r = .004$) results in biased estimates.

Direct physiological measurements found the relative resistance to toppling torque at the ankle, K, to be as high as 91% on average [22] when the average magnitude of disturbance is small. Another study estimated relative resistance to be around 64% when disturbances were larger [23]. Conversely, the chosen value of r involves a conjecture about perceptual sensitivity that is specific to this model and has not been measured directly.

Table 1. Parameters of the IPC model with units and descriptions.

Fixed / Observed		Unit
I	Inertia	$(kgm)^2$
m	Body mass	(kg)
h	Distance of center of mass from the ankle	(m)
g	Acceleration from gravity	(m/s^2)
Estimated		
K	Intrinsic upright stiffness	% (of total Nm/rad)
B	Joint friction	Nms/rad
P	Active response force	Nm/rad
D	Active response damping	Nms/rad

a	Percentage of phase space active	%
r	Insensitivity radius	rad
τ	Feedback delay	s
σ	Process noise variance	Nm
ϵ	Measurement error variance	rad

Tietäväinen et al. [18] obtained a value of τ around 300 ms, while other methods of assessment have produced estimates including 125 ms [24] and 200 ms [21]. Direct mea- surements of ankle response, however, found response to start at 30 ms with maximal dis- placement around 120 ms [25]. If feedback delay is too long, then intermittent periods of acceleration due to gravity or muscle feedback will be consequently prolonged even as the state enters unstable regions of the phase space. One result is overcompensation for error, in which the fast oscillations found in sway are more amplified than would be the case with shorter delays. Alternatively, if the value of a is too high, then delayed feedback may bypass the unstable manifold and activate at inappropriate locations in the phase space, potentially amplifying slower oscillations over time. Long feedback delays can therefore contribute to instability, sway amplification, and higher risk of falling, but the exact kinds of error are determined by the joint behavior of several parameters, including a, r, and disturbance mag- nitude σ [1].

Estimation

The above equations represent a Stochastic Delay Differential Equation (SDDE). The Kalman-Bucy filter provides minimum-variance unbiased estimates of the state of a stochastic process when both measurement and process noise are present and can be used to estimate the parameters of continuous-time differential equations from noisy data [26]. However, two challenges arise when estimating the parameters of an SDDE, including the lag interval τ and the lagged position and velocity coefficients, P and D. First, interpolation of the lagged states must be used to allow a continuous domain of possible values for τ. Second, backward extrapolation must be used to estimate the unmeasured interval of lagged states preceding initial state $\mathbf{x}_0$.

Last, we address problems that occur when the discrete switching conditions are toggled between measured instances. For most intervals between measures, the dynamics are linear and the prediction is exact, but state predictions that traverse the condition thresholds will sys- tematically introduce bias to the linear dynamics unless the correct ratio of active and inactive dynamics within each traversal is estimated. We detail an algorithm to resolve this bias by adjusting the prediction according to each of the possible threshold-traversal scenarios.

Optimal filtering. The state-space equation for the time-lagged IPC system is given as

$$\dot{\mathbf{x}}_t = \mathbf{A}\mathbf{x}_t + \mathbf{A}_\tau \mathbf{x}_{t-\tau} + \mathbf{Q}, \qquad (5)$$

$$\mathbf{y}_t = \mathbf{H}\mathbf{x}_t + \mu + \mathbf{R}, \qquad (6)$$

where Q is the process noise covariance matrix, H is the measurement matrix, μ is the estimated origin about which the COM oscillates, and R is the covariance matrix of measurement error. The contemporaneous and lagged state vectors are

$$\mathbf{x}_t = \begin{bmatrix} x_t \\ \dot{x}_t \end{bmatrix}, \quad \dot{\mathbf{x}}_t = \begin{bmatrix} \dot{x}_t \\ \ddot{x}_t \end{bmatrix}, \quad \mathbf{x}_{t-\tau} = \begin{bmatrix} x_{t-\tau} \\ \dot{x}_{t-\tau} \end{bmatrix},$$

and the state transition matrices are

$$\mathbf{A} = \begin{bmatrix} 0 & 1 \\ mgh(1-K)/I & -B/I \end{bmatrix}, \quad \mathbf{A}_\tau = \begin{bmatrix} 0 & 0 \\ -mghP/I & -D/I \end{bmatrix}, \quad \mathbf{Q} = \begin{bmatrix} 0 & 0 \\ 0 & \sigma^2 \end{bmatrix},$$

Matrix A contains the parameters of the passive, instantaneous forces, while A_τ contains the conditional parameters of active feedback. When the conditions given in Eq 3 evaluate to false, $A_\tau = 0$.

The measurement matrices simply attribute the observed COM to the state position with estimated origin μ and measurement error variance ϵ:

$$\mathbf{H} = \begin{bmatrix} 1 & 0 \end{bmatrix} \quad \mathbf{R} = [\epsilon]$$

The complete algebra for the prediction and correction steps of Kalman Filtering is excluded, as its derivation can be found in many resources [27] and remains largely unchanged for this model. However, the key difference in this case is that the prediction step is altered to include the delayed term. Using the following matrix discretizations,

$$\mathbf{A}^d = e^{\mathbf{A}\Delta t}, \tag{7}$$

$$\mathbf{A}_\tau^d = \mathbf{A}^{-1}(\mathbf{A}^d - \mathbf{I})\mathbf{A}_\tau \tag{8}$$

$$\mathbf{Q}^d = \int_{\delta=0}^{\Delta t} e^{\mathbf{A}\delta} \mathbf{Q} e^{\mathbf{A}\delta \, \mathrm{T}} d\delta \tag{9}$$

we can then provide the prediction equations for the state mean and covariance as follows:

$$\hat{\mathbf{x}}_{t+\Delta t} = \mathbf{A}^d \bar{\mathbf{x}}_t + \mathbf{A}_\tau^d \bar{\mathbf{x}}_{t-\tau+\Delta t} \tag{10}$$

$$\hat{\mathbf{P}}_{t+\Delta t} = \mathbf{A}^d \bar{\mathbf{P}}_t \mathbf{A}^{d\,\mathrm{T}} + \mathbf{A}_\tau^d \bar{\mathbf{P}}_t \mathbf{A}_\tau^{d\,\mathrm{T}} + \mathbf{Q}^d \tag{11}$$

For stationary series with large number of observations, $\mathbf{P}_t \approx \mathbf{P}\infty$. For convenience, we use $\mathbf{P}_{t-1}$ as an approximation to $\mathbf{P}_{t-\tau}$. Note that Eq 8 does not work if $K = 1$, making $\mathbf{A}$ singular. However, small, numerically viable deviations from $K = 1$ will not substantially impact solution topology. Point singularities will also not impede derivative-free optimization methods.

Estimation of feedback delay. Linear interpolation To obtain estimates of the state at time lags that do not fall on measurement instances, we use linear interpolation of the state:

$$\lambda = \frac{\tau}{\Delta t}, \tag{12}$$

$$\hat{\mathbf{x}}_{t-\tau} = \mathbf{x}_{i-\lfloor\lambda\rfloor} + (\mathbf{x}_{i-\lfloor\lambda\rfloor} - \mathbf{x}_{i-\lceil\lambda\rceil})(\lambda - \lfloor\lambda\rfloor), \tag{13}$$

λ is the conversion of the time delay to the number of measured occasions comprising that interval. The ceiling and floor functions thus give valid measurement indices and are used to give a combination of measurements falling to either side of λ, weighted proportionally. If $\tau = 0$, then the second term of Eq 13 can be neglected.

Backward extrapolation of initial values By introducing an initial value parameter for acceleration, we can estimate a quadratic extrapolation backward from t_0 to $t_0 - \tau$, allowing the influence of lagged states and switching conditions to be respected within the first λ iterations of filtering:

$$\text{If } t \leq \tau \begin{cases} \hat{x}_{t-\tau} = x_0 + \dot{x}_0(t-\tau) + \ddot{x}_0(t-\tau)^2, \\ \dot{\hat{x}}_{t-\tau} = \dot{x}_0 + 2\ddot{x}_0(t-\tau), \end{cases} \tag{14}$$

Constrained interpolation of dynamic switching points. To avoid bias due to missing transitional information between measures that straddle the threshold of the conditions given by Eq 3, we explicitly detect each case, interpolate the state falling on the condition threshold, and predict its traversal in two steps. For convenience, take the shortened terms u and v as the delayed states leading up to, and away from the condition threshold:

$$\begin{aligned} u &:= x_{t-\tau}, & \dot{u} &:= s\dot{x}_{t-\tau} \\ v &:= x_{t-\tau+\Delta t}, & \dot{v} &:= s\dot{x}_{t-\tau+\Delta t}, \end{aligned} \tag{15}$$

Where s is the seconds constant, such that v, u, $\dot{v}$, and u_- are measured in radians. For use later, we note here that the slope between the two points is $m = \frac{\dot{v}-\dot{u}}{v-u}$.

Conditions for switching off:

$$\text{If } [u(\dot{u} - \alpha u) > 0 \text{ and } u^2 + \dot{u}^2 > r^2] \text{ and } [v(\dot{v} - \alpha v) \leq 0 \text{ or } v^2 + \dot{v}^2 \leq r^2], \tag{16}$$

then A_τ is switching off. If this holds true, then the following conditions further apply:

$$\begin{aligned} &\text{If } v^2 + \dot{v}^2 > r^2 \\ &\text{and } ![(v > 0 \text{ and } \dot{v} < 0 \text{ and } u < 0 \text{ and } \dot{u} < 0) \end{aligned} \tag{17}$$

or $(v < 0$ and $\dot{v} > 0$ and $u > 0$ and $\dot{u} > 0)]$,

then the lagged state is traversing the line $\dot{x} = \alpha x$ outside of the slack radius and not traversing $u = 0$. The interpolated point $(\hat{u}, \hat{\dot{u}})$ falls on the line, and is calculated as

$$\hat{u} = -\frac{mv - \dot{v}}{\alpha - m}, \qquad \hat{\dot{u}} = m\hat{u} - mv + \dot{v}, \tag{18}$$

If $v^2 + \dot{v}^2 \leq r^2$, then the lagged state is traversing into the slack radius, and the interpolated point is

$$\hat{u} = \frac{r^2 + v^2 + 2mv\dot{v} - \dot{v}^2}{2(v + m\dot{v})}, \qquad \hat{\dot{u}} = m\hat{u} - mv + \dot{v}, \tag{19}$$

In all other cases in which (16) holds true, u is traversing the axis at $u = 0$.

$$\hat{u} = 0, \qquad \hat{\dot{u}} = -mv + \dot{v}, \tag{20}$$

Conditions for switching on: For cases where the delayed feedback is switching on, the roles of u and v are simply traded. The interpolated point is calculated identically under each set of conditions analogous to those for switching off.

$$\text{If } [u(\dot{u} - \alpha u) \leq 0 \text{ or } u^2 + \dot{u}^2 \leq r^2] \text{ and } [v(\dot{v} - \alpha v) > 0 \text{ and } v^2 + \dot{v}^2 > r^2], \tag{21}$$

then $\mathbf{A}_\tau$ is switching on. If this holds true, then the following conditions further apply:

$$\begin{aligned} &\text{If } v^2 + \dot{v}^2 < r^2 \\ &\text{and } ![(u > 0 \text{ and } \dot{u} < 0 \text{ and } v < 0 \text{ and } \dot{v} < 0) \\ &\text{or } (u < 0 \text{ and } \dot{u} > 0 \text{ and } v > 0 \text{ and } \dot{v} > 0)], \end{aligned} \tag{22}$$

then the lagged state is traversing the line $\dot{x} = \alpha x$ αx outside of the slack radius and not traversing $u = 0$, and the interpolated point is calculated as Eq (18). If $u^2 + \dot{u}^2 \leq r^2$, then the lagged state is traversing the slack radius from within, and the interpolated point is calculated with Eq (19). In all other cases in which Eq (21) holds true, u is traversing the axis at $u = 0$ and the interpolated point is calculated as Eq (20).

Prediction for threshold traversal: The time for u to reach the switching threshold, Δt^-, and the time to reach the next observation after the threshold, Δt^+, can be calculated from the interpolated state at the threshold and its neighboring states, u and v:

$$\Delta t^- = \frac{\|\mathbf{u} - \hat{\mathbf{u}}\|}{\|\mathbf{v} - \mathbf{u}\|} \Delta t, \qquad \Delta t^+ = \frac{\|\mathbf{v} - \hat{\mathbf{u}}\|}{\|\mathbf{v} - \mathbf{u}\|} \Delta t, \tag{23}$$

In the first step, **A**, **A**$_\tau$, and **Q** are discretized for the interval Δt^-, and the prediction is given as:

$$\hat{\mathbf{x}}_{t+\Delta t^-} = \mathbf{A}^d \bar{\mathbf{x}}_t + \mathbf{A}^d_\tau \mathbf{u} \tag{24}$$

$$\hat{\mathbf{P}}_{t+\Delta t^-} = \mathbf{A}^d \bar{\mathbf{P}}_t \mathbf{A}^{d\,\mathrm{T}} + \mathbf{A}^d_\tau \mathbf{P}_t \mathbf{A}^{d\,\mathrm{T}}_\tau + \mathbf{Q}^d. \tag{25}$$

In the second step, **A**, **A**$_\tau$, and **Q** are discretized for the interval Δt^+, and the prediction is computed from time $t + \Delta t^-$ as:

$$\hat{\mathbf{x}}_{t+\Delta t} = \mathbf{A}^d \hat{\mathbf{x}}_{t+\Delta t^-} + \mathbf{A}^d_\tau \hat{\mathbf{u}} \tag{26}$$

$$\hat{\mathbf{P}}_{t+\Delta t} = \mathbf{A}^d \hat{\mathbf{P}}_{t+\Delta t^-} \mathbf{A}^{d\,\mathrm{T}} + \mathbf{A}^d_\tau \hat{\mathbf{P}}_{t+\Delta t^-} \mathbf{A}^{d\,\mathrm{T}}_\tau + \mathbf{Q}^d. \tag{27}$$

For either step, $\mathbf{A}^d_\tau = 0$, depending on whether the active feedback is switching on or off.

Optimization. The toggling of active feedback is not a smooth process and results in dis- continuities in the space of a cost function for fitting the model, though these are greatly miti- gated by the interpolation measures described above. The complexity of the model nonetheless gives rise to multiple local solutions, and attempts to find optimal parameters using local methods such as gradient descent and Nelder-Mead reliably fail. Instead, we recommend using a method of global, derivative-free optimization such as Differential Evolution (DE) [28]. The optimization parameters that we chose are listed below.

- Strategy: DE / rand / 1 / bin with per-vector-dither
- Iterations = 15000
- Population size = 30
- Crossover Probability (CR) = .95
- F = .15
- Weighting of successful members (c) = 0
- Step tolerance: 500
- Relative tolerance: 1e-10

We chose a high crossover probability (CR) due to high dependence between parameters of the model and used simulations to confirm reasonable convergence given the chosen popula- tion size, iterations, and F value. DE does not require initial values for parameter estimation, but instead populates a region within explicit bounds. The bounds used here for simulation and data analysis are given in Table 2. Parameter bounds were generally restricted to poten- tially stable and theoretically meaningful ranges, such as for K, P, r, and τ. Theoretical interpre- tations of parameters B and D were less certain and were therefore allowed to vary beyond boundaries imposed under any particular physiological definition. τ was constrained to the extremes of the empirical distribution of neural delay given the results from Peterka [21].

Otherwise, bounds were made extreme enough to capture all reasonable possibilities without unnecessarily slowing convergence.

Software. All analyses used R statistical programming environment [29]. Differential Evolution was provided by the R package DEoptim [30]. The IPC model was implemented in C++ using R packages Rcpp [31] and RcppArmadillo [32], and compiled to the open-source R package IPC-model. The package includes the following functions:

- `ipcModel()`: C++ Kalman Filter with delayed terms and switching conditions that returns a -2Log-likelihood value for optimization.

- `ipcSimulate()`: C++ numerical integrator that generates simulated data for the IPC model.

- `ipcMultiGroup()`: R wrapper for ipcModel() that incorporates physical constants, parameter algebras, and enables the estimation of both within and between-series parameters.

- `kalmanIntegrate()`: C++ helper function that accepts continuous-time state-space matrices and returns discretized matrices for Kalman-Bucy filtering.

Simulations

Two simulations were used: the first to check model specification, and the second to evaluate the accuracy and precision of parameter estimates. The first simulation used noiseless (i.e. deterministic trajectories) with perfect measurement to check for systematic bias due to the estimation strategy. The second simulation used data simulated to include both process and measurement noise according to the possible properties of real data recorded by a force plate.

Table 2. Optimization bounds for all parameters.

Par.	Domain
K	[0, 1]
B	[-1000, 1000]
P	[0, 2]
D	[-1000, 1000]
a	[0, 1]
r	[0, 2]
τ	[0.15, 0.25]
σ_w	[0, 5]
σ_v	[0, 1]
$x_{0,i}$	[-10, 10]
$\dot{x}_{0;i}$	[-50, 50]
$\ddot{x}_{0;i}$	[-100, 100]
μ_i	[-20, 20]

Solutions for both deterministic and noiseless simulations were generated in linearized steps of size 10^{-5}s then downsampled according to the design of each simulation. This procedure ensured both numerical accuracy of solutions and simulated real world mapping of analogue processes to discrete measurements. The statistical properties of simulated series were expected to be invariant to downsampling due to the fractal property of continuous random walks (i.e., Wiener processes) where $\Delta t \sim N(0, \Delta t)$.

Parameter sets

Six sets of simulated parameters were defined to test the model's estimation capability over a variety of possible behaviors and are shown in Table 3. The first set replicates the simulated data for Model 4 by Asai et al. [1] and is named accordingly. The second and third sets ("Low Noise" and "High Noise") respectively decrease and increase the variance of process noise to examine its effect on other parameters. The fourth and fifth("Active Control" and "Passive Control") sets respectively increase and decrease the ratio of active to passive control, repre- senting different plausible configurations for stability. The sixth set ("Rambling and Trem- bling") represents a stationary random-walk series that diverges markedly from the underlying theory but is nonetheless a stable and plausible configuration.

Sim 1: Noiseless series

To test for improper model specification and systematic sources of bias, noiseless series were generated to span 20s, with a step size of 10^{-5}s, then downsampled to an observation every 0.01s and again to every 0.1s. The noiseless series used in the first simulation are shown in Fig.1. Only one series per set and per downsample rate was used, as there were no sources of sam- pling error. To ensure that estimates converged to a high degree of precision, 3000 iterations of optimization were used.

Table 4 contains the parameter estimates for these simulations, with sampling rate shown to the left. Only the velocity coefficients B and D exhibited substantial bias all throughout, and the "High Active" set incurred the greatest bias over nearly all parameters. Most parameter estimates given 100Hz sampling were exact to at least 3-5 decimal places. Reducing sampling resolution by a factor of ten increased biases to parameters B and D by a factor of ten to fifteen, but much less so for K and P. The nonlinear parameters a, r, and τ exhibited the least bias for all sets.

Table 3. Parameter sets for generating simulated data.

	K	B	P	D	a	r	τ	σ	ξ
Asai et al.	0.80	4.00	0.25	10.00	0.62	0.40	0.20	0.20	1E-04
Low Noise	0.80	4.00	0.25	10.00	0.62	0.40	0.20	0.05	1E-04
High Noise	0.80	4.00	0.25	10.00	0.62	0.40	0.20	1.00	1E-04
Passive Control	0.95	4.00	0.15	10.00	0.50	0.70	0.20	0.20	1E-04
Active Control	0.75	4.00	0.70	120.00	0.80	0.20	0.20	0.20	1E-04
Rambling and Trembling	0.98	500.00	0.20	-50.00	0.45	0.05	0.20	2.00	1E-04

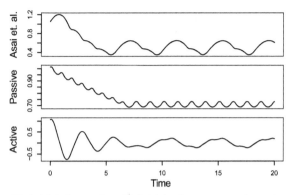

Fig 1. Noiseless series with initial values $x_0 = 1, \dot{x}_0 = 0$.

Table 4. Parameter point estimates for simulated noiseless series. Downsampling rate in Hz is shown in the left column. See Table 3 for the true, data-generating param- eter values of each set.

Hz	Set	K	B	P	D	a	r	τ	σ
100	Asai et al.	0.80003	4.10640	0.24996	10.62579	0.62000	0.40000	0.20001	0.00000
	Passive Control	0.95000	4.08142	0.14998	10.46709	0.50000	0.70000	0.20000	0.00000
	Active Control	0.75086	4.09833	0.69372	122.07313	0.79987	0.19990	0.20020	0.00000
10	Asai et al.	0.80024	5.65678	0.24975	15.62635	0.62030	0.39914	0.20000	0.00000
	Passive Control	0.95001	5.40384	0.14939	19.03988	0.49999	0.70016	0.19978	0.00001
	Active Control	0.74853	4.17357	0.63636	140.82292	0.80120	0.19951	0.20097	0.00022

The small biases to K, B, P, and D most likely occur as a result of the approximate, linear interpolation methods and inability to account for process noise before t_0 in the quadratic backward extrapolation. Biases may be further mitigated using polynomial interpolation of the lagged state. However, the exact accuracy of the estimated τ indicates that bias from linear interpolation is probably trivial in this case.

A second source of bias may be the limits of numerical precision. When no noise is present in the system, the state only occupies a small area of the phase space where certain values of B and D may have nearly unobservable effects on the solution. We show later that relatively unbiased estimates of B and D can indeed be obtained as a function of the other parameters, including the process noise variance σ.

Sim 2 Estimation from noisy data

To test the precision and accuracy of IPC parameter estimates given the dimensions and expected structure of the data from Santos et al. [20], one-hundred individuals were simulated for each parameter set in Table 3, with examples series shown in Fig 2. Each individual con- sisted of three trials, and each trial consisted of a 60s series downsampled to 100 Hz. The same parameters were estimated for all three trials, making for a total of 18,000 observations per individual model.

Figs 3 through 8 show the sampling variation and bias for each parameter set. Boxplots are grouped by common axis scale. Table 5 gives the means and standard deviations of each parameter for each set.

Variance and baises across all parameters were highly interdependent. Estimates of both process noise (σ) and measurement error were precise and close to their true values, indicating successful filtering of the state. The precision of active and passive and active control parame- ters depended on the true values of parameters and resulting behavior of the process. For the Asai et al. replication and the sets with low and high process noise, most control parameters had only small bias and high precision, while others were less reliable under particular condi- tions. The greatest apparent contrast may be the insensitivity radius r, which was not estimable for the Rambling and Trembling set in which its true value was small, and much less reliable in the increased noise set where its value matched Asai et al.

The B and D parameters were the least reliable, and are possibly empirically unidentified without a sufficiently high process noise variance. This is evident from the increased noise set (Fig 6) and the Rambling and Trembling set (Fig 4). The active control set (Fig 8) also showed successful estimation of the B parameter, and improvements in estimating D over the the Asai et al. set, low noise, and passive control.

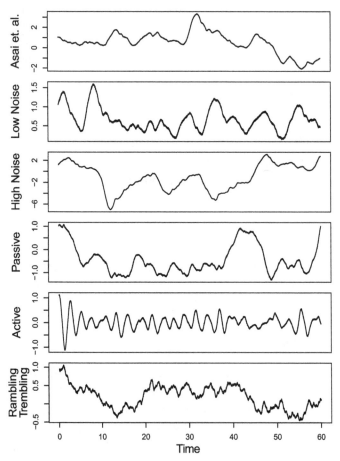

Fig 2. Examples of simulated series from six parameter sets.

From the variation in results across sets, it can be inferred that a parameter can only be estimated reliably when the state occurs for a sufficient amount of time in the portions of the phase space for which that parameter has an influence. For instance, the insensitivity radius will not be estimable if the state tends to bypass it entirely. This may be due to a large variance of process noise, or for large values of *B* that distort the saddle shape of the passive attractor space, causing an orbital path that never intersects the origin. Likewise *B* and *D* cannot be esti- mated reliably if the process does not frequently visit the extremes of the phase space where their influence is most apparent.

Empirical under-identification of some parameters is not necessarily problematic for the others, and does not imply the unreliable parameters should be fixed to some value or excluded. Two solutions to empirical under-identification are to increase the length and reso- lution of the sam- ple to increase the chances of observing informative behavior, and perhaps to introduce small interventions or disturbances such that subjects express the full range of rele- vant dynamic behaviors.

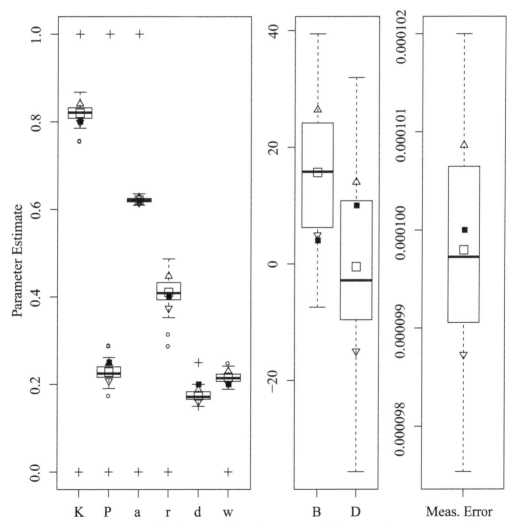

Fig 3. Parameter recovery results for the Asai et al. replication set. Black squares: data-generating value; Empty squares: Estimate mean; Triangles: Upper and lower std. dev.; Circles: Outliers; Crosses: Optimization boundaries.

Standing apart from the other parameters is the feedback delay τ. Despite perfect accuracy in the noiseless case, it tended to bias downwards when estimated from noisy data. It is unclear from our simulations why the bias occurs and whether it accounts for bias to other parameters. However, the estimates were not generally boundary cases, and the sampling variability was small. If the bias is consistent, the delay parameter should still be comparable between persons, with the caveat that the estimate is understated by 20-40ms.

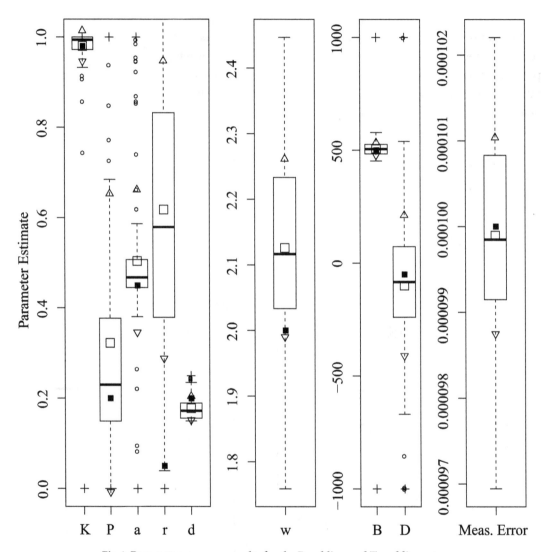

Fig 4. Parameter recovery results for the Rambling and Trembling set.

Data analysis

The IPC model was fit to empirical postural control data to 1) estimate the multivariate distributions of each parameter, 2) test for expected effects from age and visual feedback, 3) test the consistency of parameters within-person, 4) compare the proposed model to simpler alternatives. COM data were obtained from the data set published for public use by Santos et al. [20]

and included 49 individuals at 100Hz for 60 seconds per trial. Three trials were conducted with eyes open, and three with eyes closed. Only trials tested with a rigid floor were used for our analyses. Height and weight were provided for each individual and included as the con- stants h and m in the model, scaled to units of meters and kilograms respectively. Height was scaled by 0.51, the approximate ratio of vertical COM to total height in upright standing (cal- culated from Table 1, p.7 of [23]). By visual inspection of the sample, it was found that the first and last several seconds of many series contained large, sudden changes in position likely relating to movement during the initiation and termination of the trial period. To ensure that only the stationary dy- namics of interest were modeled, 500 occasions were trimmed from the beginning and end of each series, leaving 5000 occasions or 50 seconds of data per trial, and 30,000 measurements in total per individual.

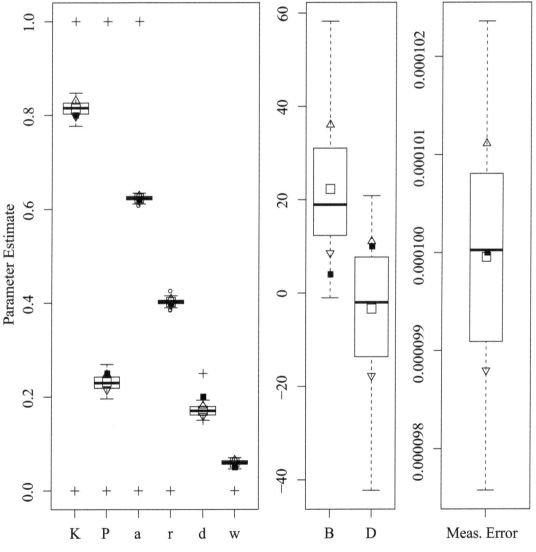

Fig 5. Parameter recovery results for the Low Noise set.

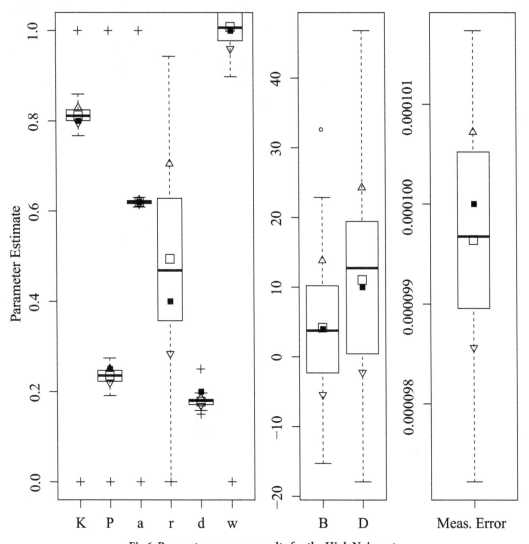

Fig 6. Parameter recovery results for the High Noise set.

Models

Three models were fit to each of three trials per individual to examine the statistical signifi- cance of the parameters involved in intermittent activation and delayed feedback. The models includ- ed, in descending order of complexity, the complete intermittent stochastic delay differ- ential equation (ISDDE),

$$I\ddot{\theta}_t = mgh(1-K)\theta_t + B\dot{\theta}_t + mghf_P(\theta_{t-\tau}) + f_D(\dot{\theta}_{t-\tau}) + \sigma w_t, \tag{28}$$

a stochastic delay differential equation (SDDE) with delayed feedback but no intermittent switching conditions,

$$I\ddot{\theta}_t = mgh(1-K)\theta_t + B\dot{\theta}_t + mghP\theta_{t-\tau} + D\dot{\theta}_{t-\tau} + \sigma w_t, \tag{29}$$

and a stochastic differential equation (SDE) containing only instantaneous, continuous PID control terms:

$$I\ddot{\theta}_t = mgh(1-K)\theta_t + B\dot{\theta}_t + \sigma w_t. \tag{30}$$

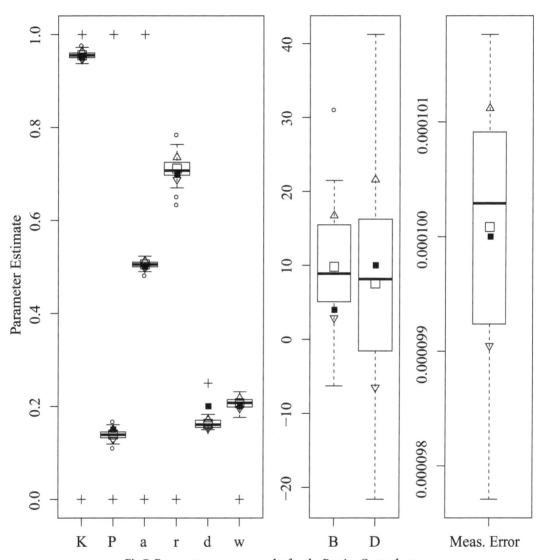

Fig 7. Parameter recovery results for the Passive Control set.

All models included trial-specific initial conditions $x_{0,i}$ and $\dot{x}_{0,i}$ and sway oriins μ_i for $i \in [1, 2, 3]$. The ISDDE and SDDE both included trial-specific estimation of $\ddot{x}_{0,i}$ for backward extrapolation. All models included measurement error variance σ_ϵ. Parameter boundaries, shown in Table 6 reflected both theoretical and analytic roles of each parameter. For example, B could not be less than zero in the ISDDE because it is conjectured to represent ankle stiff- ness, and stability is required to come from values of P and D in the given domains. In the SDE, stable solutions must rely on only instantaneous feedback with coefficients K and B. In the absence of other theoretical mechanisms, the same physiological interpretations of K and B could not be assumed and thus the same theoretical constraints were not applied.

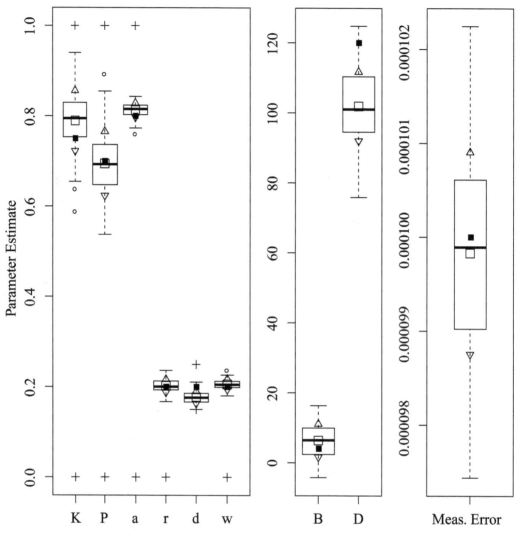

Fig 8. Parameter recovery results for the Active Control set.

Table 5. Simulation results: Means (μ) and standard deviations (σ) of estimated parameters over 100 iterations of simulation for six parameter sets. True values are given in Table 3, and parameter descriptions are given in Table 1.

Par.		Asai et al.	Low Noise	High Noise	Low Active	High Active	Ramb./Tremb.
K	μ	0.820	0.814	0.812	0.955	0.789	0.978
	σ	0.021	0.017	0.017	0.008	0.067	0.025
B	μ	15.654	22.278	4.203	9.811	6.350	504.441
	σ	10.727	13.743	9.631	6.922	4.743	28.426
P	μ	0.228	0.231	0.235	0.139	0.694	0.301
	σ	0.020	0.017	0.016	0.011	0.071	0.211

D	μ	-0.511	-3.337	11.039	7.556	101.869	-138.941
	σ	14.490	14.398	13.258	14.042	9.850	281.247
a	μ	0.621	0.623	0.620	0.505	0.813	0.481
	σ	0.006	0.005	0.004	0.008	0.016	0.072
r	μ	0.410	0.402	0.494	0.712	0.202	0.609
	σ	0.037	0.006	0.210	0.024	0.014	0.322
τ	μ	0.174	0.170	0.178	0.162	0.177	0.176
	σ	0.013	0.011	0.009	0.010	0.013	0.023
σ	μ	0.216	0.060	1.008	0.207	0.206	2.117
	σ	0.014	0.005	0.049	0.011	0.011	0.134
ϵ	μ	9.98E-05	1.00E-04	9.96E-05	1.00E-04	9.98E-05	9.99E-05
	σ	1.07E-06	1.16E-06	1.08E-06	1.04E-06	1.08E-06	1.15E-06

Multiple regression was used to test the association between each parameter, visual feed-back, and age, accounting for height and mass as covariates. Pearson correlation was used to estimate the correlation between parameter estimates during trials with eyes open and trials with eyes closed. Maximum likelihood estimation was used to fit each model, assuming the multivariate normality of measurement and process noise.

Table 6. Optimization boundaries [lower, upper] for each parameter, under each model.

Par.	ISDDE	SDDE	SDE
K	[0, 1]	[0, 1]	[-10, 10]
B	[0, 2000]	[0, 2000]	[-2000, 2000]
P	[0, 2]	[-2, 2]	
D	[-2000, 2000]	[-2000, 2000]	
a	[0, 1]		
r	[0, .1]		
τ	[0, 1]	[0, 1]	
σ	[0, 5]	[0, 5]	[0, 100]
ϵ	[0, .03]	[0, .03]	[0, 10]
x_o	[-5, 5]	[-5, 5]	[-5, 5]
$\dot{x}_0$	[-150, 150]	[-150, 150]	[-150, 150]
ϵ_0	[-500, 500]	[-500, 500]	
Origin	[-1, 1]	[-1, 1]	[-1, 1]

Table 7. Summary statistics, effects of age and vision accounting for height and mass, and person-level intraclass correlations of parameter estimates under each model. Bonferroni adjusted $\alpha = .0029$.

		$\hat{m}$	$\hat{s}$ (Trimmed)	Median	β_{Vision}	CI	p	β_{Age}	CI	p	ρ_{ICC}
ISDDE	K	0.920	0.021 (0.019)	0.920	0.002	(0, 0.004)	0.133	7.6e-5	(1.9e-5, 1.3e-4)	*0.009	0.912
	B	3.057	10.223 (5.2)	0.000	1.019	(-1.314, 3.352)	0.393	-0.078	(-0.144, -0.012)	*0.023	0.191
	P	0.174	0.272 (0.066)	0.116	0.048	(-0.015, 0.111)	0.132	-0.002	(-0.004, 0)	0.074	-0.025
	D	154.789	160.153 (47)	131.123	-20.307	(-57.37, 16.76)	0.284	-0.043	(-1.097, 1.011)	0.936	0.040
	a	0.730	0.259 (0.26)	0.853	-0.055	(-0.114, 0.004)	0.069	0.001	(-0.001, 0.003)	0.275	0.049
	r	0.002	0.003 (0.0017)	0.002	0	(-0.001, 0.001)	0.572	0	(-2.2e-5, 2.2e-5)	0.971	-0.112
	τ	0.302	0.124 (0.095)	0.284	0.018	(-0.01, 0.046)	0.212	-0.001	(-0.002, 0)	*0.049	0.291
	σ_w	0.047	0.006 (4.2e-4)	0.046	-0.001	(-0.002, 0)	0.086	4e-06	(0, 8e-6)	*0.033	0.490

SDDE	K	0.918	0.019	0.919	0	(-0.001, 0.001)	0.923	8.3e-5	(4.6e-5, 1.2e-4)	** 1e-5	0.999
	B	0.838	3.508	0.000	0.163	(-0.647, 0.973)	0.693	-0.006	(-0.029, 0.017)	0.582	0.233
	P	0.162	0.071	0.146	-0.012	(-0.026, 0.002)	0.071	0.001	(0.001, 0.001)	**0.002	0.580
	D	68.818	53.280	50.384	-3.668	(-15.80, 8.47)	0.554	0.135	(-0.21, 0.48)	0.442	0.386
	τ	0.478	0.031	0.494	0.006	(-0.001, 0.013)	0.099	-2.4e-4	(-4e-4, -5e-5)	*0.015	0.034
	σ_w	0.047	0.006	0.046	-0.001	(-0.002, 0)	*0.044	5e-6	(1e-6, 9e-6)	*0.012	0.494
SDE	K	0.931	0.020	0.930	0.004	(0.001, 0.007)	*0.007	1.8e-5	(-5.8e-5, 9.4e-5)	0.641	0.804
	B	17.475	13.231	14.616	2.462	(-0.475, 5.399)	0.102	-0.137	(-0.221, -0.053)	**0.001	0.608
	σ_w	0.048	0.006	0.048	-0.002	(-0.003, -0.001)	*0.011	6e-6	(2e-6, 1e-5)	**0.002	0.544

*Significant at unadjusted $\alpha = .05$
**Significant at adjusted $\alpha = .0029$

The estimated means $\hat{\mu}$, standard deviations $\hat{\sigma}$, and medians of each estimated parameter across all trials × participants × visual feedback conditions, are given in Table 7. The estimated individual-level intraclass correlations (ρ_{ICC}), effect sizes, and p-values for age and visual feedback are also given for each model. Measurement error estimates were generally small ($\sigma_\epsilon < 1e-3$) and were omitted from the tables. Minor, trial-specific "nuisance" parameters including sway origins and initial values were also omitted. Mean sway origin was estimated to be 0.217, with a standard deviation of 0.11 and a median of 0.226.

The estimated parameters of the ISDDE fell within the expected domains. Several parameters of the ISDDE had outliers that substantially inflated estimates of their standard deviations. Trimmed standard deviations in which the highest 15 values were excluded are given in paren- theses in Table 7. The marginal distributions of each parameter with these trimmed means and standard deviations are shown in Fig 9. B, P, D, and r in particular were skewed upward by outli- ers but otherwise had relatively precise distributions about their medians, with similar precision to those of the SDDE. K had consistent values around .91 to .93 in all three models. B was close to zero for most series but skewed upward by outliers as high as 80. In the SDE, B was allowed to take negative values but had a mean around 17. All values of B in the SDE were positive and greater than zero, with a minimum of .82. a was generally high, representing active control over 75-85% of the phase space. Similarly, r was 50% smaller on average than values used in previous studies. τ had a median of 284 ms and was distributed between 200 to 400 ms. If the bias found in simulations is consistent and proportional, then the true median delay was closer to 240 ms. The SDDE estimated much longer delays on average at 470-490 ms but much lower values of D. Process noise standard deviation σ_w estimates were distributed identically between models.

No significant effects of visual feedback were observed in the parameters of any of the three models. The lowest p-values were for $a(p = .069)$ and $\sigma_w(p = .086)$. Both the SDDE and SDE showed effects on σ with $p < .05$, the alpha level before adjusting for the 17 tests in total.

In the SDDE, both passive ankle stiffness K and active control coefficient P were shown to significantly increase with age. The effects were detected given the adjusted alpha level, with $p < .0029$. Both B and σ_w in the SDE showed significant trends with age as well. Other non-sig- nificant effects with $p < .05$ were K and σ_w in both the ISDDE and SDDE, B in the ISDDE, and τ in the SDDE. Effect estimates of σ_w were consistent across models.

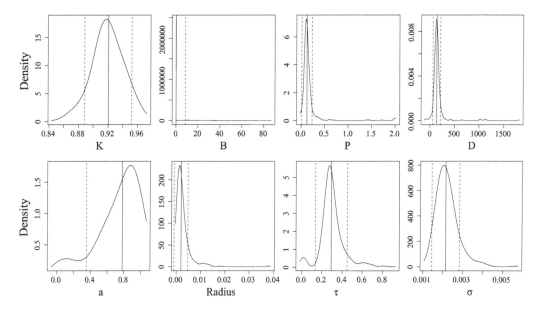

Fig 9. Marginal distributions of the ISDDE parameter estimates. Solid lines are means and dashed lines are standard deviations, both trimmed for the 15 highest values.

Overall, parameters tended to be more consistent within person for the simpler models. The highest intra-class correlation for all parameters in all models was K, with extreme reliability ($\rho =$ .999) in the SDDE. σ generally correlated around .5 for each model. The ISDDE had the least consistent parameters with intraclass correlations near zero for P, D, a, and r. The SDDE and SDE intraclass correlations were moderate to high for all except feedback delay, τ, which was near zero.

Akaike's Information Criterion (AIC) [33], as $-2 \ln(\hat{\mathcal{L}}) + 2k$ where k is the number of estimated parameters, was used to compare overall model fit for every individual. For each trial, the model with the lowest value of the AIC was selected as the best fitting option. In total, the ISDDE was selected for 227 trials, SDDE for 62, and SDE for 0. No significant associations were found between model selections over trials within person or by visual feedback condition.

Discussion

Simulation

Simulation studies were used to determine whether the parameters of the intermittent activation feedback control model proposed by Asai et al. [1] can be estimated using a Kalman Filtering-based framework with delayed proportional and derivative terms and discrete activation-thresholds. The results of the simulations show that the parameters of the model can be estimated with relatively low bias and high precision if the behaviors for which they are influential are sufficiently expressed in the data (i.e., empirically identified). Every parameter of the model was successfully recovered in at least one of the parameter configurations tested, though no single configuration of parameters resulted in a completely unbiased set. The set of results shown in Fig 8 comes close, with downward bias only to the active derivative controller. We can also see by comparing Fig 6 to Figs 5 and 3 that an increased variance of process noise allowed the identification of the B and D parameters, but with large standard errors. The deriv- ative coefficients were likely biased and unreliable when the trajectories did not frequent the extremes of position and velocity where the directional effects of derivative terms could be distinguished from other sources. Fig 4 shows that with process noise of a standard deviation much greater than the insensitivity radius and a weak attraction to point equilibria (K 1), the state is prone to drifting away from the origin where it will rarely traverse the insensitivity radius or switching boundary. If the data can be optimally explained without the use of the switching parameters, then they are said to be empirically unidentified. For this reason, both a and r do not contribute crucial information and converge to precise solutions when the data are optimally described by other parameter values characteristic of rambling and trembling. However, estimates of the derivative controllers B and D in that case were unbiased.

Across configurations, some iterations of model fitting resulted in negative values of B. In continual PID controllers, this would result in amplification and instability over time. In the ISDDE and SDDE, the stability of the system given a value of B depends on the corresponding values of P, D, and a, as the instantaneous proportional and derivative terms do not control the complete periodic behavior on their own. Negative values of B will promote further insta- bility in the already unstable manifold of instantaneous feedback but will be counteracted when the state reaches the stable manifold determined by active feedback. It is informative that solutions occasionally involved negative values of B that breach its theoretical interpreta- tion as joint friction. Solutions that did better identify B and D only did so when their effects were much larger than physically plausible, a priori values of joint friction. In simpler PID cases, estimates of damping tend to be far less reliable than, for instance, the proportional coefficients, so for these reasons together it may be inadvisable to rely on postural sway data and estimation approaches to specifically determine joint friction. Similar concerns may be directed toward the active feedback damping D, though the prior ISDDE literature does not assert as specific of a definition nor necessary theoretical boundaries.

Estimates of both process noise and measurement error were very close to their true values in every case, with only small upward bias proportional to the magnitude of the estimate for certain parameter sets (Figs 3 and 4). The standard deviation of measurement error that we chose to simulate was $\sigma = .01$ cm, twenty times the error of the force plate used by Santos et al. [20] to obtain the data. The success of estimation despite greatly exaggerated sensor noise demonstrates the reliability of Kalman filtering and adequate technical specification of the model, and relieves researchers from the need to choose a preliminary noise reduction step such as spectral filtering. Instead, using the raw data and including measurement error in the model avoids removing fine-grained details of the signal represented in the domain of high fre- quencies typically suppressed by low-pass filtering.

The feedback delay, τ, was unbiased in noiseless simulations, and consistently biased down- ward in noisy simulations. It is not clear what causes the bias, but it did not appear to consis- tently induce bias in other parameters that depended on the correct lag interval, such as the active proportional and derivative controllers.

The results of our simulation demonstrate that the proposed method of direct, statistical estimation by Kalman filter can recover the complete set of parameters for the model. Previous estimation methods only attempted to estimate five of the eight structural parameters [17, 18]. Among those attempted, the D parameter did not converge to its true value in simulation nor to a reliable, unimodal distribution in the empirical study. Despite this setback, no discussion was given of the role of empirical identification in determining D or other parameters, whereas we have demonstrated that the precision of estimates depends on their true values and interdependence. Additionally, the accuracy of their results rests on assumed values of K, B, and r. Due to the high degree of parameter dependence in univariate models such as this, error in one parameter is expected to propagate to other parameters in a compensatory manner. It is therefore preferable to jointly estimate all uncertain model parameters when possible.

The prior studies also did not account for measurement error. We determined that additional sources of sensor noise could be filtered simultaneously with estimation of the dynamic structure. If additive noise is Gaussian, then no preprocessing steps such as spectral filtering or downsampling should be needed and the risk of obscuring important, fine-grained topological features is greatly mitigated.

Computationally, the use of global optimization to maximize the likelihood function provided an efficient alternative to Bayesian MCMC methods as no data simulation procedures, prior distributions, or posterior sampling were required. An additional, unexplored benefit of maximum likelihood in this case is estimation of standard errors directly from the likelihood function. Because the model includes discrete thresholds, the likelihood function was stochastic and non-differentiable. This prevented the use of the Hessian matrix to calculate precision. However, future work may explore methods of smoothly approximating the marginal likelihood function, for instance by fitting splines to likelihood values retained from the optimization procedure.

Experimental data

The results of analyzing the empirical COM data show that the nonlinear mechanisms of feedback activation led to significant improvement in model fit over the simpler SDDE and SDE (i.e., delayed and instantaneous PID) models. It cannot be determined from statistical model comparisons alone whether the results validate the model-generating theory of posture control. To that end, we must compare the parameter estimates to their theoretical priors.

Overall, the distributions of parameters showed a feasible correspondence to the domains expected given the theory. K was consistently close to the 91% relative resistance found by Loram and Lakie [22] for all of the models tested, here showing resistance to 92% of the total gravitational toppling torque on average. Conversely, in the ISDDE and SDDE, B most often converged to zero and was not likely to play a critical role in the model behavior. Perhaps coincidentally, the mean of B was near its proposed value of 4 Nms/rad. It is possible that statistical power at the individual level was insufficient to identify small effects due to B, and the expected value would be recovered if it were estimated across the total data set. Active feedback was generally weaker than hypothesized but still sufficient for stability. Estimates of P were closer to .1 than the proposed .25 [1], likely due in part to the greater resistance to toppling forces from values of K closer to the high end of their theoretical distribution. D played a large role in the dynamics of active control and resulted in non-negligible damping in many individuals. Val- ues of a and r reflected greater control sensitivity than expected. a values around 75% to 80% assign a larger share of the phase space to active feedback, while smaller values of r indicate less tolerance to falling at the origin of sway. The mean estimate of a was higher than found by Tietäväinen et al. [18], which reported a control space closer to 64% in accordance with the analysis by Asai et al. [1]. We found a nearly identical distribution of the feedback delay, τ, to Tietäväinen et al. [18], ranging from 200 to 400 ms with a mean around 300 ms. Estimates of σ_w were an order of magnitude smaller than expected by Asai et al. [1], and about half of those found by Tietäväinen et al. [18].

A graphical vignette of these results is provided in Fig 10, which shows six raw data series with their respective intermittent activation conditions estimated by the model. The horizontal axis is the tilt angle and the vertical axis is the tilt angle's velocity. The shaded region represents behavior where P and D are equal to zero. In the unshaded region, all parameters are active with their non-zero values. Fig 10a shows two cases that resemble the theorized structure with combinations of stable and unstable manifolds in nearly equal proportion. In Fig 10b, the val- ues of B and a are sufficiently large to minimize the influence of the unstable manifold. The result is behavior that closely resembles harmonic oscillation around a single equilibrium. The opposite trend is shown in Fig 10c, where the unstable manifold is not influential, but a high ratio of the derivative coefficients to proportional coefficients results in continual suppression of velocity. This pattern results in wandering oscillations without clear equilibria.

The optimal solution for the SDE model had a much larger, positive value for B than the other models while maintaining a theoretically plausible value of K less than 1. Furthermore, all values of B in the SDE were positive, as we might expect given that negative values would result in in- stability. When a linear system is strongly overdamped (in this case, a high value of B in the SDE, or either B or D in the SDDE) with relatively weak proportional feedback, as the SDE, then it exhibits non-equilibrium Langevin dynamics. These dynamics have convention- ally described the random walk of a large molecule due to its collisions with a many smaller molecules in a sol- vent. The resulting trajectories can appear locally stationary by chance and exhibit short intervals of oscillation. Previous studies have modeled posture control in the con- text of Langevin dy- namics [34–37]. Our simplest model of COM movement, the SDE, resem- bled a model of COP proposed by Bosek et al. [34] that describes trajectories as a second order SDE with no propor- tional feedback and a large derivative coefficient B. Fig 11 shows how the theoretical model and Langevin dynamics differ markedly in their mechanistic parameteriza- tion and observed phase portraits, yet they share many notable features. In both, high-fre- quency oscillations move grad- ually across the sample space in a "rambling" pattern. By chance, the Langevin equation in Fig 11b can result in concentrated oscillations around a few apparent equilibria, but no equilibrium mechanism is present in the model. The parsimony of generating these patterns with only three parameters poses a challenge to the specificity of evi- dence for the theoretical ISDDE. Visual in- spection of the complete results showed that trials ranged between the two extremes of theoret- ical misspecification, from harmonic oscillation to Langevin dynamics. The expected topology involves a mix of features from both, sometimes showing adherence to the principles of feedback switching with occasional deviations into Langevin-type random walk.

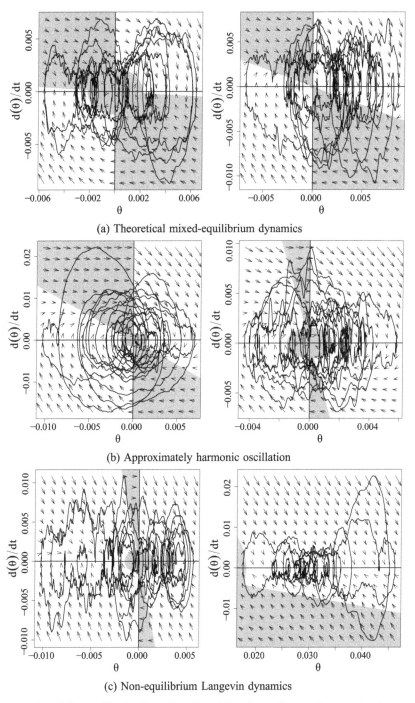

(a) Theoretical mixed-equilibrium dynamics

(b) Approximately harmonic oscillation

(c) Non-equilibrium Langevin dynamics

Fig 10. Phase portraits of observed body tilt angle series with estimated intermittent activation structures and vector fields. The horizontal axis is the COM or tilt angle position and the vertical axis is its velocity. The shaded region represents behavior where P and D are equal to zero. In (a), the estimated parameters match theoretical expectations, showing mixed equilibrium behaviors. In (b) and (c), the nonlinear mechanisms are fit in unique ways that deviate from their theoretically expected function.

Regardless of the true form of the underlying process, we might expect that if the parameters represent underlying physiological mechanisms, they should exhibit some degree of trait-like stability within-person. The intraclass correlations in Table 7 show that the nonlinear switching parameters were generally unreliable within-person. The correlations for the remaining parameters increase as the model is simplified to the SDDE and SDE. The higher consistency of the simpler models' parameters does not necessarily imply that they are more "real" than those of the ISDDE. It is expected that reliance on fewer parameters to explain the variance of sway results in fewer competing configurations of those parameters. Any consis- tency of topological features within-person will be reflected in similarity of the model solu- tions. The lack of consistency in the more complex ISDDE is, however, a challenge to the trait-like stability and actuality of its parameters.

Though no specific connections between visual feedback and the theoretical mechanisms of control were hypothesized for the present study, we expected one or more parameters to be significantly influenced over trials in which eyes were closed in correspondence with previ- ously observed effects on summary statistics. By modeling center of pressure variation with Langevin dynamics, Bosek et al. [34] found that the process noise distribution was influenced by visual feedback. The same finding was replicated with further connections to Parkinson's disease [35]. Vieira et al. [38] found associations of visual feedback with stabilogram measures of sway. All three models models tested here had lower p-values for σ_w than for other parame- ters, suggesting that effects may be discernible given a larger sample or improvement in model specification.

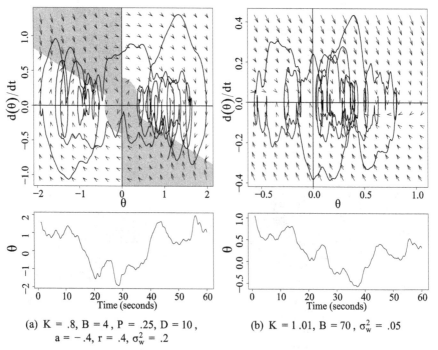

(a) $K = .8$, $B = 4$, $P = .25$, $D = 10$, $a = -.4$, $r = .4$, $\sigma_w^2 = .2$

(b) $K = 1.01$, $B = 70$, $\sigma_w^2 = .05$

Fig 11. Phase portraits and time series from two models generated from the same vector of noise (scaled by σ_w). In (a) data were simulated from the ISDDE model with the theoretical priors. In (b) data were simulated from a 3-parameter SDE. The large ratio of derivative (B) to proportional (K) force results in non-equilibrium Langevin dynamics that may exhibit similar features to the ISDDE for limited periods of time. Characteristic features distinguish (a) from (b), such as sharp changes in velocity localized to quadrants II and IV, slow change in velocity associated with quadrants I and III, and higher density at two spatial equilibria.

Age has been previously associated with more general metrics of sway, such as path length [39], frequency band [38, 40] and mean velocity [40], though findings vary and few effects have been consistently reproduced in COP and COM data. Significant effects of age were observed in the present results, including ankle stiffness and active feedback force in the SDDE and process noise and ankle viscosity in the SDE. Interpretation of effects on the SDE is more difficult because the parameters of the SDE do not correspond to specific explanatory mecha- nisms in this case. The consistent positive associations of all models with noise magnitude σ_w with age may be linked to previously observed associations of stabilogram-based diffusion metrics with age [41]. The significant associations of ankle stiffness and active proportional feedback in the SDDE found here may reflect previously observed increases in stiffness and damping with age estimated from a simpler PID model [42].

Finally, the model concerns an abstract notion of body tilt angle, though there are many ways to represent this using the full kinematic data. For simplicity and consistency with past studies, we chose to represent tilt angle by the COM. Preliminary tests using alternative mea- sures included COP and the average angle of both ankle joints. The results were found to differ markedly from both our current results and those previously obtained with the COM, but a complete compari- son of alternative measures is too complex to discuss here. We leave detailed examination of this question with regard to the feasibility of this model to future study.

Conclusions

We designed and implemented an Extended Kalman Filter-based estimation model of inter- mittent, delayed feedback control in postural sway and demonstrated that for a variety of stable configurations, parameters can be recovered accurately given adequate empirical identifica- tion. Application of the model to experimental data resulted in distributions of the parameters the correspond well to previous findings and suggest that physiologically informative and clinically useful attributes of human balance may be extracted directly from COM data. While the model replicates previous findings, the conjectured parameters of feedback activation were not reliable within-person or strongly associated with visual feedback and age. Further com- parisons with alternative mechanistic theories and model parameterizations are warranted. Beyond postural control, the model stands as a framework for estimating parameters of sto- chastic delay differ- ential equation models controlled by discrete activation thresholds.

Author Contributions Conceptualization: Kevin L. McKee. **Formal analysis:** Kevin L. McKee.

Funding acquisition: Michael C. Neale.

Investigation: Kevin L. McKee.

Methodology: Kevin L. McKee.

Project administration: Michael C. Neale.

Resources: Michael C. Neale.

Software: Kevin L. McKee, Michael C. Neale.

Supervision: Michael C. Neale.

Writing – original draft: Kevin L. McKee.

Writing – review & editing: Kevin L. McKee, Michael C. Neale.

References

1. Asai Y, Tasaka Y, Nomura K, Nomura T, Casadio M, Morasso P. A Model of Postural Control in Quiet Standing: Robust Compensation of Delay-Induced Instability Using Intermittent Activation of Feedback Control (Intermittent Postural Control). PLoS ONE. 2009; 4(7). https://doi.org/10.1371/annotation/ 96e08e7f-22f0-445d-8fb3-fe7b071d0a3a

2. Milton JG, Insperger T, Cook W, Harris DM, Stepan G. Microchaos in human postural balance: Sensory dead zones and sampled time-delayed feedback. Physical review E. 2018; 98(2-1). https://doi.org/10. 1103/PhysRevE.98.022223 PMID: 30253531

3. Collins JJ, De Luca CJ. Random walking during quiet standing. Physical review letters. 1994; 73(5). https://doi.org/10.1103/PhysRevLett.73.764

4. Lafond D, Duarte M, Prince F. Comparison of three methods to estimate the center of mass during bal- ance assessment. Journal of Biomechanics. 2004; 37(9):1421–1426. https://doi.org/10.1016/S0021- 9290(03)00251-3 PMID: 15275850

5. Collins JJ, De Luca CJ. Open-loop and closed-loop control of posture: A random-walk analysis of cen- ter-of-pressure trajectories. Experimental Brain Research. 1993; 95(2):308–318. https://doi.org/10. 1007/bf00229788 PMID: 8224055

6. Yamamoto T, Smith CE, Suzuki Y, Kiyono K, Tanahashi T, Sakoda S, et al. Universal and individual characteristics of postural sway during quiet standing in healthy young adults. Physiological Reports. 2015; 3(3):n/a–n/a. https://doi.org/10.14814/phy2.12329

7. Zatsiorsky VM, Duarte M. Rambling and trembling in quiet standing. Motor control. 2000; 4(2). https:// doi.org/10.1123/mcj.4.2.185 PMID: 11500575

8. Maurer C, Peterka R. A New Interpretation of Spontaneous Sway Measures Based on a Simple Model of Human Postural Control. Journal of Neurophysiology. 2005; 93(1):189–200. https://doi.org/10.1152/ jn.00221.2004 PMID: 15331614

9. Gawthrop P, Loram I, Lakie M, Gollee H. Intermittent control: a computational theory of human control. Biological Cybernetics. 2011; 104(1):31–51. https://doi.org/10.1007/s00422-010-0416-4 PMID: 21327829

10. Milton J, Meyer R, Zhvanetsky M, Ridge S, Insperger T. Control at stability's edge minimizes energetic costs: expert stick balancing. Journal of the Royal Society, Interface. 2016; 13(119). https://doi.org/10. 1098/rsif.2016.0212 PMID: 27278361

11. Michimoto K, Suzuki Y, Kiyono K, Kobayashi Y, Morasso P, Nomura T. Reinforcement learning for sta- bilizing an inverted pendulum naturally leads to intermittent feedback control as in human quiet stand- ing. Conference proceedings: Annual International Conference of the IEEE Engineering in Medicine and Biology Society IEEE Engineering in Medicine and Biology Society Annual Conference. 2016;2016:37–40.

12. Gawthrop PJ, Wang L. Intermittent model predictive control. Proceedings of the Institution of

Mechani- cal Engineers, Part I: Journal of Systems and Control Engineering. 2007; 221(7):1007–1018.

13. Gawthrop P, Loram I, Gollee H, Lakie M. Intermittent control models of human standing: similarities and differences. Biological cybernetics. 2014; 108(2). https://doi.org/10.1007/s00422-014-0587-5 PMID: 24500616

14. Eurich, Milton. Noise-induced transitions in human postural sway. Physical review E, Statistical physics, plasmas, fluids, and related interdisciplinary topics. 1996; 54(6). https://doi.org/10.1103/physreve.54. 6681 PMID: 9965894

15. Bottaro A, Yasutake Y, Nomura T, Casadio M, Morasso P. Bounded stability of the quiet standing pos- ture: An intermittent control model. Human Movement Science. 2008; 27(3):473–495. https://doi.org/ 10.1016/j.humov.2007.11.005 PMID: 18342382

16. Nomura T, Oshikawa S, Suzuki Y, Kiyono K, Morasso P. Modeling human postural sway using an inter- mittent control and hemodynamic perturbations. Mathematical Biosciences. 2013; 245(1):86–95. https://doi.org/10.1016/j.mbs.2013.02.002 PMID: 23435118

17. Wang H, Li J. Adaptive Gaussian Process Approximation for Bayesian Inference with Expensive Likeli- hood Functions. Neural Computation. 2018; 30(11):3072–3094. https://doi.org/10.1162/neco_a_01127

18. Tietä̈vä̈inen A, Gutmann MU, Keski-Vakkuri E, Corander J, Hæggströ̈m E. Bayesian inference of physi- ologically meaningful parameters from body sway measurements. Scientific reports. 2017; 7(1). https:// doi.org/10.1038/s41598-017-02372-1 PMID: 28630413

19. Aandahl RZ, Stadler T, Sisson SA, Tanaka MM. Exact vs. approximate computation: reconciling differ- ent estimates of Mycobacterium tuberculosis epidemiological parameters. Genetics. 2014; 196(4). https://doi.org/10.1534/genetics.113.158808 PMID: 24496011

20. Santos DAD, Fukuchi CA, Fukuchi RK, Duarte M. A data set with kinematic and ground reaction forces of human balance. PeerJ. 2017; 5(7).

21. Peterka RJ. Sensorimotor integration in human postural control. Journal of neurophysiology. 2002; 88 (3):1097–1118. https://doi.org/10.1152/jn.2002.88.3.1097 PMID: 12205132

22. Loram ID, Lakie M. Direct measurement of human ankle stiffness during quiet standing: the intrinsic mechanical stiffness is insufficient for stability. Journal of Physiology. 2002; 545(3):1041–1053. https:// doi.org/10.1113/jphysiol.2002.025049 PMID: 12482906

23. Casadio M, Morasso PG, Sanguineti V. Direct measurement of ankle stiffness during quiet standing: implications for control modelling and clinical application. Gait & Posture. 2005; 21(4):410–424. https:// doi.org/10.1016/j.gaitpost.2004.05.005

24. Li Y, Levine WS, Loeb GE. A Two-Joint Human Posture Control Model With Realistic Neural Delays. IEEE Transactions on Neural Systems and Rehabilitation Engineering. 2012; 20(5):738–748. https:// doi.org/10.1109/TNSRE.2012.2199333 PMID: 22692939

25. Woollacott M, Hosten C, Rö̈sblad B. Relation between muscle response onset and body segmental movements during postural perturbations in humans. Experimental Brain Research. 1988; 72(3):593–604. https://doi.org/10.1007/bf00250604 PMID: 3234505

26. Kalman RE, Bucy RS. New Results in Linear Filtering and Prediction Theory. Journal of Basic Engineer- ing. 1961; 83(1).

27. Wei WWS. Time series analysis: univariate and multivariate methods. 2nd ed. Boston: Pearson Addi- son Wesley; 2006.

28. Price KV, Storn RM, Lampinen JA. Differential Evolution—A Practical Approach to Global Optimization. Natural Computing. Springer-Verlag; 2006.

29. R Core Team. R: A Language and Environment for Statistical Computing; 2018. Available from: https:// www.R-project.org/.

30. Mullen KM, Ardia D, Gil DL, Windover D, Cline J. DEoptim: An R Package for Global Optimization by Differential Evolution. Journal of Statistical Software. 2011; 40(6). https://doi.org/10.18637/jss.v040.i06

31. Eddelbuettel D, Balamuta JJ. Extending extitR with extitC++: A Brief Introduction to extitRcpp. PeerJ Preprints. 2017; 5:e3188v1.

32. Eddelbuettel D, Sanderson C. RcppArmadillo: Accelerating R with high-performance C++ linear alge- bra. Computational Statistics and Data Analysis. 2014; 71:1054–1063. https://doi.org/10.1016/j.csda. 2013.02.005

33. Akaike H. A new look at the statistical model identification. IEEE Transactions on Automatic Control. 1974; 19(6):716–723. https://doi.org/10.1109/TAC.1974.1100705

34. Bosek M, Grzegorzewski B, Kowalczyk A. Two-dimensional Langevin approach to the human stabilo- gram. Human Movement Science. 2004; 22(6):649–660. https://doi.org/10.1016/j.humov.2004.02.005 PMID: 15063046

35. Bosek M, Grzegorzewski B, Kowalczyk A, Lubiński I. Degradation of postural control system as a con- sequence of Parkinson's disease and ageing. Neuroscience letters. 2005; 376(3). https://doi.org/10. 1016/j.neulet.2004.11.056 PMID: 15721224

36. Gottschall J, Peinke J, Lippens V, Nagel V. Exploring the dynamics of balance data—movement vari- ability in terms of drift and diffusion. Physics Letters A. 2009; 373(8):811–816. https://doi.org/10.1016/j. physleta.2008.12.026

37. Lauk M, Chow CC, Pavlik AE, Collins JJ. Human Balance out of Equilibrium: Nonequilibrium Statistical Mechanics in Posture Control. Physical Review Letters. 1998; 80(2):413–416. https://doi.org/10.1103/ PhysRevLett.80.413

38. Vieira TdMM, Oliveira LFd, Nadal J. An overview of age-related changes in postural control during quiet standing tasks using classical and modern stabilometric descriptors. Journal of Electromyography and Kinesiology. 2009; 19(6):e513–e519. https://doi.org/10.1016/j.jelekin.2008.10.007

39. Hageman PA, Leibowitz JM, Blanke D. Age and gender effects on postural control measures. Archives of physical medicine and rehabilitation. 1995; 76(10). https://doi.org/10.1016/s0003-9993(95)80075-1 PMID: 7487439

40. Lin D, Seol H, Nussbaum MA, Madigan ML. Reliability of COP-based postural sway measures and age- related differences. Gait & Posture. 2008; 28(2):337–342. https://doi.org/10.1016/j.gaitpost.2008.01. 005

41. Collins JJ, De Luca CJ, Burrows A, Lipsitz LA. Age-related changes in open-loop and closed-loop pos- tural control mechanisms. Experimental brain research. 1995; 104(3). https://doi.org/10.1007/ bf00231982 PMID: 7589299

42. Cenciarini M, Loughlin PJ, Sparto PJ, Redfern MS. Stiffness and Damping in Postural Control Increase With Age. IEEE Transactions on Biomedical Engineering. 2010; 57(2):267–275. https://doi.org/10. 1109/TBME.2009.2031874 PMID: 19770083

A novel power-driven fractional accumulated grey model and its application in forecasting wind energy consumption of China

Peng Zhang 1,2, Xin Ma2,3, Kun She1*

1 School of Information and Software Engineering, University of Electronic Science and Technology of China, Chengdu, China, 2 School of Science, Southwest University of Science and Technology, Mianyang, China, 3 State Key Laboratory of Oil and Gas Reservoir Geology and Exploitation, Southwest Petroleum University, Chengdu, China

* kun@uestc.edu.cn

Editor: Wei Yao, Huazhong University of Science and Technology, CHINA

Funding: This work is funded by National Natural Science Foundation of China (No. 71901184, No. 11872323, No. 61672136), Humanities and Social Science Project of Ministry of Education of China (No. 19YJCZH119), the Open Fund (PLN201710) of State Key Laboratory of Oil and Gas Reservoir Geology and Exploitation (Southwest Petroleum University), the Doctoral Research Foundation of Southwest University of Science and Technology (No. 16zx7140), and National Statistical Scientific Research Project (2018LY42). The funders had no role in study design, data collection and analysis, decision to publish, or preparation of the manuscript.

Competing interests: The authors have declared that no competing interests exist.

Abstract

Wind energy is one of the most important renewable resources and plays a vital role in reducing carbon emission and solving global warming problem. Every country has made a corresponding energy policy to stimulate wind energy industry development based on wind energy production, consumption, and distribution. In this paper, we focus on forecasting wind energy consumption from a macro perspective. A novel power-driven fractional accu- mulated grey model (PFAGM) is proposed to solve the wind energy consumption prediction problem with historic annual consumption of the past ten years. PFAGM model optimizes the grey input of the classic fractional grey model with an exponential term of time. For boosting prediction performance, a heuristic intelligent algorithm WOA is used to search the optimal order of PFAGM model. Its linear parameters are estimated by using the least- square method. Then validation experiments on real-life data sets have been conducted to verify the superior prediction accuracy of PFAGM model compared with other three well- known grey models. Finally, the PFAGM model is applied to predict China's wind energy consumption in the next three years.

Introduction

Wind energy is one of the important vital resources of renewable energy, which is widely distributed with large reserves. Wind energy and other renewables will play a vitally important role in solving global warming issues and reducing carbon emission in future decades. Interna- tional energy agency (IEA) has optimistically estimated that renewables will account for 39% share of total electricity generation by 2050. In light of Global Wind Report 2018 of the GWEC, it noticed that new wind installed capacity had overtaken new fossil fuel capacity for the first time in many developing or mature markets. The global total cumulative wind installed capacity has brought up to 591GW at the end of 2018 with new installations 51.3GW. In China, the total wind installed capacity has reached 211GW in 2018, which indicates that the wind energy target of Five-Year-Plan 2016-2020 has been achieved two years ahead of schedule. Along with the rapid development of the energy industry, it also brings a great chal- lenge to make energy policy upgrade energy structure. According to the definition in the litera- ture [1], the energy policy of an entity (especially a government) is used to solve the problems of energy production, consumption, and distribution [2]. Most scholars mainly focused on forecasting the state of wind energy, including wind speed forecasting and wind power fore- casting, which is expected to provide a reference for formulating energy production planning from micro-perspective. These forecasting approaches mainly include deterministic forecast- ing and uncertainty analysis [3]. For these approaches, forecasting wind energy is considered as a prediction process of stochastic time series. Usually, the future values are predicted by these data-driven algorithms based on the historical wind speed or wind energy sequence or other related data. Many scholars applied Artificial Neural Network (ANN) to forecast wind energy. Numerous variants of ANN, such as Back Propagation Neural Network (BPNN) [4, 5], Radial Basis Function Neural Network (RBFNN) [6], Generalized Regression Neural Network (GRNN) [7] and Wavelet Neural Network (WNN) [8, 9], were also proposed to predict the future wind energy. Besides, some scholars utilized Support Vector Machine (SVM) [10] or its variants [11, 12] to forecast wind energy. Deep learning approaches, such as Autoencoder (AE), Deep Boltzmann Machine (DBM) [13], Convolutional Neural Network (CNN) [14], Recurrent Neural Networks (RNNs) [15] and so on, were also adopted to forecast the future wind energy. Though these approaches can be used to forecasting wind energy, they usually need a larger dataset, including historical data and some exogenous data, to train the predictors for better prediction performance. And they were often used to predict hours or minutes ahead of wind energy or speed for making the production plan of a company or energy farm in a local region [16, 17]. Besides, many scholars utilized ARMA, ARIMA, and their successors to predict wind energy or wind speed [18, 19]. However, the study of wind energy consumption prediction from the macro view is very few at present. Previously, many scholars mainly concentrated in study of forecasting electricity [20–29], natural gas [30,

31], oil [32, 33], nuclear [34, 35], renewable [36, 37] energy consumption and so on. Forecasting wind energy consumption is also an important task for making energy policy and plan for the government. Grey prediction model is an effective prediction approach, which is one of the best choices for solving the wind energy consumption prediction problem with small samples.

Recently, scholars have conducted numerous studies to improve prediction performance and enlarge the application range of grey models. Lots of novel or improved grey models have been put forward to solve prediction problems with partially know or unknown information effectively. On the one hand, many scholars devoted their efforts to optimize the initial GM (1,1) model. Wang et al. proposed a rolling grey model with PSO and data cleaning techniques to predict Beijing's tertiary industry effectively [38]. Xia et al. [39] proposed an improved grey model with grey input of time power based on the new information priority. Ding et al. over- came the fixed structure and poor adaptability of GM(1,1) model and put forward an opti- mized grey model called NOGM(1,1) to forecast China's electricity consumption [20]. Xu et al. proposed IRGM(1,1) model with optimizing the initial condition of time response func- tion [21]. Wang et al. proposed SGM(1,1) model to forecast the seasonal time series [23]. Besides, other improved grey models such as WBGM(1,1) [40], GARGM(1,1) [41], Nash NBGM [42], SIGM [43], GRA-IGSA [44] and so on were proposed and obtained satisfactory prediction accuracy. But these models have the same insufficiency of traditional GM(1,1) model because they are essentially grey models with one order and one variable. For the multi- variable time series, Wang et al [45] proposed an improved GM(1,n) model which considered the effects of the dependent variables. Zeng et al. improved the GM(1,n) model with a dynamic background parameter [46]. Soon afterwards Zeng et al. [47] presented a new multi-variable grey model to enhance the structure compatibility of grey model. Ma et al. [48] proposed a parameter optimization method for CGM(1,n) model to promote its prediction accuracy. These efforts not only enriched the theory of grey system but also improved the prediction per- formance of grey prediction model.

On the other hand, fractional order accumulating operator was imported to decrease the stochasticity and uncertainty of raw sequence to boost the performance of grey model. Wu et al. initially improved traditional GM(1,1) with fractional order accumulated operator called FAGM(1,1), which guarantees the priority of new information under small fractional order and boosts the prediction accuracy compared with traditional GM(1,1) model [49]. Yang et al. generalized the traditional GM(1,1) models with fractional calculus of which fractional-order derivative is profoundly determined the accuracy of prediction and can be optimized by intelli- gent algorithms [27]. Mao et al. proposed a fractional grey model based on fractional order derivative called FGM(q,1) of which the whitening equation is a fractional order differential equation, and forecasted the gross national income per capita accurately [50]. Some research- ers optimized the grey action quantity of FAGM(1,1) and achieved better prediction accuracy. Ma et al. optimized the grey input of the original FAGM(1,1) model with a fractional time delay term and applied it to predict the gas consumption and coal consumption [51]. Wu et al. proposed a fractional FAGM(1,1,k) model with linear grey input of time instead of constant grey input in initial FAGM(1,1) model and optimized it with optimal linear parameters and optimal order [34]. Then Wu et al. proposed a fractional accumulated Bernoulli grey model and adopted an intelligent optimization algorithm to seek optimal fractional order of this model [37]. Optimization of fractional order accumulation is another improvement for the traditional FAGM(1,1) model. Ma et al. firstly proposed a CFGM(1,1) model in which the computational complexity of accumulation is lower than that of traditional fractional accumu- lated operator [52]. Zeng proposed a self-adaptive intelligent fractional grey model called as SAIGMFO model and predicted the electricity consumption of China [53]. Besides, the frac- tional multivariate grey model is another kind of grey model to deal with multivariate time series. Ma et al. proposed a discrete multivariate grey prediction model (FDGM) and mathe- matically proved that it is an unbiased grey model that was applied in four real-life applications and achieved better accuracy compared with other well-known grey models [54]. Though these fractional grey models have gained better performance, boosting prediction accuracy of fractional grey models is still worth studying.

In this paper, we focus on forecasting wind energy consumption by using grey models to provide reference information for formulating and adjusting energy policy. In order to boost the prediction accuracy, we improved the classic fractional grey model with an exponential grey input term. There are two aspects of the contribution. On the one hand, we proposed a novel power-driven fractional accumulated grey model called PFAGM of which optimal order is sought out by WOA algorithm. Meanwhile, PFAGM model can be easily reduced into the classical grey model and fractional grey model. To some extends, PFAGM model has better adaptability than classical grey models. On the other hand, PFAGM model is applied to fore- cast China's wind energy consumption in the next three years.

The rest of this paper is structured as follows. In section, we will introduce the fractional order accumulation and the classic fractional accumulated grey model in detail. In section, a power-driven fractional accumulated grey model (PFAGM) with optimization of grey action quantity is proposed. In section, we validate the prediction accuracy of PFAGM model on several real-life datasets compared with those of three well-known grey models. In section, PFAGM model is utilized to forecast the total wind energy consumption of China. In the last section, several conclusions are drawn.

Fractional accumulated operation and fractional accumulated grey model

The fractional accumulating operation plays a very important role in grey prediction applications. It can be used to decrease the randomness of raw sequence data and boost the performance of grey models. The detail of fractional accumulate operation and its inverse operation are introduced as follows.

Fractional accumulated operation

For an original data sequence $\overline{X^{(0)}} = (x^{(0)}(1), x^{(0)}(2), \cdots, x^{(0)}(n))$, the r-order accumulated operation sequence can be defined as follow.

Definition 1 *The r-order fractional accumulated operation sequence of raw data is defined as:*

$$X^{(r)} = (x^{(r)}(1), x^{(r)}(2), \cdots, x^{(r)}(n)). \tag{1}$$

where $x^{(r)}(k) = \sum_{i=1}^{k} \frac{(k-i+r-1)!}{(r-1)!(k-i)!} x^{(0)}(i), k = 1, 2, \ldots, n$.

Definition 2 *The inverse r-order fractional accumulated operation sequence is defined as*:

$$X^{(-r)} = (x^{(-r)}(1), x^{(-r)}(2), \cdots, x^{(-r)}(n)). \tag{2}$$

where $x^{(-r)}(k) = \sum_{i=1}^{k} \frac{(k-i-r-1)!}{(-r-1)!(k-i)!} x^{(0)}(i), k = 1, 2, \ldots, n.$

Methodology of fractional order accumulated grey model

The fractional order accumulated grey model abbreviated as FAGM(1,1) was firstly proposed by Wu et al. in 2013 [49]. The whitening differential equation of FAGM(1,1) model is defined as:

$$\frac{dx^{(r)}(t)}{dt} + ax^{(r)}(t) = b. \tag{3}$$

where a is the development coefficient, b is the grey input, and r is the fractional order of grey model. Integrating both side of Eq (3) within the interval $[k-1, k]$, the discrete form of FAGM model is obtained as follow:

$$x^{(r)}(k) - x^{(r)}(k-1) + az^{(r)}(k) = b. \tag{4}$$

where $z^{(r)}(k) = 0.5(x^{(r)}(k) + x^{(r)}(k))$.

In order to estimate the parameters of FAGM(1,1) model, the least-square method is used to solve the problem with the objective of minimizing the errors of simulation under the assumption that the order is given. So, the parameters a and b can be calculated as follow:

$$[a, b]^T = (A^T A)^{-1} A^T Y. \tag{5}$$

where

$$Y_u = \begin{bmatrix} x^{(r)}(2) - x^{(r)}(1) \\ x^{(r)}(3) - x^{(r)}(2) \\ \vdots \\ x^{(r)}(u) - x^{(r)}(u) \end{bmatrix}, A_u = \begin{bmatrix} z^{(r)}(2) & 1 \\ z^{(r)}(3) & 1 \\ \vdots & \vdots \\ z^{(r)}(u) & 1 \end{bmatrix}. \tag{6}$$

in which u is the number of in-samples used to build model. Then we solve the differential Eq (3) and have

$$x^{(r)}(t) = Ce^{-at} + \frac{b}{a}. \tag{7}$$

Substituting the initial condition into Eq (7) and setting $t = k$, the time response function of FAGM model is obtained as follow:

$$x_p^{(r)}(k) = \left(x^{(0)}(1) - \frac{b}{a}\right)e^{-a(k-1)} + \frac{b}{a}. \tag{8}$$

where $k = 2, 3, \ldots, n$. However, the time response sequence is only an intermediate result and needs to be restored to restored values by using inverse r-order fractional accumulated operation. The restored value can be represented as:

$$x_p^{(0)}(k) = \sum_{i=1}^{k} \frac{(k-i-r-1)!}{(-r-1)!(k-i)!} x_p^{(r)}(i). \tag{9}$$

where $k = 1, 2, \ldots, n$.

The proposed fractional accumulated grey model

In this section, we prove that the grey action quantity of fractional grey model built by various subsequences with the same length changes with time in a homogeneous exponential sequence. Then a novel power-driven fractional accumulated grey model is proposed, which optimizes the classical FAGM model with an exponential grey action quantity. It makes the grey action quantity change from a constant term to an exponential term of time.

The basis of grey action quantity optimization

The optimization of grey input is a remarkable approach to improve classical grey models and promote their prediction accuracy. The basis of grey action quantity optimization is that the grey input of the grey model changes with time in the real-world grey system. Xu et al. [55] presented a theory to illustrate that the grey action quantity of the classical grey model would change with time. The theory is represented as follow:

Theorem 1 *Assuming that the time series* $X^{(0)}(k) = Ae^{\lambda(k-1)}$, $k = 1, 2, \cdots$ *is raw sequence, there are two subsequence* $X_1^{(0)}(k) = Ae^{\lambda(k-1)}, k = 1; 2; \ldots; n$ *and* $X_2^{(0)}(k) = Ae^{\lambda(k+t-1)}, k = 1; 2; \ldots; n$ *with the same number of samples. The parameters* a_1 *and* b_1 *are respectively the development coefficient and grey input of the GM(1,1) model built by* $X_1^{(0)}$. *The parameters* a_2 *and* b_2 *are respectively the development coefficient and grey input of the GM(1,1) model built by* $X_2^{(0)}$. *Then,* $a_1 = a_2$ *and* $b_2 = b_1 e^{\lambda t}$.

Theorem 1 indicates that the grey action quantity of GM(1,1) changes with time. If the first order term and constant term of Maclaurin's series of $e\lambda t$ are only remained, the SAIGM(1,1) model is obtained with the grey input term $bt + c$. If the first order term of Maclaurin's series of $e^{\lambda t}$ is only remained, the NGM(1,1) model can be obtained with the grey input term bt. Obviously, the grey inputs of SAIGM and NGM are linear time functions. In fact, the grey input of the grey system is nonlinear. Notably, the EOGM(1,1) model has been presented when the term $be\lambda t$ is directly used as its grey action quantity.

For the fractional accumulation grey model, its grey action quantity should vary with time according to the above-mentioned optimization of the classical GM(1,1) model. This proposition can also be illustrated with the theorem as follow:

Theorem 2 *Assuming that the sequence $Y^{(0)} = \{y^{(0)}(k) = Ae^{\lambda}(k-1) | k = 1, 2,\}$ is raw data, there are two subsequence $Y_1^{(0)} = \{y_1^{(0)}(k) = Ae^{\lambda(k-1)} | k = 1, 2, \ldots, n\}$ and $Y_2^{(0)} = \{y_2^{(0)}(k) = Ae^{\lambda(k+t-1)} | k = 1, 2, \ldots, n\}$ with the same number of samples. The parameters a_1 and b_1 are respectively the development coefficient and grey input of the FAGM(1,1) model built by $Y_1^{(0)}$. The parameters a_2 and b_2 are respectively the development coefficient and grey input of the FAGM(1,1) model built by $Y_2^{(0)}$. Then, $a_1 = a_2$ and $b_2 = b_1 e^{\lambda t}$.*

Proof 1 *According to the modeling procedure of FAGM(1,1) model, the fractional order accumulation generated sequence of $Y_1^{(0)}$ is represented as*

$$Y_1^{(r)} = \left\{y_1^{(r)}(k) = \sum_{i=1}^{k} \frac{(k-i-r-1)!}{(-r-1)!(k-i)!} y_1^{(0)}(i) \Big| k = 1, 2, \ldots, n \right\}.$$

The fractional order accumulation generated sequence of $Y_2^{(0)}$ is represented as

$$Y_2^{(r)} = \left\{y_2^{(r)}(k) = \sum_{i=1}^{k} \frac{(k-i-r-1)!}{(-r-1)!(k-i)!} y_2^{(0)}(i) \Big| k = 1, 2, \ldots, n \right\}.$$

The grey differential equation of FAGM(1,1) can be built by using $Y_1^{(0)}$ as follow:

$$y_1^{(r)}(k) - y_1^{(r)}(k-1) + a_1 z_1^{(r)}(k) = b_1, \tag{10}$$

where $z_1^{(r)}(k) = 0.5(y_1^{(r)}(k) + y_1^{(r)}(k-1)), k = 2, 3, \ldots, n$. The grey differential equation of FAGM(1,1) can be built by using $Y_2^{(0)}$ as follow:

$$y_2^{(r)}(k) - y_2^{(r)}(k-1) + a_2 z_2^{(r)}(k) = b_2, \tag{11}$$

where $z_2^{(r)}(k) = 0.5(y_2^{(r)}(k) + y_2^{(r)}(k-1)), k = 2, 3, \ldots, n$. According to the Eqs (5) and (6), the grey parameters a_1 and b_1 can be obtained as follow:

$$a_1 = -\frac{(n-1)\sum_{i=2}^{n} z_1^{(r)}(i)(y_1^{(r)}(i) - y_1^{(r)}(i-1))}{(n-1)\sum_{i=2}^{n}(z_1^{(r)}(i))^2 - \left(\sum_{i=2}^{n} z_1^{(r)}(i)\right)^2} + \frac{\sum_{i=2}^{n} z_1^{(r)}(i) \sum_{i=2}^{n}(y_1^{(r)}(i) - y_1^{(r)}(i-1))}{(n-1)\sum_{i=2}^{n}(z_1^{(r)}(i))^2 - \left(\sum_{i=2}^{n} z_1^{(r)}(i)\right)^2} \tag{12}$$

$$b_1 = -\frac{\sum_{i=2}^{n} z_1^{(r)}(i) \sum_{i=2}^{n} z_1^{(r)}(i)(y_1^{(r)}(i) - y_1^{(r)}(i-1))}{(n-1)\sum_{i=2}^{n}(z_1^{(r)}(i))^2 - \left(\sum_{i=2}^{n} z_1^{(r)}(i)\right)^2}$$

$$+ \frac{\sum_{i=2}^{n}(z_1^{(r)}(i))^2 \sum_{i=2}^{n}(y_1^{(r)}(i) - y_1^{(r)}(i-1))}{(n-1)\sum_{i=2}^{n}(z_1^{(r)}(i))^2 - \left(\sum_{i=2}^{n} z_1^{(r)}(i)\right)^2} \tag{13}$$

In a similar way, the grey parameters a_2 and b_2 can be obtained as:

$$a_2 = -\frac{(n-1)\sum_{i=2}^{n} z_2^{(r)}(i)(y_2^{(r)}(i) - y_2^{(r)}(i-1))}{(n-1)\sum_{i=2}^{n}(z_2^{(r)}(i))^2 - \left(\sum_{i=2}^{n} z_2^{(r)}(i)\right)^2}$$

$$+ \frac{\sum_{i=2}^{n} z_2^{(r)}(i) \sum_{i=2}^{n}(y_2^{(r)}(i) - y_2^{(r)}(i-1))}{(n-1)\sum_{i=2}^{n}(z_2^{(r)}(i))^2 - \left(\sum_{i=2}^{n} z_2^{(r)}(i)\right)^2} \tag{14}$$

$$b_2 = -\frac{\sum_{i=2}^{n} z_2^{(r)}(i) \sum_{i=2}^{n} z_2^{(r)}(i)(y_2^{(r)}(i) - y_2^{(r)}(i-1))}{(n-1)\sum_{i=2}^{n}(z_2^{(r)}(i))^2 - \left(\sum_{i=2}^{n} z_2^{(r)}(i)\right)^2}$$

$$+ \frac{\sum_{i=2}^{n}(z_2^{(r)}(i))^2 \sum_{i=2}^{n}(y_2^{(r)}(i) - y_2^{(r)}(i-1))}{(n-1)\sum_{i=2}^{n}(z_2^{(r)}(i))^2 - \left(\sum_{i=2}^{n} z_2^{(r)}(i)\right)^2} \tag{15}$$

Based on the relation between the sequence $Y_1^{(0)}$ and $Y_2^{(0)}$Þ, the equation between the accumulated sequence $Y_1^{(r)}$ and $Y_2^{(r)}$Þ are obtained as follow:

$$y_2^{(r)}(k) = \sum_{i=1}^{k} \frac{(k-i-r-1)!}{(-r-1)!(k-i)!} y_1^{(0)}(i) e^{\lambda t} = e^{\lambda t} y_1^{(r)}(k) \tag{16}$$

The equation between the sequence $z_1^{(r)}(k)$ and $z_2^{(r)}(k)$ are obtained as follow:

$$z_2^{(r)}(k) = 0.5(y_1^{(r)}(k)e^{\lambda t} + y_1^{(r)}(k-1)e^{\lambda t}) = e^{\lambda t}z_1^{(r)}(k) \qquad (17)$$

Substituting Eqs (16) and (17) into Eqs (14) and (15), the equations can be obtained as follow:

$$a_2 = -\frac{(n-1)\sum_{i=2}^{n}z_1^{(r)}(i)e^{\lambda t}(y_1^{(r)}(i)e^{\lambda t} - y_1^{(r)}(i-1)e^{\lambda t})}{(n-1)\sum_{i=2}^{n}(z_1^{(r)}(i)e^{\lambda t})^2 - \left(\sum_{i=2}^{n}z_1^{(r)}(i)e^{\lambda t}\right)^2}$$

$$+\frac{\sum_{i=2}^{n}z_1^{(r)}(i)e^{\lambda t}\sum_{i=2}^{n}(y_1^{(r)}(i)e^{\lambda t} - y_1^{(r)}(i-1)e^{\lambda t})}{(n-1)\sum_{i=2}^{n}(z_1^{(r)}(i)e^{\lambda t})^2 - \left(\sum_{i=2}^{n}z_1^{(r)}(i)e^{\lambda t}\right)^2} = a_1 \qquad (18)$$

$$b_2 = -\frac{\sum_{i=2}^{n}z_1^{(r)}(i)e^{\lambda t}\sum_{i=2}^{n}z_1^{(r)}(i)e^{\lambda t}(y_1^{(r)}(i)e^{\lambda t} - y_1^{(r)}(i-1)e^{\lambda t})}{(n-1)\sum_{i=2}^{n}(z_1^{(r)}(i)e^{\lambda t})^2 - \left(\sum_{i=2}^{n}z_1^{(r)}(i)e^{\lambda t}\right)^2}$$

$$+\frac{\sum_{i=2}^{n}(z_1^{(r)}(i)e^{\lambda t})^2\sum_{i=2}^{n}(y_1^{(r)}(i)e^{\lambda t} - y_1^{(r)}(i-1)e^{\lambda t})}{(n-1)\sum_{i=2}^{n}(z_1^{(r)}(i)e^{\lambda t})^2 - \left(\sum_{i=2}^{n}z_1^{(r)}(i)e^{\lambda t}\right)^2} = b_1 e^{\lambda t} \qquad (19)$$

This proof is completed.

From Theorem 2, it can be noticed that the grey input varies with time, and the develop- ment coefficient remains unchanged if two different subsequences with the same number of samples employed to construct models for a homogeneous exponential sequence. When the grey input linearly changes with time, the typical fractional grey model is the FAGM(1,1,k) model [34] with the linear grey input term $bt + c$. Moreover, the input of grey system is non- linear in many other cases. The typical fractional grey model is FTDGM model [51] with non- linear grey input ty. Though these models have obtained better prediction performance, the optimization of classical grey models is still required to enhance their adaptability and applicability.

The power-driven fractional accumulated grey model

Optimization of grey action quantity is an effective and common method to improve the grey model. Theorem 2 shows that the grey input should not be a constant while it should vary with time. Meanwhile, Theorem 2 is derived based on homogeneous exponential time series. In fact, more sequences have the non-homogeneous exponential characteristics in the real world. Therefore, the term $be^{rt} + c$ is considered as the grey action quantity of the proposed fractional accumulated grey model. The definition of the proposed model is represented as follows.

Definition 3 *The differential equation*

$$\frac{dx^{(r)}(t)}{dt} + ax^{(r)}(t) = be^{rt} + c. \tag{20}$$

is the whitening equation of power-driven fractional order accumulated grey model abbreviated as PFAGM, in which a is defined the same in FAGM model, bert + c is the power-driven grey action quantity, and r is the fractional order of this grey model.

Integrating the whitening Eq (20) within $[k - 1, k]$, we have

$$\int_{k-1}^{k} \frac{dx^{(r)}(t)}{dt} dt + a \int_{k-1}^{k} x^{(r)}(t) dt = \int_{k-1}^{k} (be^{rt} + c) dt. \tag{21}$$

Then we have

$$x^{(r)}(k) - x^{(r)}(k-1) + a \int_{k-1}^{k} x^{(r)}(t) dt = br^{-1}(e^r - 1)e^{r(k-1)} + c. \tag{22}$$

Substituting the background value $z^{(r)}(k) = \int_{k-1}^{k} x^{(r)}(t)\, dt$ into Eq (22), the discrete form of PFAGM model can be obtained as follow.

Definition 4 *The discrete differential equation of PFAGM model is defined as:*

$$x^{(r)}(k) - x^{(r)}(k-1) + az^{(r)}(k) = br^{-1}(e^r - 1)e^{r(k-1)} + c. \tag{23}$$

where $z^{(r)}$ is the background value, and $z^{(r)}(k) = 0.5x^{(r)}(k) + 0.5x^{(r)}(k-1)$.

When the parameter b of PFAGM model is set to 0, PFAGM model can be reduced into FAGM(1,1) model. When the parameter b is set to 0 and the fractional order is set to 1, PFAGM model can be degenerated into GM(1,1) model. Because $e^{\gamma t} \approx 1 + rt$, PFAGM model can be degenerated into FAGM(1,1,k) model if the term $1 + rt$ replaces the term $e^{\gamma t}$ of the grey input in PFAGM model.

Parameter estimation of power-driven factional grey model

To realize the prediction of the fractional grey model, one of the most important problems is parameter estimation for building a model. In fact, it is effective to boost the prediction performance of grey model by using suitable parameters. For PFAGM model, it needs to determine three linear parameters and search out the optimal value of its fractional order. The optimal linear parameters can be gained by using the least-square method under the condition of the given fractional order of PFAGM model. In this subsection, we mainly introduce the principle of linear parameters' estimation while the methodology of seeking out the optimal fractional order is presented in subsection.

Once the order of PFAGM model is given, the linear parameters of the PFAGM model can mathematically be calculated as:

$$(a, b, c)^T = (B_u^T B_u)^{-1} B_u^T Y_u. \tag{24}$$

where

$$Y_u = \begin{bmatrix} x^{(r)}(2) - x^{(r)}(1) \\ x^{(r)}(3) - x^{(r)}(2) \\ \vdots \\ x^{(r)}(u) - x^{(r)}(u-1) \end{bmatrix}, B_u = \begin{bmatrix} -z^{(r)}(2) & r^{-1}(e^r - 1)e^r & 1 \\ -z^{(r)}(3) & r^{-1}(e^r - 1)e^{2r} & 1 \\ \vdots & \vdots & \vdots \\ -z^{(r)}(u) & r^{-1}(e^r - 1)e^{(u-1)r} & 1 \end{bmatrix}. \tag{25}$$

in which u is the number of samples for fitting. The proof process of parameter estimation is omitted here because it is like that of traditional FAGM model.

The time response function and restored values

After linear parameters of PFAGM model are calculated, the time response function and the restored values of PFAGM can be got by solving Eq (20). Firstly, we consider the homogeneous differential equation:

$$\frac{dx^{(r)}(t)}{dt} + ax^{(r)}(t) = 0. \tag{26}$$

which is corresponding to the whitening equation of PFAGM model and solve Eq (26). Then the solution of Eq (26) can be obtained as:

$$x^{(r)}(t) = Ae^{-at}. \tag{27}$$

where A is a constant which is determined by initial condition. Let $x^{(r)}(t) = A(t)e^{-at}$ be the solution of Eq (20). And substituting it into Eq (20), we have

$$\frac{dA(t)}{dt} e^{-at} = be^{rt} + c. \tag{28}$$

Then the solution of Eq (28) can be easily obtained as:

$$A(t) = \frac{b}{a+r} e^{(a+r)t} + \frac{c}{a} e^{at} + C. \tag{29}$$

So, the solution of the whitening equation of PFAGM model can be represented as:

$$x^{(r)}(t) = Ce^{-at} + \frac{b}{a+r}e^{rt} + \frac{c}{a}. \tag{30}$$

Substituting the initial condition $x^{(r)}(1) = x^{(0)}(1)$ into Eq (30), the arbitrary constant can be easily determined. And then the time response function of PFAGM model can be derived and represented as:

$$x_p^{(r)}(k) = \left(x^{(0)}(1) - \frac{b}{a+r}e^r - \frac{c}{a}\right)e^{-a(k-1)} + \frac{b}{a+r}e^{rk} + \frac{c}{a}. \tag{31}$$

By using inverse r-order accumulated operation, the restored value $x_p^{(0)}(k)$ can be obtained as:

$$x_p^{(0)}(k) = (x_p^{(r)}(k))^{(-r)}. \tag{32}$$

where $k = 1, 2, \ldots, n$.

Searching the optimal order of power-driven fractional accumulated grey model

In subsection, the linear parameters are estimated by the least-square method under the assumption that the order of fractional power-driven grey model has been given. In fact, the order of PFAGM model needs to be sought out and plays a vital role in boosting its prediction performance effectively. In order to search for an optimal value of the order, we establish a constrained optimization problem of which the objective is minimizing the mean absolute per- centage errors for building the grey model. The constraints of PFAGM model have been derived in the above-mentioned modeling process. In summary, the optimization problem can be represented as follow:

$$\min MAPE(r) = \frac{1}{u-1}\sum_{i=2}^{u}\left|\frac{x_p^{(0)}(i) - x^{(0)}(i)}{x^{(0)}(i)}\right| \times 100\%.$$

$$s.t. \begin{cases} (a,b,c)^T = (B_u^T B_u)^{-1} B_u^T Y_u. \\ x_p^{(r)}(k) = \left(x^{(0)}(1) - \frac{b}{a+r}e^r - \frac{c}{a}\right)e^{-a(k-1)} + \frac{b}{a+r}e^{rk} + \frac{c}{a}. \\ x^{(0)}(k) = (x_p^{(r)})^{(-r)}, k = 2, 3, \ldots, u. \end{cases} \tag{33}$$

where B_u and Y_u are defined as Eq (25). It can be clearly noticed that the optimization problem Eq (33) is a complex nonlinear programming problem with nonlinear objective function and a few nonlinear constraints. Obviously, it is difficult to derive the exact solution for optimal-fractional order. However, the optimal value of fractional order plays a very important roles in boosting the prediction accuracy of existing grey models dramatically. For example, Ma et al. established a similar optimization problem to optimize the fractional order of FTDGM model and applied it to forecast energy consumption accurately [51]. Wu et al. also build a similar optimization problem to obtain the optimal fractional order and power index of FANGBM and forecasted the renewable energy consumption successfully [37]. These facts have shown that the optimal fractional order can be used to boost the prediction performance of grey models.

In order to obtain the optimal order of PFAGM, we adopted a nature-based heuristic intel- ligent method called Whale Optimization Algorithm (WOA) to solve the nonlinear program- ming problem Eq (33). WOA algorithm was firstly proposed by Mirjalili et al. in 2016 [56].

The inspiration of WOA is from the social behaviors of whales. It mimics the bubble-net feeding strategy including shrinking encircling, spiral updating position and randomly hunting behavior. Due to the simple rules and local optimization performance of WOA, it is widely used in many domains such as feature selection, clustering, classification, image processing and so on [57]. In this paper, it is assumed by WOA that there is a population with 30 hump- back whales as search agents in search space. $P(k)$ indicates the position vector of search agents at iteration k. $P^*(k)$ represents the candidate solution which is the best one or near to the optimum. The procedures of searching optimal fractional order with WOA are depicted as follow:

Step 1: Initialize the initial position $P(k)$ of search agent randomly in search space at $k = 1$ in search space. And set maximum iterations T_m to be 300. The value of each agent's position indicates a possible fractional order of PFAGM model.

Step 2: Compute the fitness of each agent and obtain the candidate solution $P^*(k)$ at $k = 1$.

According to Eq 33, the fitness function is defined as:

$$Fitness = \frac{1}{u-1} \sum_{i=2}^{u} \left| \frac{x_p^{(0)}(i) - x^{(0)}(i)}{x^{(0)}(i)} \right| \times 100\%. \tag{34}$$

The position of search agent with minimum fitness is considered as the candidate solution at first iteration.

Step 3: Update the position vectors of humpback whales. When humpback whales forage the prey, they usually move around the prey by shrinking encircling and spiral updating posi- tion simultaneously. In order to imitate this simultaneous behavior, we assume that there is a 50% probability to shrink encircling or update position spirally in the iteration of optimi- zation. In each iteration, it generates a random number in [0, 1] and set it to the parameter p. If $p \leq 0.5$, humpback whales select to shrink encircling. Mathematically, the behavior of shrinking encir- cling can be represented as follows:

$$\vec{A} = 2g(k) \cdot \vec{r} - g(k). \tag{35}$$

$$\vec{B} = \left| 2 \cdot \vec{r} \cdot \vec{P}^*(k) - \vec{P}(k) \right|. \tag{36}$$

$$\vec{P}(k+1) = \vec{P}^*(k) - \vec{A} \cdot \vec{B}. \tag{37}$$

$$g(k) = 2 - 2k/T_m. \tag{38}$$

where $\vec{r}$ is a random vector in interval [0, 1]. If $p > 0.5$, the humpback whales will select spi- ral updating position to imitate the helix-shaped moving behavior in process of optimiza- tion. The behavior is shown mathematically as follow:

$$\vec{P}(k+1) = \left|\vec{P}^*(k) - \vec{P}(k)\right| \cdot e^{\omega l} \cdot \cos(2\pi l) + \vec{P}^*(k). \tag{39}$$

where l denotes a random value in [−1, 1], ω is a constant. In fact, the humpback whales search for the prey randomly. If $|\vec{A}| > 1$, WOA algorithm imitates the behavior and performs a global search. Mathematically, the model is formulated as follows:

$$\vec{C} = |2 \cdot r \cdot \vec{P}_{rand}(k) - \vec{P}(k)|. \tag{40}$$

$$\vec{P}(k+1) = \vec{P}_{rand}(k) - \vec{A} \cdot \vec{C}. \tag{41}$$

where $\vec{P}_{rand}(k)$ is the position of a random search agent.

Step 4: Calculate the fitness of each humpback whales. If there is a better candidate with mini- mum fitness value, it is needed to update the optimal solution $\vec{P}^*$.

Step 5: If the ending condition is satisfied, the optimal fractional order is obtained, otherwise $k = k + 1$.

Based on the process of building PFAGM model and optimization of the fractional order, the detailed modeling procedure of power-driven fractional accumulated grey model can be shown in Fig 1. For enhancing the prediction performance of PFAGM model, the key is to obtain the optimal values of linear parameters and fractional order. The optimal value of frac- tional order can be sought out through solving the established optimization problem by WOA algorithm. Moreover, the linear parameters are estimated by using the least-square method. Then the estab- lished PFAGM model is used to predict future values in different applications.

Applicability analysis of power-driven fractional grey model

Though the classical grey models are effective approaches to predict future values with small samples, each of these models has a certain scope of application. Similarly, the proposed pow- er-driven fractional grey model also has its scope of application. Therefore, we construct a se- ries of sequences, including two kinds of non-homogeneous exponential sequences and some other sequences with special shapes, to study the applicability of the proposed grey model. There are three cases studied and compared with GM(1,1), FAGM(1,1), and FAGM (1,1,k) as follows.

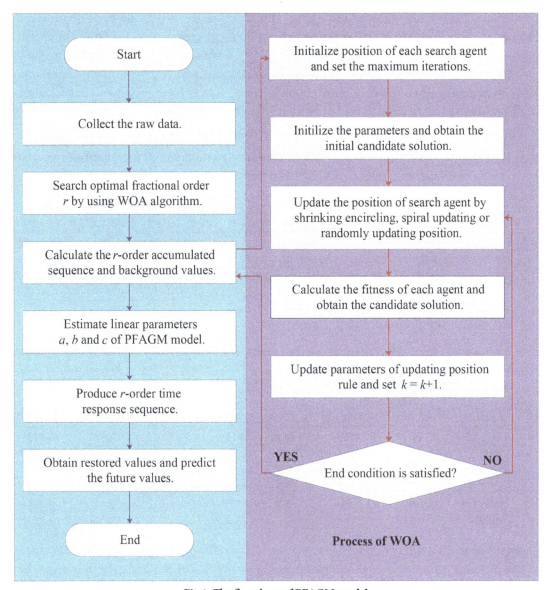

Fig 1. The flowchart of PFAGM model.

Firstly, we construct a series of non-homogeneous exponential sequences with the larger values of the development coefficient. These sequences are represented as $x(0)(k) = Ae\lambda t + B$ ($k = 1, 2, \ldots, 9$) in which A is set to 1, B is set to 10, and a is set to 0.7, 1.0, 1.3, 1.5, 1.7 or 1.8. The first six digits of each sequence are used to train the grey models. The other three digits of each sequence are employed to examine the prediction performance of the proposed grey model. The MAPEs of different grey models for fitting and prediction are filled in Table 1. It can be noticed that PFAGM model obtains a significant advantage of prediction accuracy over the other three classical grey models. Meanwhile, PFAGM model has significant advantages in fitting. Secondly, we construct the other kind of non-homogeneous exponential sequences rep- resented as $x^{(0)}(k) = Ae^{\lambda t} + B(k-1)$ ($k = 1, 2, \ldots, 9$) in which A is set to 1, B is set to 5, and a is set to 0.7, 1.0, 1.3, 1.5, 1.7 or 1.8. Similarly, the first six digits of each sequence are used to train the grey models. The other three digits of each sequence are employed to examine the prediction performance of the proposed grey model. Table 2 shows the MAPEs of different grey models for fitting and prediction. The results indicate that PFAGM model has a significant advantage of fitting ability over the other three classical grey models. PFAGM model also has a significant superiority of prediction accuracy, especially when the development coeffi- cient is large. From the above analysis results, it indicates that PFAGM model has better adapt- ability and applicability for the two kinds of non-homogeneous exponential sequences with the larger values of the development coefficient.

Table 1. Comparison of different grey models for non-homogeneous exponential sequences with various development coefficients ($X^{(o)}(k) = Ae^{\lambda(k-1)} + B$, $k = 1, 2, ., n$).

λ	MAPE of Fitting				MAPE of Prediction			
	GM(1,1)	FAGM(1,1)	FAGM(1,1,k)	PFAGM	GM(1,1)	FAGM(1,1)	FAGM(1,1,k)	PFAGM
0.7	8.0214	1.4179	2.0788	0.3766	43.4858	5.1797	26.5822	3.9745
1	69.4463	4.6622	5.2342	3.38E-07	88.7597	19.0313	27.3230	1.61E-06
1.3	188.7666	10.4601	11.3918	1.0518	180.0536	46.3293	45.6655	0.9277
1.5	237.1642	15.7192	17.0864	1.9898	193.0147	59.6113	59.0422	2.3112
1.7	249.5205	22.4229	24.1564	3.0308	171.0404	71.1620	70.8141	3.9040
1.8	246.3506	26.4090	28.2787	3.5722	156.4693	76.0759	75.8225	4.7317

Table 2. Comparison of different grey models for non-homogeneous exponential sequences with various development coefficients ($X^{(o)}(k) = Ae^{\lambda(k-1)} + B(k-1)$, $k = 1, 2, ., n$).

λ	MAPE of Fitting				MAPE of Prediction			
	GM(1,1)	FAGM(1,1)	FAGM(1,1,k)	PFAGM	GM(1,1)	FAGM(1,1)	FAGM(1,1,k)	PFAGM
0.7	5.4555	2.7347	2.4818	0.8435	14.2598	22.5767	14.4327	4.6936
1	26.9863	13.4895	9.2376	2.4296	54.5762	44.7105	36.2111	23.0307
1.3	111.0393	10.5668	9.9856	3.6362	119.3470	29.8071	29.0046	15.3936
1.5	163.5235	9.7733	9.6019	3.9296	144.4912	45.4235	45.0679	9.4585
1.7	190.9126	13.5743	13.7557	3.8576	142.1207	58.5609	58.0279	4.6953
1.8	195.8534	16.2803	16.0250	3.7240	135.6079	72.4797	72.4032	3.0292

Finally, we generate a series of non-homogeneous exponential sequences with different characteristics randomly. The raw data of each sequence is tabulated in Table 3. Fig 2 exhibits the characteristics of these generated sequences. These sequences are divided into two groups. The first group, including the first eight digits of each sequence, is used to build grey models. The second group, including the other four digits of each sequence, is utilized to validate the prediction performance of the proposed grey model. The MAPEs of the grey models for fitting and prediction are filled in Table 4. It can be noticed that the prediction accuracy of the pro- posed grey model is better than those of the other three grey models. The fitting errors of the proposed grey model are the lowest or very approximate to the lowest.

From the above validations and analysis, it can be drawn that PFAGM model has a significant superiority to deal with the non-homogeneous exponential sequences (formu- lated as $X^{(0)}(k) = Ae^{\lambda(k-1)} + B$) or $X^{(0)}(k) = Ae^{\lambda(k-1)} + B(k-1)$) compared with the classical GM(1,1), FAGM(1,1) and FAGM(1,1,k) model. Meanwhile, PFAGM model has a certain advantage to handle the above-mentioned special sequences. Therefore, PFAGM model has better adaptability and applicability than the other classical grey models for non-homoge- neous exponential sequences with larger development coefficients or special shape characteristics.

Table 3. Raw data of special non-homogeneous exponential sequences.

Index	S1	S2	S3	S4	S5	S6	S7	S8	S9
1	10	10	10	10	10	10	1	1	1
2	13.6138	13.7664	16.7975	8.9671	7.8421	5.9371	1.32435	1.6487	6.6487
3	16.4143	13.4857	19.1673	8.4053	6.9623	4.7812	1.85915	2.7183	12.7183
4	18.7775	12.6666	19.7109	8.0436	6.5948	4.359	2.74085	4.4817	19.4817
5	20.8251	12.0434	19.4351	7.8582	6.5571	4.2515	4.19455	7.3891	27.3891
6	22.6213	11.7834	18.7967	7.8306	6.7771	4.3335	6.59125	12.1825	37.1825
7	24.2093	11.9347	18.0058	7.9513	7.2224	4.5737	10.54275	20.0855	50.0855
8	25.6219	12.5487	17.153	8.2224	7.8808	4.9824	17.05775	33.1155	68.1155
9	26.8853	13.7188	16.2686	8.658	8.7532	5.5959	27.7991	54.5982	94.5982
10	28.0212	15.5993	15.3514	9.2834	9.85	6.4732	45.50855	90.0171	135.0171
11	29.0477	18.4248	14.3833	10.1355	11.19	7.6988	74.7066	148.4132	198.4132
12	29.9801	22.5364	13.3348	11.2646	12.7997	9.3895	122.846	244.6919	299.6919

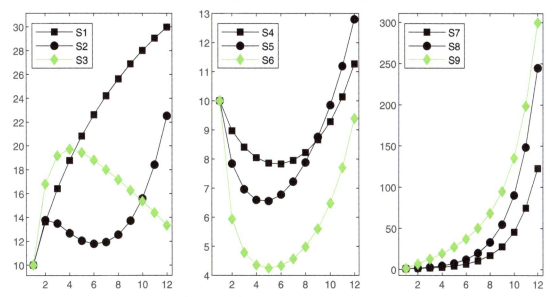

Fig 2. Different non-homogeneous exponential sequences.

Validation of the power-driven grey model

Performance metrics and comparative grey models

In order to evaluate the accuracy of grey prediction models, four performance metrics are adopted in numerical validation and case studies, including residual percentage error (RPE), absolute percentage error (APE), mean absolute percentage error (MAPE) and Correlation coefficient (R)

[58, 59]. The metrics RPE and APE are used to validate the prediction accuracy of a grey model for a single data point. Mathematically, they are represented as follows:

$$RPE(k) = \frac{x_p^{(0)}(k) - x^{(0)}(k)}{x^{(0)}(k)} \times 100\%. \tag{42}$$

$$APE(k) = |RPE(k)|. \tag{43}$$

Table 4. Comparison of different grey models for special non-homogeneous exponential sequences.

a	MAPE of Fitting				MAPE of Prediction			
	GM(1,1)	FAGM(1,1)	FAGM(1,1,k)	PFAGM	GM(1,1)	FAGM(1,1)	FAGM(1,1,k)	PFAGM
S1	3.1505	0.0471	0.0311	0.0022	18.4623	0.2208	0.2560	**0.0221**
S2	2.7926	2.0018	1.4507	0.4060	34.0582	30.6134	25.5486	**6.6880**
S3	4.4282	0.2048	0.1991	0.1987	22.2005	2.2671	1.0236	**0.5911**
S4	2.5212	0.4465	0.0444	0.0236	22.5855	28.3267	1.9441	**1.2779**
S5	5.2952	0.6137	0.0218	0.1080	30.1755	16.2743	1.7891	**0.6995**
S6	7.7229	0.5409	0.1851	0.2357	41.5405	3.4184	10.1779	**1.8713**
S7	15.0169	1.5770	2.2206	0.5789	25.0665	5.1843	8.6351	**2.3063**
S8	4.7913	1.8546	2.3858	0.2575	10.6270	5.3084	7.8210	**0.7552**
S9	9.0649	2.1075	1.0262	0.5335	5.8343	19.3921	6.9563	**4.4094**

where $x^{(0)}(k)$ is raw data, $x_p^{(0)}$ is the value produced by a grey model. Meanwhile, MAPE is used as a general metric to evaluate the accuracy of a prediction model. The lower value of MAPE indicates that the model has a better performance. Mathematically, MAPE is repre sented as follow:

$$MAPE = \frac{1}{n}\sum_{k=1}^{n}|APE(k)|. \tag{44}$$

where n is the number of samples. For a raw sequence with n samples, the metric FMAPE is defined as the simulation performance metric while PMAPE is defined as a prediction performance metric. Mathematically, they can be formulated as:

$$FMAPE = \frac{1}{v}\sum_{k=1}^{v}|APE(k)|. \tag{45}$$

$$PMAPE = \frac{1}{n-v}\sum_{k=v+1}^{n}|APE(k)|. \tag{46}$$

where v is the number of samples used to build a model while the rest of the raw sequence is used to examine the prediction accuracy of the model. The total MAPE is used to evaluate the whole performance of a model and abbreviated as TMAPE. The correlation coefficient (R) is used to describe the relationship between raw sequence $X^{(0)}$ and the sequence $X_p^{(0)}$ produced by grey models. The mathematical definition of R is represented as:

$$R = \frac{Cov(X^{(0)}, X_p^{(0)})}{\sqrt{Var(X^{(0)})Var(X_p^{(0)})}}. \tag{47}$$

where $Cov(s, t)$ denotes the covariance of sequence s and t, $Var(s)$ denotes the variances of sequence s.

For investigating its superiority of performance, PFAGM model compares with three existing grey models, including traditional integral-order grey model (GM(1, 1)), fractional accumulated grey model (FAGM(1,1)) and the improved fractional accumulated grey model (FAGM(1,1,k)). The whitening equation of integral-order GM(1,1) model [38] is represented as:

$$\frac{dx^{(1)}(t)}{dt} + ax^{(1)}(t) = b. \tag{48}$$

The whitening equation of FAGM(1,1,k) [34] is written as:

$$\frac{dx^{(r)}(t)}{dt} + ax^{(r)}(t) = bt + c. \tag{49}$$

The details of FAGM(1,1) model are introduced in section. All grey prediction models, including the new proposed model and the three contrast grey models, are implemented in MATLAB and performed on the platform MATLAB 2018. In particular, it needs to highlight that the orders of the two comparative fractional grey models are also sought out by the optimization algorithm WOA which also is applied in our proposed PFAGM model. In the following subsections, the validation experiments on some real-world data sets are conducted to illustrate the advantages of PFAGM compared with the other three existing grey models.

Table 5. Raw data in Example A.

Year	NEC	Year	NEC	Year	NEC
2006	12.4	2010	16.7	2014	30
2007	14.1	2011	19.5	2015	38.6
2008	15.5	2012	22	2016	48.3
2009	15.9	2013	25.3	2017	56.1

Example A: Forecasting China's nuclear energy consumption

In this numerical validation, the raw sequence containing Chinese nuclear energy consumption (NEC) from 2006 to 2017 is obtained from section of reference [34]. The dataset is tabulated in Table 5. The consumption from 2006 to 2012 is used to build a model respectively for PFAGM model and the other three contrast grey models, while the remainder of the dataset is used to validate the prediction performance of grey models. Firstly, we obtain the optimal orders of the three fractional grey models by using WOA algorithm. Then the linear parame- ters of all grey models are estimated by using the least-square method. The optimal values of linear and nonlinear parameters for each model are filled in Table 6. From Fig 3, it can be noticed that WOA algorithm converges rapidly into a stable status after a few iterations. And the optimal order of PFAGM model is obtained and equal to 0.24794. The results produced by these established grey models are also filled in Table 7. FMAPE, PMAPE, and TMAPE of PFAGM model are respectively 1.2024, 2.8591, and 1.8927. As can be noticed from Fig 4 and Table 7, all evaluation metrics' values of PFAGM model are lowest compared with those of the other three contrast grey models. Fig 5 shows the detailed analysis between the raw data and the values produced by the four grey models. It can be clearly found that the linear regression line obtained from PFAGM model almost coincides with the equal line of which the point denotes that raw data is equal to the value produced by the grey model. It indicates that PFAGM model has a better prediction performance than the other three contrast models.

Example B: Forecasting cumulative oil field production

In this subsection, we consider forecasting the cumulative oil field production of which the raw data is collected from the paper [60] as a validation example. The raw data contains the cumulative oil field production from 1999 to 2012 of the RQ block of Huabei oil field company in China. The raw data is tabulated in Table 8. Then the dataset is divided into two subsets.

Table 6. The optimal parameters of different grey models in Example A.

Parameters	GM(1,1)	FAGM(1,1)	FAGM(1,1,k)	PFAGM
a	-0.08907	-0.50681	0.07111	0.80372
b	11.94888	-6.15082	0.67748	3.86999
c	-	-	1.99915	11.19314
γ	1	-0.11349	-0.2949	0.24794

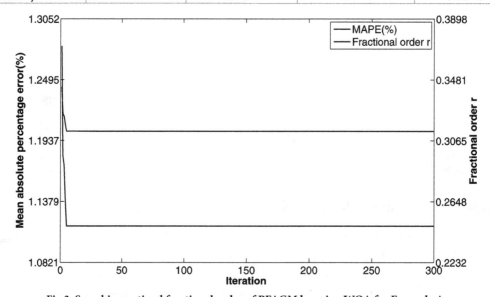

Fig 3. Searching optimal fractional order of PFAGM by using WOA for Example A.

The first subset, including the first 11 samples, is used to build a model for each grey model. The other subset, including the last 3 samples, is utilized to examine the accuracy of grey mod- els. In the stage of building model, the optimal orders of all fractional grey models are obtained by using WOA algorithm. From Fig 6, it can be obviously noticed that the optimal order of PFAGM is sought out after a few iterations of WOA. And then the linear parameters of these models are also obtained by using the least-square method after their orders are determined. The parameters of the integral order grey model can be directly estimated by using the least- square method. These optimal parameters are filled in Table 9. Then we utilize these different established grey models to calculate the fitted production from 1999 to 2009 and the predicted production from 2010 to 2012, which are filled in Table 10. As can be noticed from Table 10 and Fig 7, the proposed PFAGM model exhibits the most excellent prediction performance compared with those of the other contrast grey models. Though FMAPE of PFAGM model is not best, its PMAPE and TMAPE are superior to the other three contrast models. R of FMAPE is also best among those of these grey models. Meanwhile, it is shown that results produced by PFAGM almost approximate to the real value, and the regression line is almost coincident with the equal line for raw sequence and the produced sequence by the model in Fig 8. Above all, PFAGM model is slightly superior to the other three grey models in the aspect of prediction accuracy.

Table 7. The results produced by proposed model and other comparative grey models in Example A.

Year	Raw data	GM(1,1)		FAGM(1,1)		FAGM(1,1,k)		PFAGM	
		Value	RPE	Value	RPE	Value	RPE	Value	RPE
2006	12.4	12.4	0	12.4	0	12.4	0	12.4	0
2007	14.1	13.6523	3.1752	13.9812	0.8422	14.2607	-1.1397	14.1	0
2008	15.5	14.9241	3.7156	15.0733	2.7527	15.0798	2.7112	15.0037	3.2022
2009	15.9	16.3143	-2.6059	16.1486	-1.5633	15.9525	-0.3303	15.9486	-0.3059
2010	16.7	17.8341	-6.7911	17.4546	-4.5184	17.2798	-3.4716	17.3112	-3.6598
2011	19.5	19.4955	0.0233	19.2689	1.1854	19.259	1.2359	19.2864	1.0955
2012	22	21.3116	3.1293	22	0	22	0	22.0337	-0.1531
2013	25.3	23.2969	7.9176	26.3009	-3.9563	25.5662	-1.0521	25.736	-1.7235
2014	30	25.4671	15.1097	33.2414	-10.8046	29.9935	0.0215	30.6283	-2.0943
2015	38.6	27.8395	27.877	44.5882	-15.5136	35.3004	8.5482	37.0175	4.0998
2016	48.3	30.4329	36.992	63.269	-30.9917	41.4932	14.0928	45.3032	6.2045
2017	56.1	33.2679	40.699	94.1392	-67.806	48.5701	13.4223	56.0027	0.1734
FMAPE		2.7772		1.5517		1.2698		1.2024	
PMAPE		25.7191		25.8144		7.4274		**2.8591**	
TMAPE		12.3363		11.6612		3.8355		**1.8927**	
R		0.9642		0.9761		0.9959		**0.998**	

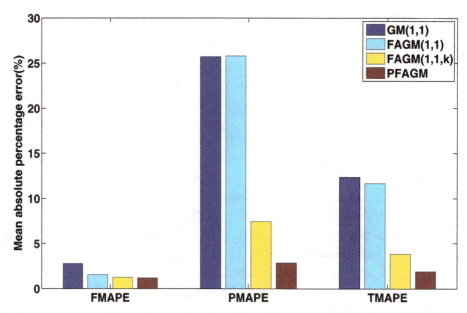

Fig 4. Performance comparison of the proposed model and other comparative grey models in Example A.

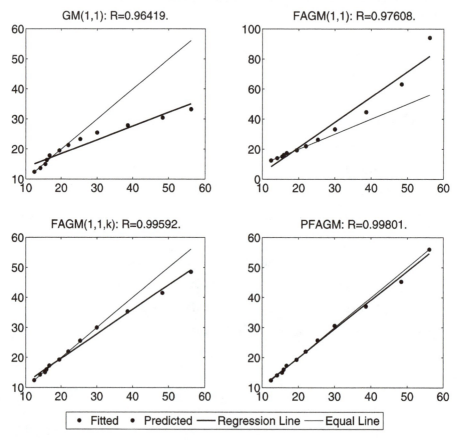

Fig 5. Analysis of detailed results obtained by using the proposed model and other comparative grey models in Example A.

Table 8. Raw data of cumulative oil field production in example B.

Year	Oil production	Year	Oil production	Year	Oil production
1999	73.8217	2004	342.6394	2009	519.8508
2000	136.8817	2005	382.4312	2010	552.6569
2001	195.059	2006	420.0399	2011	581.6092
2002	247.8547	2007	454.043	2012	608.1863
2003	297.0902	2008	485.1171		

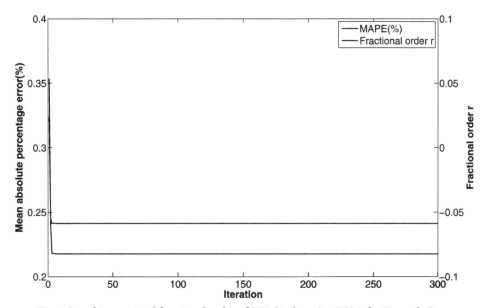

Fig 6. Searching optimal fractional order of PFAGM by using WOA for Example B.

Table 9. The optimal parameters of different grey models in Example B.

Parameters	GM(1,1)	FAGM(1,1)	FAGM(1,1,k)	PFAGM
a	-0.1142	0.0689	0.0513	0.3618
b	180.4262	73.4323	81.0772	-172.2099
c	-	-	31.5459	246.4802
γ	1	0.04046	1.12273	-0.08235

Table 10. The results produced by proposed model and other comparative grey models in Example B.

Year	Raw data	GM(1,1)		FAGM(1,1)		FAGM(1,1,k)		PFAGM	
		Value	RPE	Value	RPE	Value	RPE	Value	RPE
1999	73.8217	73.8217	0	73.8217	0	73.8217	0	73.8217	0
2000	136.8817	200.0595	46.155	136.8817	0	136.8817	0	136.8817	0
2001	195.059	224.2578	14.9692	194.4272	-0.3239	195.7859	0.3727	194.6208	-0.2247
2002	247.8547	251.3831	1.4236	247.2619	-0.2392	248.754	0.3628	247.4707	-0.1549
2003	297.0902	281.7893	-5.1502	295.9235	-0.3927	297.0427	-0.016	295.9658	-0.3785
2004	342.6394	315.8733	-7.8117	340.8246	-0.5296	341.3715	-0.37	340.6058	-0.5935
2005	382.4312	354.08	-7.4134	382.305	-0.033	382.2394	-0.0501	381.8222	-0.1592

2006	420.0399	396.908	-5.5071	420.656	0.1467	420.025	-0.0036	419.9784	-0.0146
2007	454.043	444.9163	-2.0101	456.1335	0.4604	455.0314	0.2177	455.3797	0.2944
2008	485.1171	498.7315	2.8064	488.9656	0.7933	487.5107	0.4934	488.2846	0.6529
2009	519.8508	559.0559	7.5416	519.3578	-0.0948	517.6779	-0.418	518.9141	-0.1802
2010	552.6569	626.6769	13.3935	547.4965	-0.9337	545.7198	-1.2552	547.4598	-0.9404
2011	581.6092	702.477	20.7816	573.5516	-1.3854	571.8011	-1.6864	574.0897	-1.2929
2012	608.1863	787.4456	29.4744	597.6785	-1.7277	596.0686	-1.9924	598.9529	-1.5182
FMAPE		9.1626		0.274		**0.2095**		0.2412	
PMAPE		21.2165		1.349		1.6447		**1.2505**	
TMAPE		11.7456		0.5043		0.517		**0.4575**	
R		0.9657		0.9998		0.9998		**0.9999**	

Example C: Forecasting foundation settlement

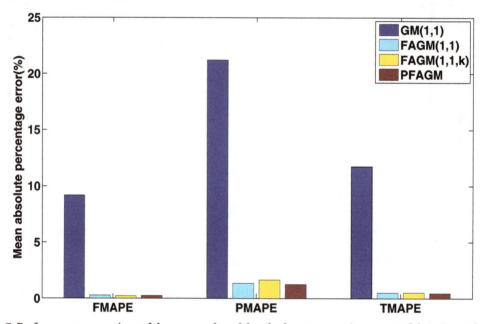

Fig 7. Performance comparison of the proposed model and other comparative grey models in Example B.

In this validation, the raw data of foundation settlement during engineering construction is obtained from the paper [61], which is tabulated in Table 11. According to the previous prac- tices, we divide the raw into two groups. The first one consists of the first eight digits of foun- dation settlement, which are used to build grey models for the proposed model and the contrast grey models. The second one consists of the last two digits used to validate the predic- tion accuracy of each grey model. Firstly, the optimal orders of the fractional accumulated grey models are found out by using WOA algorithm. Then their linear parameters are estimated by using the least-square method under their determined fractional order. The parameters of GM (1,1) model are obtained by the least-square method directly. These optimal parameters are listed in Table 12. In Fig 9, it is shown that the convergence curve of WOA declines rapidly and then stays a constant stably after about 20 iterations. The results produced by these grey models are also tabulated in Table 13. From Table 13 and Fig 10, it can be apparently found that PFAGM model has the lowest PMAPE and TMAPE compared with the other three con- trast grey models and has the second-lowest FMAPE which is almost approximate to the low- est FMAPE of FAGM(1,1,k) model. Meanwhile, it can be clearly seen from Fig 11 that the regression line between raw data and calculated values is almost coincident with their equal line. In brief, it can be concluded that PFAGM model has better prediction performance than the other three grey models in this example.

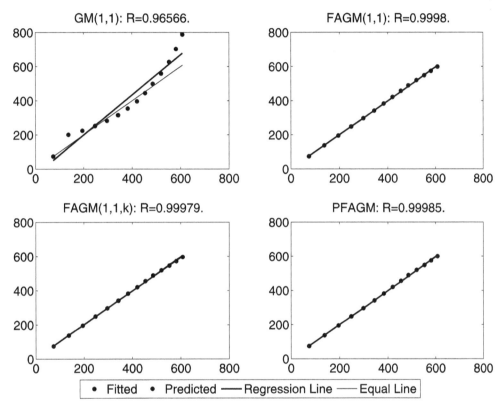

Fig 8. Analysis of detailed results obtained by using the proposed model and other comparative grey models in Example B.

Table 11. Raw data of foundation settlement in Example C.

Index	Observed value	Index	Observed value	Index	Observed value
1	43.19	6	99.73	9	112.19
2	58.73	7	105.08	10	113.45
3	70.87	8	109.73		
4	83.71	5	92.91		

Table 12. The optimal parameters of different grey models in Example C.

Parameters	GM(1,1)	FAGM(1,1)	FAGM(1,1,k)	PFAGM
a	-0.0920	0.1274	0.0421	0.0126
b	59.2681	30.1302	-1.3201	42.9745
c	-	-	15.6763	-25.9805
γ	0.95251	0.99867	0.99978	0.99980

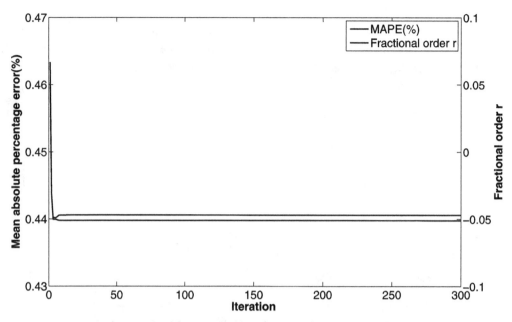

Fig 9. Searching optimal fractional order of PFAGM by using WOA for Example C.

Table 13. The results produced by proposed model and other comparative grey models in Example C.

Index	Raw data	GM(1,1)		FAGM(1,1)		FAGM(1,1,k)		PFAGM	
		Value	RPE	Value	RPE	Value	RPE	Value	RPE
1	43.19	43.19	0	43.19	0	43.19	0	43.19	0
2	58.73	66.243	12.7925	58.8407	0.1884	58.73	0	58.6616	-0.1164
3	70.87	72.6275	2.4798	72.1069	1.7453	71.7718	1.2724	71.8027	1.3161
4	83.71	79.6272	-4.8773	82.9702	-0.8838	82.8814	-0.9899	82.9569	-0.8997
5	92.91	87.3016	-6.0364	91.8531	-1.1376	92.1827	-0.7828	92.2435	-0.7173
6	99.73	95.7156	-4.0253	99.1141	-0.6176	99.7299	-0.0001	99.7491	0.0191
7	105.08	104.9405	-0.1327	105.0387	-0.0393	105.5627	0.4593	105.5524	0.4496
8	109.73	115.0546	4.8524	109.8553	0.1141	109.7181	-0.0108	109.73	0
9	112.19	126.1434	12.4373	113.749	1.3896	112.2341	0.0393	112.3567	0.1486
10	113.45	138.3009	21.9048	116.8717	3.0161	113.1498	-0.2646	113.5059	0.0493
FMAPE		4.3996		0.5908		**0.4394**		0.4398	
PMAPE		17.171		2.2029		0.152		**0.0989**	
TMAPE		6.9539		0.9132		0.3819		**0.3716**	
R		0.9525		0.9987		0.9998		**0.9999**	

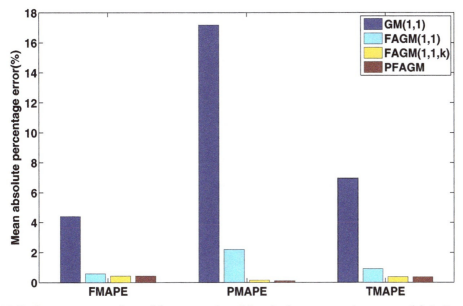

Fig 10. Performance comparison of the proposed model and other comparative grey models in Example C.

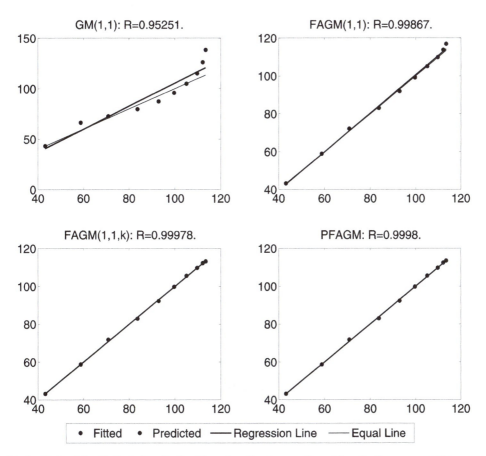

Fig 11. Analysis of detailed results obtained by using the proposed model and other comparative grey models in Example C.

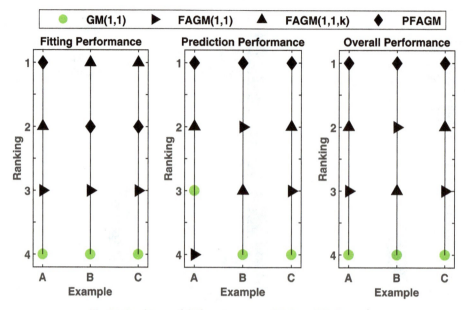

Fig 12. Rankings of different grey models in validations.

Analysis and discussion

According to the results of the above three validation experiments, we can conclude that PFAGM model has a better prediction accuracy than the other three contrast grey model. From an overall perspective, the prediction performances of fractional grey models are supe- rior to the classical GM(1,1) model. In Example A, the MAPE of PFAGM model for fitting is only slightly lower than the other grey models. However, the MAPE of PFAGM for prediction reaches 1.8927%, which is significantly lower than those of the other models. In Example B, the MAPE of PFAGM model for fitting is very closed to the lowest value though it is not the lowest. Moreover, the prediction performance of PFAGM model is slightly better than FAGM (1,1) and FAGM(1,1,k) model, and significantly better than GM(1,1) model. In Example C, PFAGM model obtains the secondary lowest MAPE of fitting which is also very closed to the lowest one among the four models. PFAGM model also achieves the lowest MAPE of predic- tion. Meanwhile, the overall MAPEs and correlation coefficients of PFAGM model in the three validation are lowest among all grey models. Such indicates that PFAGM model has bet- ter stability and adaptability. Fig 12 shows the rankings of various grey models for fitting and prediction performance. It can be clearly noticed that the prediction performance and overall performance of PFAGM is the best in the three validations. Moreover, the nuclear energy con- sumption of China is used to validate the performance of PFAGM model which obtains better accuracy in Example A. It indicates that PFAGM model can be applied in forecasting renew- able energy consumption. In the next section, we apply the novel PFAGM model to predict Chinese wind energy consumption to provide a reference for corresponding decision departments.

Application

Wind energy is safe, renewable, green, and economical. It is worth developing in all countries all over the world. The 13th Five-Year Plan (2016-2020) of China pointed out that the target of total wind installed capacity will be over 210GW at the end of 2020. However, the cumulative wind installed capacity of China has reached 211GW at the end of 2018, according to Global Wind Report 2018 of the GWEC. This means that the target of wind energy was achieved two years ahead of schedule.

Table 14. Raw data of China's wind energy consumption (million tonnes oil equivalent).

Year	Wind energy consumption	Year	Wind energy consumption
2009	6.2	2014	35.3
2010	10.1	2015	42
2011	15.9	2016	53.6
2012	21.7	2017	66.8
2013	31.9	2018	82.8

There is little room for wind energy growth in 2019 and 2020. To provide reference data for government, accurately predicting the wind energy is significant and necessary for formulating or adjusting corresponding wind energy policy. In this section, we focus on forecasting the annual wind energy consumption of China from macro perspectives. Because of small samples and some uncertainty, the grey prediction model is selected to apply in this application. According to the validation of the previous section, PFAGM model has a competitive edge compared with the other three grey models and can be used to forecast wind energy consumption. In this application, we collect the raw wind energy consumption of China from the BP Statistical Review of World Energy 2019. The wind energy consumption from 2009 to 2018, which are tabulated in Table 14, is used to forecast the consumption of the next three years. From Table 14, it can be apparently noticed that wind energy consumption increases rapidly. In 2018, the total wind energy consumption reached 82.8 million tonnes oil equivalent, which is 13.4 times as much as ten years ago. Meanwhile, wind energy is one of the cheapest forms of energy in many countries and is a kind of renewable energy used widely. Therefore, accurately forecasting wind energy consumption plays a vital role in reducing energy expenditure and carbon emissions. Firstly, we partition the raw sequence into two groups to build a model and test the model. The first group, including the consumption from 2009 to 2017, is used to build models for the four grey models separately. The second group, including wind energy consumption in 2018, is used to verify the prediction accuracy of these grey models. The linear parameters of GM(1,1) model can be estimated by using the least-square method. They are -0.22512 and 11.19559. The time response function (TRF) of GM (1,1) model can be represented as:

$$x^{(1)}(k) = 55.9317 e^{-0.2251(k-1)} - 49.7317 \qquad (50)$$

For fractional grey models, their fractional orders should be determined to obtain optimal values. The optimal orders of FAGM(1,1), FAGM(1,1,k), and PFAGM model are searched by using

WOA algorithm, which are filled in Table 15. Fig 13 shows than the optimal order of PFAGM model is obtained after a few iterations of WOA algorithm. Then the linear parame- ters of each fractional grey model can be calculated by using the least-square method, which also are listed in Table 15. The time response function of the three fractional grey models can be respectively represented as follows. The TRF of FAGM(1,1) model is

$$x^{(0.36871)}(k) = 37.55924 e^{0.17072(k-1)} - 31.35924. \tag{51}$$

Table 15. The optimal parameters of different grey models for forecasting Chinese wind energy consumption.

Parameters	GM(1,1)	FAGM(1,1)	FAGM(1,1,k)	PFAGM
a	-0.22512	-0.17072	-0.13851	0.08124
b	11.19559	5.35365	4.21723	5.11976
c	-	-	3.69298	-0.23365
γ	1	0.36871	1.13366	0.17874

The TRF of FAGM(1,1,k) model is

$$x^{(1.13366)}(k) = 283.1282 e^{0.13851(k-1)} - 30.4471 k - 246.4811. \tag{52}$$

The TRF of PFAGM model is

$$x^{(0.17874)}(k) = 19.6961 e^{0.17874 k} - 14.4727 e^{-0.08124(k-1)} - 2.8781. \tag{53}$$

By using inverse fractional or integral order accumulated operation, the restored results of the four grey models can be easily obtained and tabulated in Table 16. From Table 16 and Fig 14, FMAPE, PMAPE and TMAEP of PFAGM model are 3.1100, 2.1456 and 3.0136 respectively while those of GM(1,1) model are 8.3600, 3.2696 and 7.8510, those of FAGM(1,1) model are 3.1559, 3.3757 and 3.1779, and those of FAGM(1,1,k) model are 3.1901, 6.0252 and 3.4736. It can be clearly noticed that the total, fitting and validation MAPE of PFAGM are lowest. Mean- while, Fig 15 shows that the regression line almost coincides with the equal line. The fitted points of PFAGM model almost are on the regression line, while more fitted points of other grey models deviate from the regression line. R of PFAGM model is also best among the four grey models. Above all, all the above-mentioned evidences suggest that the prediction perfor- mance of PFAGM model is better than the other three grey models, and PFAGM model can be used to forecast the wind energy consumption of the future. Then we utilize these estab- lished grey models to forecast the wind energy consumption in the next three years. The pre- dicted wind energy consumptions from 2019 to 2021 are tabulated in Table 17. The predicted results of PFAGM model are respectively 98.0472, 118.2918, and 142.4003 in the next three years.

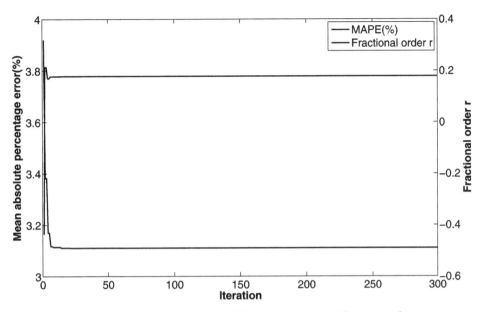

Fig 13. Searching optimal order of PFAGM by using WOA for forecasting Chinese wind energy consumption.

Table 16. The results produced by different grey models for Chinese wind energy consumption.

Year	Raw data	GM(1,1)		FAGM(1,1)		FAGM(1,1,k)		PFAGM	
		Value	RPE	Value	RPE	Value	RPE	Value	RPE
2009	6.2	6.2	0	6.2	0	6.2	0	6.2	0
2010	10.1	14.1212	39.8135	10.906	7.9799	10.7861	6.7928	10.8293	7.2208
2011	15.9	17.6863	11.2346	15.9	0	15.9521	0.328	15.9	0
2012	21.7	22.1515	2.0808	21.4735	-1.0437	21.7572	0.2637	21.5801	-0.5525
2013	31.9	27.7441	-13.028	27.8494	-12.6977	28.3244	-11.2087	28.0626	-12.0295
2014	35.3	34.7486	-1.5622	35.2399	-0.1702	35.7865	1.3782	35.5569	0.7278
2015	42	43.5214	3.6225	43.8725	4.4584	44.2901	5.4525	44.3002	5.4767
2016	53.6	54.5092	1.6963	54.0042	0.7541	53.9998	0.746	54.5674	1.8049
2017	66.8	68.271	2.2021	65.9324	-1.2989	65.1026	-2.541	66.6812	-0.1778
2018	82.8	85.5073	3.2696	80.0049	-3.3757	77.8112	-6.0252	81.0234	-2.1456
FMAPE		8.36		3.1559		3.1901		**3.11**	
PMAPE		3.2696		3.3757		6.0252		**2.1456**	
TMAPE		7.851		3.1779		3.4736		**3.0136**	
R		0.9964		0.9978		0.9971		**0.9979**	

Conclusion

In this paper, a novel fractional grey model called PFAGM is put forward based on the grey action quantity optimization of the classic fractional grey model with an exponential term. For fractional order accumulated grey models, it is the key to seek out their optimal orders to obtain the best prediction accuracy.

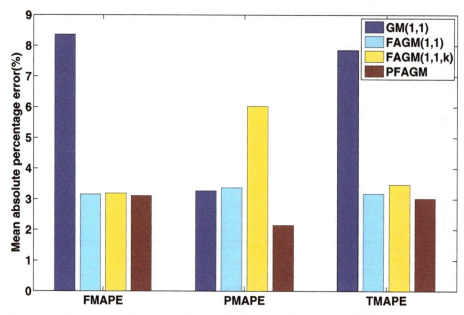

Fig 14. Performance comparison of the proposed and other comparative grey models for forecasting Chinese wind energy consumption.

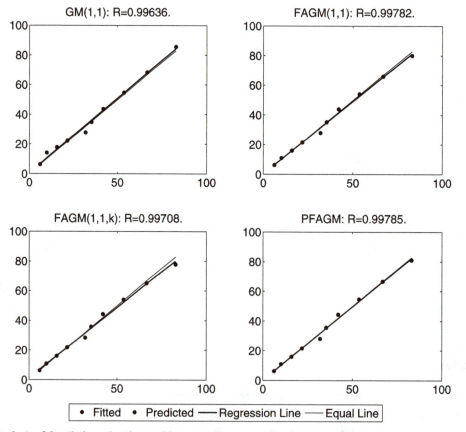

Fig 15. Analysis of detailed results obtained by using the proposed model and other comparative grey models for forecasting Chinese wind energy consumption.

In PFAGM model, WOA algorithm is adopted to search its optimal order. Moreover, its linear parameters are estimated by using the least-square method on the basis of the optimal order. Meanwhile, there are some various structures of the grey model behind PFAGM model, which can be reduced into GM(1,1) and FAGM(1,1) easily. It maybe indicates that PFAGM model has more predictive power than traditional grey mod- els. The results of validation experiments on real-life datasets show that PFAGM model has better prediction performance than the other three grey models. So PFAGM model is used to predict Chinese wind energy consumption. The predicted values of GM(1,1) in the next 3 years are 107.0951, 134.1331, 167.9974 respectively while those of FAGM(1,1) are 96.6314, 116.2954, 139.5689 and those of FAGM(1,1,k) are 92.3688, 109.0540, 128.1861 respectively. It indicates that the results of GM(1,1), FAGM(1,1,k) maybe have more some deviation from the truth. Results of PFAGM model are in the middle of the other grey models' results. So, its predicted results can be as reference data of adjusting wind energy policy.

Table 17. China's wind energy consumptions from 2019 to 2021 predicted by different grey models.

Year	GM(1,1)	FAGM(1,1)	FAGM(1,1,k)	PFAGM
2019	107.0951	96.6314	92.3688	**98.0472**
2020	134.1331	116.2954	109.054	**118.2918**
2021	167.9974	139.5689	128.1861	**142.4003**

Above all, it can be drawn that PFAGM model is efficient to realize short term prediction for time series, especially the sequence with small samples or uncertainty. In the future, the novel PFAGM model can be applied in more applications such as carbon emission forecasting, management of the petro- leum reservoirs and so on.

Acknowledgments

This work is supported by National Natural Science Foundation of China(Nos:71901184, 11872323 and 61672136), Humanities and Social Science Project of Ministry of Education of China (No.19YJCZH119), the Open Fund of State Key Laboratory of Oil and Gas Reservoir Geology (No. PLN201710) and Exploitation (Southwest Petroleum University), the Doctoral Research Foundation of Southwest University of Science and Technology (No. 16zx7140), and National Statistical Scientific Research Project (No. 2018LY42). We also thank anonymous peer reviewers for carefully revising our manuscript and for his or her useful comments.

Author Contributions Conceptualization: Peng Zhang, Xin Ma. **Data curation:** Peng Zhang.

Formal analysis: Peng Zhang.

Funding acquisition: Peng Zhang, Xin Ma, Kun She.

Investigation: Peng Zhang. **Methodology:** Peng Zhang. **Software:** Peng Zhang.

Supervision: Kun She. **Validation:** Xin Ma. **Visualization:** Peng Zhang.

Writing – original draft: Peng Zhang.

Writing – review & editing: Peng Zhang, Xin Ma, Kun She.

References

1. Saidur R, Islam MR, Rahim NA, Solangi KH. A review on global wind energy policy. Renewable and Sustainable Energy Reviews. 2010; 14(7):1744–1762. https://doi.org/10.1016/j.rser.2010.03.007

2. Zhang P. Do energy intensity targets matter for wind energy development? Identifying their heteroge- neous effects in Chinese provinces with different wind resources. Renewable Energy. 2019; 139:968– 975. https://doi.org/10.1016/j.renene.2019.03.007

3. Liu H, Chen C, Lv X, Wu X, Liu M. Deterministic wind energy forecasting: A review of intelligent predic- tors and auxiliary methods. Energy Conversion and Management. 2019; 195(January):328–345. https://doi.org/10.1016/j.enconman.2019.05.020

4. Yang Z, Wang J. A hybrid forecasting approach applied in wind speed forecasting based on a data pro- cessing strategy and an optimized artificial intelligence algorithm. Energy. 2018; 160:87–100. https:// doi.org/10.1016/j.energy.2018.07.005

5. do Nascimento Camelo H, Lucio PS, JB VL Junior, de Carvalho PCM. A hybrid model based on time series models and neural network for forecasting wind speed in the Brazilian northeast region. Sustain- able Energy Technologies and Assessments. 2018; 28:65–72. https://doi.org/10.1016/j.seta.2018.06. 009

6. Xiao L, Qian F, Shao W. Multi-step wind speed forecasting based on a hybrid forecasting architecture and an improved bat algorithm. Energy Conversion and Management. 2017; 143:410–430. https://doi. org/10.1016/j.enconman.2017.04.012

7. Liu H, O'Connor T, Lee S, Yoon S. A process optimization strategy of a pulsed-spray fluidized bed gran- ulation process based on predictive three-stage population balance model. Powder Technology. 2018; 327:188–200. https://doi.org/10.1016/j.powtec.2017.12.070

8. Santhosh M, Venkaiah C, Vinod Kumar DM. Ensemble empirical mode decomposition based adaptive wavelet neural network method for wind speed prediction. Energy Conversion and Management. 2018; 168:482–493. https://doi.org/10.1016/j.enconman.2018.04.099

9. Du P, Wang J, Yang W, Niu T. A novel hybrid model for short-term wind power forecasting. Applied Soft Computing Journal. 2019; 80:93–106. https://doi.org/10.1016/j.asoc.2019.03.035

10. Liu D, Niu D, Wang H, Fan L. Short-term wind speed forecasting using wavelet transform and support vector machines optimized by genetic algorithm. Renewable Energy. 2014; 62:592–597. https://doi.org/ 10.1016/j.renene.2013.08.011

11. Xiao L, Dong Y, Dong Y. An improved combination approach based on Adaboost algorithm for wind speed time series forecasting. Energy Conversion and Management. 2018; 160:273–288. https://doi. org/10.1016/j.enconman.2018.01.038

12. Kong X, Liu X, Shi R, Lee KY. Wind speed prediction using reduced support vector machines with fea- ture selection. Neurocomputing. 2015; 169:449–456. https://doi.org/10.1016/j.neucom.2014.09.090

13. Zhang CY, Chen CLP, Gan M, Chen L. Predictive Deep Boltzmann Machine for Multiperiod Wind Speed Forecasting. IEEE Transactions on Sustainable Energy. 2015; 6(4):1416–1425. https://doi.org/ 10.1109/TSTE.2015.2434387

14. Liu H, Mi X, Li Y. Smart deep learning based wind speed prediction model using wavelet packet decom- position, convolutional neural network and convolutional long short term memory net-

work. Energy Con- version and Management. 2018; 166:120–131. https://doi.org/10.1016/j.enconman.2018.04.021

15. Yu C, Li Y, Bao Y, Tang H, Zhai G. A novel framework for wind speed prediction based on recurrent neu- ral networks and support vector machine. Energy Conversion and Management. 2018; 178:137–145. https://doi.org/10.1016/j.enconman.2018.10.008

16. Yang W, Wang J, Lu H, Niu T, Du P. Hybrid wind energy forecasting and analysis system based on divide and conquer scheme: A case study in China. Journal of Cleaner Production. 2019; 222:942–959. https://doi.org/10.1016/j.jclepro.2019.03.036

17. Du P, Wang J, Yang W, Niu T. A novel hybrid model for short-term wind power forecasting. Applied Soft Computing. 2019; 80:93–106. https://doi.org/10.1016/j.asoc.2019.03.035

18. Shukur OB, Lee MH. Daily wind speed forecasting through hybrid KF-ANN model based on ARIMA. Renewable Energy. 2015; 76:637–647. https://doi.org/10.1016/j.renene.2014.11.084

19. Aasim, Singh SN, Mohapatra A. Repeated wavelet transform based ARIMA model for very short-term wind speed forecasting. Renewable Energy. 2019; 136:758–768. https://doi.org/10.1016/j.renene. 2019.01.031

20. Ding S, Hipel KW, guo Dang Y. Forecasting China's electricity consumption using a new grey prediction model. Energy. 2018; 149:314–328. https://doi.org/10.1016/j.energy.2018.01.169

21. Xu N, Dang Y, Gong Y. Novel grey prediction model with nonlinear optimized time response method for forecasting of electricity consumption in China. Energy. 2017; 118:473–480. https://doi.org/10.1016/j. energy.2016.10.003

22. Bianco V, Manca O, Nardini S. Electricity consumption forecasting in Italy using linear regression mod- els. Energy. 2009; 34(9):1413–1421. https://doi.org/10.1016/j.energy.2009.06.034

23. Wang ZX, Li Q, Pei LL. A seasonal GM(1,1) model for forecasting the electricity consumption of the pri- mary economic sectors. Energy. 2018; 154:522–534.

24. da Silva FLC, Cyrino Oliveira FL, Souza RC. A bottom-up bayesian extension for long term electricity consumption forecasting. Energy. 2019; 167:198–210. https://doi.org/10.1016/j.energy.2018.10.201

25. Tang L, Wang X, Wang X, Shao C, Liu S, Tian S. Long-term electricity consumption forecasting based on expert prediction and fuzzy Bayesian theory. Energy. 2019; 167:1144–1154. https://doi.org/10.1016/ j.energy.2018.10.073

26. Bahrami S, Hooshmand RA, Parastegari M. Short term electric load forecasting by wavelet transform and grey model improved by PSO (particle swarm optimization) algorithm. Energy. 2014; 72:434–442. https://doi.org/10.1016/j.energy.2014.05.065

27. Yang Y, Xue D. Continuous fractional-order grey model and electricity prediction research based on the observation error feedback. Energy. 2016; 115:722–733. https://doi.org/10.1016/j.energy.2016. 08.097

28. Wu L, Gao X, Xiao Y, Yang Y, Chen X. Using a novel multi-variable grey model to forecast the electricity consumption of Shandong Province in China. Energy. 2018; 157(2018):327–335. https://doi.org/10. 1016/j.energy.2018.05.147

29. Hamzacebi C, Es HA. Forecasting the annual electricity consumption of Turkey using an optimized grey model. Energy. 2014; 70:165–171. https://doi.org/10.1016/j.energy.2014.03.105

30. Sen D, Günay ME, Tunç KMMM. Forecasting annual natural gas consumption using so-

cio-economic indicators for making future policies. Energy. 2019; 173:1106–1118. https://doi.org/10.1016/j.energy. 2019.02.130

31. Zhang W, Yang J. Forecasting natural gas consumption in China by Bayesian Model Averaging. Energy Reports. 2015; 1:216–220. https://doi.org/10.1016/j.egyr.2015.11.001

32. Li J, Wang R, Wang J, Li Y. Analysis and forecasting of the oil consumption in China based on combina- tion models optimized by artificial intelligence algorithms. Energy. 2018; 144:243–264. https://doi.org/ 10.1016/j.energy.2017.12.042

33. Wang Q, Song X. Forecasting China's oil consumption: A comparison of novel nonlinear-dynamic grey model (GM), linear GM, nonlinear GM and metabolism GM. Energy. 2019; 183:160–171. https://doi. org/10.1016/j.energy.2019.06.139

34. Wu W, Ma X, Zeng B, Wang Y, Cai W. Application of the novel fractional grey model FAGMO(1,1,k) to predict China's nuclear energy consumption. Energy. 2018; 165:223–234. https://doi.org/10.1016/j. energy.2018.09.155

35. Tang L, Yu L, He K. A novel data-characteristic-driven modeling methodology for nuclear energy con- sumption forecasting. Applied Energy. 2014; 128:1–14. https://doi.org/10.1016/j.apenergy.2014.04.021

36. Wang H, Lei Z, Zhang X, Zhou B, Peng J. A review of deep learning for renewable energy forecasting. Energy Conversion and Management. 2019; 198(April):111799. https://doi.org/10.1016/j.enconman. 2019.111799

37. Wu L, Zhao H. Discrete grey model with the weighted accumulation. Soft Computing. 2019; 3.

38. Wang Q, Liu L, Wang S, Wang JZ, Liu M. Predicting Beijing's tertiary industry with an improved grey model. Applied Soft Computing Journal. 2017; 57:482–494. https://doi.org/10.1016/j.asoc.2017.04.022

39. Xia J, Ma X, Wu W, Huang B, Li W. Application of a new information priority accumulated grey model with time power to predict short-term wind turbine capacity. Journal of Cleaner Production. 2019; p. 118573.

40. Liu X, Xie N. A nonlinear grey forecasting model with double shape parameters and its application. Applied Mathematics and Computation. 2019; 360:203–212. https://doi.org/10.1016/j.amc.2019.05.012

41. Hu Y. Electricity consumption prediction using a neural-network-based grey forecasting approach. Jour- nal of the Operational Research Society. 2017; 68(10):1259–1264. https://doi.org/10.1057/s41274-016-0150-y

42. Wang ZX. An optimized Nash nonlinear grey Bernoulli model for forecasting the main economic indices of high technology enterprises in China. Computers and Industrial Engineering. 2013; 64(3):780–787. https://doi.org/10.1016/j.cie.2012.12.010

43. Zeng B, Li C. Forecasting the natural gas demand in China using a self-adapting intelligent grey model. Energy. 2016; 112:810–825. https://doi.org/10.1016/j.energy.2016.06.090

44. Li L, Wang H. A VVWBO-BVO-based GM (1,1) and its parameter optimization by GRA-IGSA integration algorithm for annual power load forecasting. PLoS ONE. 2018; 13(5):e0196816. https://doi.org/10. 1371/journal.pone.0196816 PMID: 29768450

45. Wang Z, Hao P. An improved grey multivariable model for predicting industrial energy consumption in China. Applied Mathematical Modelling. 2016; 40(11-12):5745–5758. https://doi.org/10.1016/j.apm. 2016.01.012

46. Zeng B, Li C. Improved multi-variable grey forecasting model with a dynamic background-value coeffi- cient and its application. Computers and Industrial Engineering. 2018; 118(March):278–290. https://doi. org/10.1016/j.cie.2018.02.042

47. Zeng B, Duan H, Zhou Y. A new multivariable grey prediction model with structure compatibility. Applied Mathematical Modelling. 2019; 75:385–397. https://doi.org/10.1016/j.apm.2019.05.044

48. Ma X, Liu Z. The GMC (1, n) Model with Optimized Parameters and Its Application. Journal of grey sys- tem. 2017; 29(4):122–138.

49. Wu L, Liu S, Yao L, Yan S, Liu D. Grey system model with the fractional order accumulation. Communi- cations in Nonlinear Science and Numerical Simulation. 2013; 18(7):1775–1785. https://doi.org/10. 1016/j.cnsns.2012.11.017

50. Mao S, Gao M, Xiao X, Zhu M. A novel fractional grey system model and its application. Applied Mathe- matical Modelling. 2016; 40(7-8):5063–5076. https://doi.org/10.1016/j.apm.2015.12.014

51. Ma X, Mei X, Wu W, Wu X, Zeng B. A novel fractional time delayed grey model with Grey Wolf Optimizer and its applications in forecasting the natural gas and coal consumption in Chongqing China. Energy. 2019; 178:487–507. https://doi.org/10.1016/j.energy.2019.04.096

52. Ma X, Wu W, Zeng B, Wang Y, Wu X. The conformable fractional grey system model. ISA Transactions. 2019;. https://doi.org/10.1016/j.isatra.2019.07.009

53. Zeng B, Liu S. A self-adaptive intelligence gray prediction model with the optimal fractional order accumulating operator and its application. Mathematical Methods in the Applied Sciences. 2017; 40(18):7843–7857. https://doi.org/10.1002/mma.4565

54. Ma X, Xie M, Wu W, Zeng B, Wang Y, Wu X. The novel fractional discrete multivariate grey system model and its applications. Applied Mathematical Modelling. 2019; 70:402–424. https://doi.org/10.1016/ j.apm.2019.01.039

55. Xu H, Liu S, Fang Z. Optimization of grey action quantity of GM(1,1) model. Mathematics in Practice and Theory. 2010; 40(2):27–32.

56. Mirjalili S, Lewis A. The Whale Optimization Algorithm. Advances in Engineering Software. 2016; 95:51–67. https://doi.org/10.1016/j.advengsoft.2016.01.008

57. Gharehchopogh FS, Gholizadeh H. A comprehensive survey: Whale Optimization Algorithm and its applications. Swarm and Evolutionary Computation. 2019; 48(November 2018):1–24. https://doi.org/ 10.1016/j.swevo.2019.03.004

58. Shaikh F, Ji Q, Shaikh PH, Mirjat NH, Uqaili MA. Forecasting China's natural gas demand based on optimised nonlinear grey models. Energy. 2017; 140:941–951. https://doi.org/10.1016/j.energy.2017. 09.037

59. Wang Y, Zhang C, Chen T, Ma X. MODELING THE NONLINEAR FLOW FOR A MULTIPLE-FRACTURED HORIZONTAL WELL WITH MULTIPLE FINITE-CONDUCTIVITY FRACTURES IN TRIPLE MEDIA CARBONATE RESERVOIR. Journal of Porous Media. 2018; 21(12):1283–1305. https://doi. org/10.1615/JPorMedia.2018028663

60. Ma X, Liu Z. Predicting the Cumulative Oil Field Production Using the Novel Grey ENGM Model. Journal of Computational and Theoretical Nanoscience. 2016; 13(1):89–95. https://doi.org/10.1166/jctn.2016. 4773

61. Chen PY, Yu HM. Foundation Settlement Prediction Based on a Novel NGM Model. Mathematical Problems in Engineering. 2014; 2014:1–8.

Risk factors in the illness-death model: Simulation study and the partial differential equation about incidence and prevalence

Annika Hoyer 1*, Sophie Kaufmann1, Ralph Brinks1,2

1 Institute for Biometrics and Epidemiology, German Diabetes Center, Leibniz Center for Diabetes Research at Heinrich Heine University Dü sseldorf, Dü sseldorf, Germany, **2** Hiller Research Unit for Rheumatology, Heinrich Heine University Dü sseldorf, Dü sseldorf, Germany

* annika.hoyer@ddz.de

Editor: Alberto d'Onofrio, International Prevention Research Institute, FRANCE

Funding: The authors received no specific funding for this work.

Competing interests: The authors have declared that no competing interests exist.

Abstract

Recently, we developed a partial differential equation (PDE) that relates the age-specific prevalence of a chronic disease with the age-specific incidence and mortality rates in the illness-death model (IDM). With a view to planning population-wide interventions, the question arises how prevalence can be calculated if the distribution of a risk-factor in the population shifts. To study the impact of such possible interventions, it is important to deal with the resulting changes of risk-factors that affect the rates in the IDM. The aim of this work is to show how the PDE can be used to study such effects on the age-specific prevalence of a chronic disease, to demonstrate its applicability and to compare the results to a discrete event simulation (DES), a frequently used simulation technique. This is done for the first time based on the PDE which only needs data on population-wide epidemiological indices and is related to the von Foerster equation. In a simulation study, we analyse the effect of a hypothetical intervention against type 2 diabetes. We compare the age-specific prevalence obtained from a DES with the results predicted from modifying the rates in the PDE. The DES is based on 10000 subjects and estimates the effect of changes in the distributions of risk-factors. With respect to the PDE, the change of the distribution of risk factors is syn- thesized to an effective rate that can be used directly in the PDE. Both methods, DES and effective rate method (ERM) are capable of predicting the impact of the hypothetical inter- vention. The age-specific prevalences resulting from the DES and the ERM are consistent. Although DES is common in simulating effects of hypothetical interventions, the ERM is a suitable alternative. ERM fits well into the analytical theory of the IDM and the related PDE and comes with less computational effort.

Introduction

Recently, we developed a partial differential equation (PDE) that links the age-specific prevalence of a chronic disease with the age-specific incidence and mortality rates [1]. This PDE is related to the classical illness-death model (IDM) where each subject of the population under consideration is either in the state *Healthy* (with respect to the considered disease), *Ill* or *Dead*. Using the PDE, we are able to estimate age-specific prevalences in case the incidence and mor- tality rates are given. Possible epidemiological applications of the PDE are, for instance, the projection of future case-numbers of a chronic disease [1] or the estimation of incidence rates from two or more cross-sectional surveys [2, 3]. Unfortunately, the PDE is currently rather unknown in context of non-communicable diseases and has only been used in a few cases. To strengthen the use of the PDE we aim to demonstrate applicability of that approach in case only population-wide epidemiological indices, as incidence or prevalence, are available. Fur- thermore, as it is of high interest in epidemiological research to project the effect of interven- tions on prevalences, we want to show how risk-factors can be incorporated in the PDE.

In order to study the effect of interventions on the future prevalence or for the projection of case-numbers, it is important to consider risk-factors (i.e., covariates) which modify the transi- tion rates between the states in the IDM. There is a long history of estimating the impact of covariates on incidence and mortality rates, for instance, by the Cox proportional hazard model [4]. Kalbfleisch and Prentice gave an extensive introduction with example data in their popular textbook [5]. A more recent but briefer introduction is presented by Bland [6]. Typi- cally, the effect of covariates is reported in terms of risk or rate reductions. For example, a meta-analysis about the effect of physical activity on the incidence of type 2 diabetes found that 2.5 hours per week of moderate-intensity physical activity (PA) decreased the incidence of diabetes by about 30% compared to almost no PA [7]. If the research question seeks for the effect of PA on the prevalence of diabetes, a study could compare a group without intervention to a group with intervention with respect to the prevalence. However, conducting an interven- tion study is sometimes lengthy, expensive and in some cases impossible due to practical problems. Comput- er models may be used to simulate the key aspects of a (hypothetical) inter- vention and estimate the resulting effects. For this, the recently developed PDE may be helpful by comparing the prevalence in the group without intervention to the intervention group where the incidence rate is modified by the intervention. A similar comparison could be made if mortality rates would be altered by an intervention. Such a group-wise comparison between an intervention group and a 'business-as-usual' group has performed in estimating the impact of increased active travel in the urban population of Germany [8].

Comparing two groups with and without intervention has limited value in studying the effects of population-wide interventions. If health authorities decide to promote a campaign for increased PA, typically not all persons in the population will start to increase their PA. Possibly, some persons will increase their activities and some will not. Assumed that before the campaign 30% of the population were active for 2.5 hours per week with moderate intensity, after the campaign this percentage could be increased to, say, 35%. Of course, other persons might also increase their PA, but remain below the recommended weekly 2.5 hours. This means, that the campaign is likely to change the the distribution of PA in the population. If we want to compare the prevalence of type 2 diabetes before and after the campaign, we are in a different situation than comparing two groups with the intervention entirely being effective in one group while the other group remains completely unaffected.

With a view to planning population-wide interventions, the question arises how the prevalence can be calculated if the distribution of a risk factor in the population shifts. In this work, we present two different approaches for answering this question.

First, the frequently used discrete event simulation (DES) is performed. DES generates relevant events in the illness-death model (IDM) for each individual, i.e., in our situation, the onset of disease and death with or without disease [9]. Second, prevalences are calculated based on the PDE. We illustrate and compare both methods using an example from type 2 diabetes. For this, we focus on the body mass index (BMI) as a risk factor. The relation between BMI on the incidence rate of type 2 diabetes and the mortality rates is considerably well understood from epidemiological studies [10, 11]. The aim of this work is to demonstrate applicability of both methods and compare them in estimating the impact of an intervention on the prevalence of a chronic disease. This is done for the first time based on the PDE developed by Brinks and Landwehr [1] which only needs data on population-wide epidemiological indices. Such a usage of a PDE with implications on the distribution of a modifiable risk factor is not well-known and can also be shown for the famous von Foerster equation [12] which applies to all age-structured equations of population dynamics. Unfortunately, this relationship is widely unknown to a broader audience [13], implying that the von Foerster equation is rarely used in medical and epidemiological research. Therefore, the present article gives a brief overview on the related mathematical background, corresponding equations and, finally, practical applications. It can basically be seen as guideline for assessing the impact of a potential intervention in an epidemiological context. Note that our work can be seen only as one ingredient for planning population-wide interventions. For assessing the overall impact of a population-wide intervention, many more aspects like, for example, costs, logistics, side-effects, sustainability and acceptance in the population, have to be considered.

The article is organized as follows: first, we give a brief overview on the IDM. Afterwards, the considered population and its characteristics are described. Then, the two methods for estimating the age-specific prevalences, which depend on the distribution of risk factors in the population, are introduced in detail. The relation of BMI and type 2 diabetes serves as an example application. After this, the results of the two methods are compared. Finally, in the discussion and conclusion the central findings are discussed critically.

Materials and methods

First we introduce and review the IDM and a related partial differential equation. Then, we describe our hypothetical intervention and the target population and how the prevalences can be determined analytically and in a DES.

Illness-death model

The classical IDM as introduced by [14] is depicted in Fig 1. The IDM consists of three states: Healthy (with respect to the considered disease (*H*)), Ill (*I*) and Dead (*D*). The transition rates between these states are the incidence rate (λ), the mortality of the susceptibles (μ_0) and the mortality of the cases (μ_1). These rates depend on calender time *t* and age *a*.

If we denote the prevalence of the chronic disease in those persons aged *a* at time *t* with $\pi(t, a)$, the PDE [1] reads as

$$\left(\frac{\partial}{\partial t} + \frac{\partial}{\partial a}\right)\pi = (1 - \pi)(\lambda - \pi(\mu_1 - \mu_0)). \tag{1}$$

Given the rates λ, μ_0 and μ_1, the age-specific prevalence $\pi = \pi(\lambda, \mu_0, \mu_1)$ can be calculated from the PDE.

Description of the population

We simulated a birth cohort consisting of *n* = 10000 people born in 1980. Every person was followed-up from birth to death passing states presented in the IDM in Fig 1. For simplicity, we can omit the dependency on time *t* in case of a birth cohort and confine ourselves to only one time scale, age *a*, to account for temporal progression.

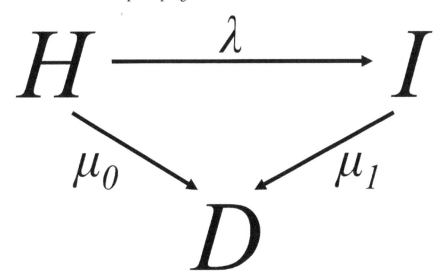

Fig 1. Illness-death model. The transition rates between the states *Healthy* (H), *Ill* (I), and *Dead* (D) are denoted by λ, μ_0, μ_1.

Although the IDM is applicable to any chronic condition, for demonstration of the methods we choose type 2 diabetes as exemplary application. As the epidemiological relationship between the BMI and type 2 diabetes is well known, we included the BMI as potential risk fac- tor. Hence, our modelled transition rates, depend only on age a and the BMI [10, 11].

For each person $j, j = 1, \ldots, n$, we assigned a BMI, denoted by z_j, which was randomly drawn from a beta-distribution with shape parameters $\phi = 3$ and $\psi = 8$. The support of the beta-distribution was chosen to be [17;47] which are conceivable BMI values. The choice of these parameters mirrors the right-skewed distribution of the BMI in the German male popu- lation [15]. As a high BMI is known as a potential risk factor for type 2 diabetes [11], we also simulated a population-wide intervention with the aim to reduce the BMI. This could be a life- style intervention as, for example, a change in eating habits or increased sporting activities.

Assuming such an intervention, we generated the BMI of each person from a beta-distribution with shape parameters $\phi = 2.8$ and $\psi = 8$. This leads to a slightly shifted distribution, meaning a change in the expected BMI from 23.9 in the group without any intervention to 23.3 in the intervention group. This seems to be a reasonable change that can be reached in a convenient amount of time as it also has been observed in the Finnish Diabetes Prevention Study [16]. Fig 2 depicts both distributions of the BMI in the population.

Discrete event simulation

Given the BMI z_j of an individual j, we simulated a possible transition to the diabetic state (incidence rate λ) or to the death state without contracting diabetes (mortality rate μ_0). The incidence rate is chosen in accordance to the incidence rate reported by Tamayo et al. [17] for the German population. As no data on μ_0 is available for Germany, we used the general mor- tality as given by the German Federal Statistical Office [18]. Both rates, λ and μ_0, depend on the BMI as well as on age a as it is shown in Eqs (2) and (3):

$$\lambda(a, Z) = \varepsilon \exp\left(\alpha a^2 + \beta a + \gamma + \delta Z\right) \qquad (2)$$

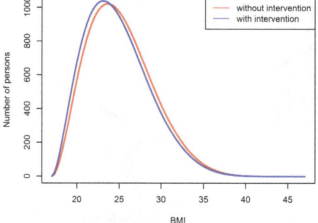

Fig 2. BMI distribution in Germany.

and

$$\mu_0(a) = \exp(\xi \mid Z - z_0 \mid + \vartheta + va), \qquad (3)$$

where z_0 is chosen to be 23.75 according to Tobias et al. [19]. For the age-specific incidence rate, we chose an exponential parabolic function. The mortality rate μ_0 was modelled using the Gompertz-Makeham law of mortality and to guarantee a j-shaped dependency on BMI, because the mortality is increasing with a high as well as with a low BMI. The parameters $\varepsilon = 0.014$, $\alpha = -0.001$, $\beta = 0.209$, $\gamma = -11.112$ and $\delta = 0.1$ for the incidence as well as $\xi = 0.02$, $\vartheta = -10.948$ and $v = 0.095$ were estimated to mimic the incidence of Tamayo et al. [17] and the general mortality in Germany, respectively.

For those individuals who contracted diabetes, we additionally simulated the mortality rate μ_1 which also depends on Z and a. As it is reasonable that the mortality of people with diabetes is elevated compared to the mortality rate of people without diabetes, μ_1 is assumed to be the double of μ_0 as it is presented in Eq (4) [20]:

$$\mu_1(a) = 2\mu_0(a). \qquad (4)$$

From Eq (3), it becomes obvious that the mortality rate increases if the BMI is below or above of 23.75. As mentioned above, this is due to the empirical finding that a very low or very high BMI is associated with increased mortality rates.

For each individual j, we simulated failure times T_{1j} of contracting diabetes or dying with- out diabetes similar to Brinks et al. [21]. If the subject indexed j contracted diabetes, a second failure time T_{2j} was simulated, which corresponds to death after contracting the disease. It should be noted that we assume no change on the BMI in case a person contracted diabetes.

We estimated the prevalence of diabetes by dividing the number of cases at age a by the total number of people who are alive at age a. All prevalences were estimated in absence and in case of the intervention that reduces the BMI.

Effective rate method

The DES as described in the previous section comes along with the disadvantage of being time-consuming. If many settings of input parameters have to be simulated, this can be quite cumbersome or even impossible [9]. As such, the question arises if the age-specific prevalence can be determined more time-saving and without the need for individual participant data. The effective rate method (ERM) which uses the partial differential Eq (1) offers such a possibility as this method is only based on the mean rates. In the following we present the mathematical background of the ERM as well as corresponding equations. It should be noted that these are consequences of well-known facts that also holds true for the von Foerster equation [12]. However, with respect to epidemiological applications it is helpful to reiterate. For this, we first need to convert the individual participant data from the birth cohort to mean rates.

The mean incidence rate, denoted by λ, is calculated by the general formula

$$\lambda^* = \int \lambda(Z) f(Z) dZ, \qquad (5)$$

where $\lambda(Z)$ is the incidence rate for the BMI value Z and $f(Z)$ is the probability density function (pdf) for the BMI. Furthermore, Eq (5) is the expected value of the incidence rate.

The same holds true for the mean mortality rates μ_0^* and μ_1^*:

$$\mu_0^* = \int \mu_0(Z) f(Z) dZ \qquad (6)$$

and

$$\mu_1^* = HR \cdot \mu_0^*, \qquad (7)$$

where HR indicates a hazard ratio. The value of HR = 2 seems to be a reasonable choice [20].

As we assume the BMI to be beta-distributed, $f(Z)$ is the pdf of this beta distribution. For the lower and upper bound of the integral, the minimum (= 17) and maximum (= 47) of the BMI in our birth cohort are chosen. Based on the mean incidence and mortality rates, we can estimate the prevalence by solving the PDE (1). In our example of a single birth cohort, we can omit the calender time t and the PDE simplifies to an ordinary differential Eq (8) [22]:

$$\frac{\partial}{\partial a}\pi = (1-\pi)(\lambda^* - \pi(\mu_1^* - \mu_0^*)) \qquad (8)$$

which can be solved using the Runge-Kutta method [23] with the initial value $p(30) = 0$. This initial value is assumed to be reasonable because the majority of people contracts type 2 diabetes after the age of 30 [17].

Results

The results of both methods are depicted in Fig 3. On the left hand side of the figure, the results for the base-case scenario (without intervention) are presented whereas the findings for the intervention scenario are presented on the right hand side. The red and blue solid lines depict the estimated prevalences based on the effective rate method. Fig 3 shows, for example, that a man aged 80 years has diabetes with a probability of approximately 25% using the DES and of 26% using the ERM in the base-case scenario. However, the prevalence of diabetes for an 80 years old man in the intervention group ranges between 24% and 25% for the DES and the ERM, respectively.

The estimated prevalences in the base-case and intervention scenario are the highest for the age group 75–90 which was also reported by Tamayo et al. [17]. However, for the population- wide intervention, the maximum prevalence is reduced from approximately 26% to 25% only by a slightly decreasing BMI. The overall prevalence of type 2 diabetes is reduced by 0.24 per- centage points, from 6.51% to 6.27%. Estimates of the ERM agree very well with the point esti- mates of the DES method, especially in lower age groups. All of the ERM-estimates are located in the confidence intervals of the DES. The confidence intervals are somewhat wider in higher age groups above 80 years which is caused by higher mortality rates and therefore a decreased number of individuals in that age classes.

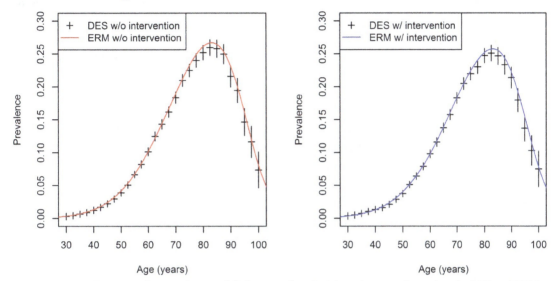

Fig 3. Estimated age-specific prevalences of diabetes with and without intervention using the DES and ERM.
Left panel: ase-case scenario, right panel: intervention scenario.

The results of the DES and the ERM agree well in the intervention case as well as in the case no intervention was assumed. Results from the DES are given as black crosses which represent the point estimates with the corresponding 95% confidence intervals. Confidence intervals for the DES result from the stochasticity of the simulation [9]. As the ERM is an exact analytical method to determine the age-specific prevalences, no confidence bounds need to be provided.

An advantageous feature of the ERM is that this method is computationally more efficient, resulting in faster simulation times. This can be seen in Table 1 where the computation times for different sample sizes, i.e. different numbers of simulated persons in the birth cohort, are given. In case of small birth cohorts of 100 persons, the DES is as fast as the ERM. However, with increasing sample sizes the DES is quite slower compared to the ERM. For example, to estimate the prevalences shown in Fig 3, the DES needs approximately 14.66 seconds whereas the ERM leads to results in just 0.36 seconds.

Table 1. Simulation time (in seconds) of the DES compared to the ERM.

Number of persons simulated	DES	ERM
100	0.36	0.36
1000	3.74	0.34
10000	14.66	0.36

Discussion

In this article, we aimed to compare two methods of including the effect of a covariate into the IDM. On the one hand, a discrete event simulation was applied that simulated the relevant event times on the individual subject level. On the other hand, we integrated the effect of the covariates into the transition rates in the IDM. For this an effective rate was calculated that takes into account the distribution of the risk factor in the population (ERM). The ERM is based only on aggregated data in form of mean mortality and incidence rates. These rates are the input data for the (partial) differential equation [22] which describes the temporal and age- specific change in prevalences. For both methods, the potential effect of a covariate is modelled by a shift of the distribution of the covariate which is caused by an intervention.

We illustrated both methods using the example of type 2 diabetes mellitus and use the BMI as a very common risk factor for that non-communicable disease. Based on empirical findings, we assumed that the BMI is beta-distributed in the German population. Depending on the BMI value, individual events (contracting diabetes, dying with and without diabetes) were sim- ulated. It becomes obvious that both methods yield similar results. This finding was confirmed in a second scenario where we changed the beta-distribution of the BMI according to a lifestyle intervention. For this work we assumed that BMI remains constant over the whole lifespan for both methods. However, in reality people will change BMI with age. Such time-varying covari- ates or interaction terms can be easily included in the ERM with regard to Eq (5) if data are available. However, in case of the DES it is more difficult to account for time-varying covari- ates or interaction terms [24]. Therefore, we decided to keep the simulation as simple as possi- ble aiming to compare the ERM to the DES.

The ERM comes with some advantages. As it is an analytical method only mean rates are needed, i.e., the incidence and mortality rates, to calculate the prevalence at several points in time. The use of such kind of aggregated data is a definite advantage compared to the DES which can be very time consuming. Moreover, the ERM has in some cases, e.g. for non-differ- ential mortality $m_0 = m_1$ or for given m_0 and m_1 [25] a solution which can be directly deter- mined. Additionally, the ERM does automatically account for competing risks as the used differential equation is based on the IDM. In case of the DES, modelling competing events is quite more difficult and should be based on the algorithm presented in [21].

Conclusion

Altogether, we have shown that the ERM could be used to model potential changes in covari- ates. Moreover, it is a time-saving alternative to the DES if we are interested in population- wide epidemiological indices like age-specific prevalences. Therefore, the ERM based on a dif- feren- tial equation is of high interest for future work whenever it is necessary to model the effect of covariates on the change of prevalences.

Supporting information

S1 File. Scripts for the statistical software R. The zip-file contains the complete R-code for analyzing the simulated data set and to reproduce the results. For detailed instructions unzip the file and refer to the readme.txt file.

(ZIP)

Author Contributions Conceptualization: Annika Hoyer, Ralph Brinks. **Data curation:** Annika Hoyer, Ralph Brinks.

Formal analysis: Annika Hoyer, Sophie Kaufmann, Ralph Brinks.

Investigation: Annika Hoyer, Ralph Brinks.

Methodology: Annika Hoyer, Sophie Kaufmann, Ralph Brinks.

Project administration: Ralph Brinks.

Resources: Annika Hoyer, Ralph Brinks.

Software: Annika Hoyer, Sophie Kaufmann, Ralph Brinks.

Supervision: Ralph Brinks.

Validation: Annika Hoyer, Ralph Brinks.

Visualization: Annika Hoyer, Sophie Kaufmann, Ralph Brinks.

Writing – original draft: Annika Hoyer.

Writing – review & editing: Annika Hoyer, Sophie Kaufmann, Ralph Brinks.

References

1. Brinks R, Landwehr S. Age-and time-dependent model of the prevalence of non-communicable dis- eases and application to dementia in Germany. Theoretical population biology. 2014; 92:62–68. https:// doi.org/10.1016/j.tpb.2013.11.006 PMID: 24333220

2. Brinks R, Landwehr S. A new relation between prevalence and incidence of a chronic disease. Mathe- matical Medicine and Biology. 2015; 32:425–435. https://doi.org/10.1093/imammb/dqu024 PMID: 25576933

3. Brinks R, Hoyer A, Landwehr S. Surveillance of the Incidence of Non-Communicable Diseases (NCDs) with Sparse Resources: A Simulation Study Using Data from a National Diabetes Registry, Denmark, 1995-2004. PloS one. 2016; 11:e0152046. https://doi.org/10.1371/journal.pone.0152046 PMID: 27023438

4. Cox DR. Regression models and life tables (with discussion). Journal of the Royal Statistical Society. 1972; 34:187–220.

5. Kalbfleisch JD, Prentice RL. The Statistical Analysis of Failure Time Data. John Wiley & Sons; 2011.

6. Bland M. An introduction to medical statistics. Oxford University Press (UK); 2015.

7. Jeon CY, Lokken RP, Hu FB, van Dam RM. Physical activity of moderate intensity and risk of type

2 dia- betes: a systematic review. Diabetes Care. 2007; 30:744–752. https://doi.org/10.2337/dc06-1842 PMID: 17327354

8. Brinks R, Hoyer A, Kuss O, Rathmann W. Projected Effect of Increased Active Travel in German Urban Regions on the Risk of Type 2 Diabetes. PLoS one. 2015; 10:e0122145. https://doi.org/10.1371/ journal.pone.0122145 PMID: 25849819

9. Law AM. Simulation Modeling & Analysis. Fourth Edition. McGraw-Hill, New York. 2007.

10. Narayan KMV, Boyle JP, Thompson TJ, Gregg EW, Williamson DF. Effect of BMI on Lifetime Risk for Diabetes in the U.S. Diabetes Care. 2007; 30:1562–1566. https://doi.org/10.2337/dc06-2544 PMID: 17372155

11. Berentzen TL, Jakobsen MU, Halkjaer J, Tjonneland A, Sorensen TI, Overvad K. Changes in waist cir- cumference and the incidence of diabetes in middle-aged men and women. PloS one. 2011; 6:e23104. https://doi.org/10.1371/journal.pone.0023104 PMID: 21829698

12. Von Foerster H. Some remarks on changing populations. In: The Kinetics of Cellular Proliferation, Stohl- man F. Jr., editor. Greene and Stratton, New York.

13. Chubb MC, Jacobsen KH. Mathematical modeling and the epidemiological research process. European Journal of Epidemiology. 2010; 25:13–19. https://doi.org/10.1007/s10654-009-9397-9 PMID: 19859816

14. Keiding N. Age-specific incidence and prevalence: a statistical perspective. Journal of the Royal Statisti- cal Society. Series A (Statistics in Society). 1991;371–412. https://doi.org/10.2307/2983150

15. Robert Koch-Institut and Destatis. Distribution of the population to groups in terms of body mass index in percent. Classification: years, Germany, age, sex, body mass index. 2003. Available at: http://www. gbe-bund.de/gbe10/trecherche.prc_them_rech?tk=5800&tk2=6000&p_uid=gast&p_aid=10171277&p_ sprache=D&cnt_ut=11&ut=6150 (Last accessed May 15, 2019).

16. Lindströ¨m J, Louheranta A, Mannelin M, Rastas M, Salminen V, Eriksson J, Uusitupa M, Tuomilehto J, Finnish Diabetes Prevention Study Group. The Finnish Diabetes Prevention Study (DPS): Lifestyle intervention and 3-year results on diet and physical activity. Diabetes Care. 2003; 26:3230–3236.

17. Tamayo T, Brinks R, Hoyer A, Kuss O, Rathmann W. The Prevalence and Incidence of Diabetes in Ger- many. Deutsches Arzteblatt International. 2016; 133:177–182.

18. Statistisches Bundesamt. Sterbetafel 2012/2014—Methoden- und Ergebnisbericht zur laufenden Berechnung von Periodensterbetafeln fu¨r Deutschland und die Bundesla¨nder. Available at: https:// www.destatis.de/DE/Themen/Gesellschaft-Umwelt/Bevoelkerung/Sterbefaelle-Lebenserwartung/_ inhalt.html#sprg233418 (Last accessed May 15, 2019).

19. Tobias DK, Pan A, Jackson CL, O'Reilly EJ, Ding EL, Willett WC, Manson JE, Hu FB. Body-mass index and mortality among adults with incident type 2 diabetes. New England Journal of Medicine. 2014; 370:233–244. https://doi.org/10.1056/NEJMoa1304501 PMID: 24428469

20. Saydah SH, Loria CM, Eberhardt MS, Brancati FL. Subclinical states of glucose intolerance and risk of death in the U.S. Diabetes Care. 2001; 24:447–453. https://doi.org/10.2337/diacare.24.3.447 PMID: 11289466

21. Brinks R, Landwehr S, Fischer-Betz R, Schneider M, Giani G. Lexis Diagram and Illness-Death Model: Simulating Populations in Chronic Disease Epidemiology. PLoS one. 2014; 9:e106043. https://doi.org/ 10.1371/journal.pone.0106043 PMID: 25215502

22. Brinks R, Landwehr S, Icks A, Koch M, Giani G. Deriving age-specific incidence from prevalence

with an ordinary differential equation. Statistics in Medicine. 2013; 32:2070–2078. https://doi.org/10.1002/ sim.5651 PMID: 23034867

23. Dahlquist G, Bjö̈rk A. Numerical Methods. Prentice-Hall Inc. Englewood Cliffs, New Jersey. 1964.

24. Brennan A, Chick SE, Davies R. A taxonomy of model structures for economic evaluation of health tech- nologies. Health Economics. 2006; 15:1295–1310. https://doi.org/10.1002/hec.1148 PMID: 16941543

25. Brinks R, Landwehr S. Change rates and prevalence of a dichotomous variable: simulations and appli- cations. PloS one. 2015; 10:e0118955. https://doi.org/10.1371/journal.pone.0118955 PMID: 25749133

Parameter identification for gompertz and logistic dynamic equations

Elvan Akın [1☯*], *Neslihan Nesliye Pelen* [2☯], *Ismail Uğur Tiryaki* [3☯], *Fusun Yalcin* [4☯]

1 Department of Mathematics and Statistics, Missouri University of Science and Technology, Rolla, Missouri, United States of America, **2** Department of Mathematics, Ondokuz Mayıs University, Arts and Science Faculty, Samsun, Turkey, **3** Department of Mathematics, Bolu Abant Izzet Baysal University, Faculty of Arts and Science, Bolu, Turkey, **4** Department of Mathematics, Faculty of Science, Akdeniz University, Antalya, Turkey

☯ These authors contributed equally to this work.

* akine@mst.edu

Editor: J. Alberto Conejero, IUMPA - Universitat Politecnica de Valencia, SPAIN

Funding: The authors received no specific funding.

Competing interests: The authors have declared that no competing interests exist.

Abstract

In this paper, we generalize and compare Gompertz and Logistic dynamic equations in order to describe the growth patterns of bacteria and tumor. First of all, we introduce two types of Gompertz equations, where the first type 4-paramater and 3-parameter Gompertz curves do not include the logarithm of the number of individuals, and then we derive 4- parameter and 3-parameter Logistic equations. We notice that Logistic curves are better in modeling bacteria whereas the growth pattern of tumor is described better by Gompertz curves. Increasing the number of parameters of Logistic curves give favorable results for bacteria while decreasing the number of parameters of Gompertz curves for tumor improves the curve fitting. Moreover, our results overshadow some of the existing results in the literature.

Introduction

Most of the growth curves are described by linear, power, parabolic, power-exponential, logistic, log-logistic, von Bertalanffy, Gompertz, and Richards curves; see [1], [2], [3], [4], [5], and [6] for the tumor, [7] for the human fetus, [8] for the human life. A recent research article [8] related with a human life modeled by Gompertz and Mirror Gompertz differential equations are

$$(\ln x)' = -\beta \ln x, \tag{1}$$

and

$$x' = -\beta(1-x)\ln(1-x), \tag{2}$$

respectively, where β is a positive parameter. The well-known logistic equation in the literature, see [9] is given by

$$x' = kx\left(1 - \frac{x}{K}\right), \tag{3}$$

where k is the proportionality constant and K is the carrying capacity.

In this study, mathematical modeling is applied to the Pseudomonas putida and mammary tumor datas given in [10, 11], respectively. Note that Pseudomonas putida is a bacterium found in most soil and water habitats, and is significant to the environment due to its complex metabolism and ability to control pollution, [12] and [13]. We model their growth patterns by continuous and discrete Gompertz and Logistic curves. To achieve our goal, we derive 4-parameter and 3-parameter Gompertz and Logistic dynamic equations. We first propose two types of Gompertz dynamic equations: The first type Gompertz dynamic equations are motivated by [14]. We contribute two first type continuous Gompertz curves to the literature. All of the discrete Gompertz curves in this type are new. 4-parameter second type Gompertz dynamic equations are motivated by [2] in which only 3-parameter discrete Gompertz curves are considered. 3-parameter second type continuous Gompertz are investigated earlier in [10]. Inspired by [15], we come up with 4-parameter Logistic dynamic equations while 3-parameter Logistic dynamic equations are constructed earlier in [16]. 4-parameter Logistic discrete curves are new. To establish both dynamic equations, we use the variation of constant formulas together with the circle dot multiplication and the circle minus substraction on time scales. We refer readers to [17] and [18] by Bohner and Peterson for the theory of time scales calculus.

The parameters of these models are estimated by NonlinearModelFit function of Wolfram Mathematica 11.0 applying Monte Carlo simulation and our comparison is based on outputs following from the p-values of parameters, adjusted R-squared, and RMSE (root mean square error), RRMSE (Relative Root Mean Square Error), MAPE (Mean Absolute Percent Error), MAE (mean absolute error), U1 (Theil inequality coefficient, Theil's U1), and U2 (Theil inequality coefficient,

Theil's U2). We use the Mathematica 11 program for the goodness of fit test of the models. Having at least three small values of each determined statistical criterion, the p value less than 0.05 for each parameter, and adjusted R-squared value close to 1 show bet- ter performance in terms of goodness of fit.

Outline of this paper is as follows: In Section, we introduce the time scales calculus together with some preliminary results. Sections are related with first and second type Gompertz dynamic equations. In each section we obtain 4-parameter and 3-parameter contin- uous and discrete Gompertz curves. In Section, Logistic dynamic equations are introduced and we explicitly cal- culate 4-parameter and 3-parameter continuous and discrete Logistic curves. In the last section, we discuss how Gompertz and Logistic curves fit the growth of Pseu- domonas putida and mammary tumor and include our conclusion.

Preliminary results

A *time scale*, $\mathbb{T}$, is an arbitrary nonempty closed subset of the real numbers $\mathbb{R}$. The theory of time scales is to introduce a new calculus so that we can unify the continuous and discrete analysis. Here, we give basic definitions and some essential results without proofs. Neverthe- less, we mainly refer readers two books [17] and [18] by Bohner and Peterson and the manu- script [16] by Akin-Bohner and Bohner.

The *forward jump operator* σ on $\mathbb{T}$ is defined as $\sigma(t):=\inf\{s > t : s \in \mathbb{T}\} \in \mathbb{T}$, for all $t \in \mathbb{T}$. For this definition we also have $\sigma(\emptyset) = \sup \mathbb{T}$. The *backward jump operator* ρ on $\mathbb{T}$ is defined by $\rho(t):= \sup\{s < t : s \in \mathbb{T}\} \in \mathbb{T}$, for all $t \in \mathbb{T}$. Here, we have $\rho(\emptyset) = \inf \mathbb{T}$. If $\sigma(t) > t$, we say t is *right-scattered*, while if $\rho(t) < t$ we say t is *left-scattered*. If $\sigma(t) = t$, we say t is *right-dense*, while if $\rho(t) = t$ we say t is *left-dense*. The *graininess* function $\mu : \mathbb{T} \mapsto [0, \infty)$ is defined by $\mu(t) := \sigma(t) - t$. It is apparent that for $\mathbb{T} = \mathbb{Z}$, $\sigma(t) = t + 1$, $\rho(t) = t - 1$ and for $\mathbb{T} = \mathbb{R}$, $\sigma(t) = t$, $\rho(t) = t$. The set $\mathbb{T}^\kappa$ is derived from $\mathbb{T}$: If $\mathbb{T}$ has left-scattered maximum m, then $\mathbb{T}^\kappa = \mathbb{T} - \{m\}$. Otherwise, $\mathbb{T}^\kappa = \mathbb{T}$. The following notations are also useful: $f^\sigma(t) = f(\sigma(t))$. Note that

$$t \in [t_0, \infty)_\mathbb{T} = [t_0, \infty) \cap \mathbb{T}.$$

Assume $f : \mathbb{T} \mapsto \mathbb{R}$ and let $t \in \mathbb{T}^\kappa$, then we define $f^\Delta(t)$ to be the number (provided it exists) with the property that given any $\epsilon > 0$, there is a neighborhood U of t such that

$$| [f(\sigma(t)) - f(s)] - f^\Delta(t)[\sigma(t) - s] | \leq \epsilon | \sigma(t) - s |,$$

for all $s \in U$. $f^\Delta(t)$ is called the *delta derivative* of $f(t)$ at t. Note that the delta-derivative turns out to be the usual derivative when $\mathbb{T} = \mathbb{R}$ while it is the forward difference operator when $\mathbb{T} = \mathbb{Z}$. If f is differentiable at t, then f is continuous at t. If f is continuous at t and t is right-scattered, then f is differentiable at t with

$$f^\Delta(t) = \frac{f(\sigma(t)) - f(t)}{\mu(t)}.$$

If f is differentiable and t is right-dense, then

$$f^\Delta(t) = \lim_{s \to t} \frac{f(t) - f(s)}{t - s}.$$

If f is differentiable at t, then

$$f^\sigma(t) = f(t) + \mu(t)f^\Delta(t). \tag{4}$$

If $f, g : \mathbb{T} \mapsto \mathbb{R}$ are differentiable at $t \in \mathbb{T}^k$, then the product $fg : \mathbb{T} \mapsto \mathbb{R}$ is also differentiable at t with

$$(fg)^\Delta(t) = f^\Delta(t)g(t) + f(\sigma(t))g^\Delta(t).$$

If f is continuous at each right-dense point $t \in \mathbb{T}$ and whenever $t \in \mathbb{T}$ is left-dense $\lim_{s \to t^-} f(s)$ exists as a finite number, then we say that $f : \mathbb{T} \mapsto \mathbb{R}$ is *rd-continuous*. A function $F : \mathbb{T}^\kappa \mapsto \mathbb{R}$ is called an *antiderivative* of $f : \mathbb{T} \mapsto \mathbb{R}$ provided $F^\Delta(t) = f(t)$ holds for all $t \in \mathbb{T}^k$. In this case, we define the integral of f by

$$\int_a^t f(s)\Delta s = F(t) - F(a) \text{ for } t \in \mathbb{T}. \tag{5}$$

If $1 + \mu(t)p(t) \neq 0$ for all $t \in \mathbb{T}^\kappa$, $p : \mathbb{T}^\kappa \mapsto \mathbb{R}$ is called *regressive*. The set of all regressive and rd-continuous functions is denoted by R. If $1 + \mu(t)p(t) > 0$ for all $t \in \mathbb{T}^\kappa$, $p : \mathbb{T}^\kappa \mapsto \mathbb{R}$ is called *positively regressive*. The set of all positively regressive and rd-continuous functions is denoted by R^+.

If $p, \epsilon 2 R$ and α is a constant, then we define

$$\ominus p(t) = -\frac{p(t)}{1 + \mu(t)p(t)}, \qquad p(t) \ominus q(t) = \frac{p(t) - q(t)}{1 + \mu(t)q(t)}, \tag{6}$$

and

$$(p \oplus q)(t) = p(t) + q(t) + \mu(t)p(t)q(t)$$

for all $t \in \mathbb{T}^k$. Finding a simple formula of the derivative of any power of a function yields to the introduction of a circle dot multiplication. A circle dot multiplication $\odot$ is defined in [16] as

$$(\alpha \odot p)(t) = \alpha p(t) \int_0^1 (1 + h\mu(t)p(t))^{\alpha-1} dh.$$

Note that $\ominus p = -p$, $p \oplus q = p + q$ and $\alpha \odot p = \alpha p$ for the continuous case. If p is regressive, then we define the exponential function by

$$e_p(t, s) = \exp\left(\int_s^t \xi_\mu(p(\tau))\Delta\tau\right) \quad \text{for } s, t \in \mathbb{T}, \tag{7}$$

where $\xi_h(z) = \frac{1}{h}\text{Log}(1+hz), h > 0$ is the cylinder transformation such that $\xi_0(z) = z$. If $p : \mathbb{T}^\kappa \mapsto \mathbb{R}$ is rd-continuous and regressive, then the *exponential function* $e_p(t, t_0)$ is the unique solution of the IVP

$$x^\Delta = p(t)x, \quad x(t_0) = 1$$

on $\mathbb{T}$ for each fixed $t_0 \in \mathbb{T}^k$. For data analysis we need to calculate exponential functions

$$e_\beta(t, t_0) = e^{\beta(t-t_0)}, \quad e_{\ominus\beta}(t, t_0) = e^{-\beta(t-t_0)} \quad \text{when } \mathbb{T} = \mathbb{R} \tag{8}$$

$$e_\beta(t, t_0) = (1+\beta)^{t-t_0}, \quad e_{\ominus\beta}(t, t_0) = (1+\beta)^{-(t-t_0)} \quad \text{when } \mathbb{T} = \mathbb{Z} \tag{9}$$

for a regressive constant β, see Table 2.4 in [17].

We use the following properties of the exponential function $e_p(t, s), t; s \in \mathbb{T}$.

Theorem 0.1. *If p, q are regressive and $t_0 \in \mathbb{T}$, then*

1. $e_p(t, t) \equiv$ *and* $e_0(t, s) \equiv 1$;
2. $e_p(\sigma(t), s) = (1 + \mu(t)p(t))e_p(t, s)$;
3. $\dfrac{1}{e_p(t, s)} = e_{\ominus p}(t, s) = e_p(s, t)$;
4. $\dfrac{e_p(t, s)}{e_q(t, s)} = e_{p \ominus q}(t, s)$;
5. $e_p(t, s)e_q(t, s) = e_{p \oplus q}(t, s)$;
6. *if $p > 0$ for all $t \in \mathbb{T}$, then $e_p(t, t_0) > 0$ for all $t \in \mathbb{T}$;*
7. *if $p \in R^+$, then $e_p(t, t_0) > 0$ for all $t \in \mathbb{T}$.*

In addition, two of the useful formulas for a circle dot are

$$e_{\alpha \odot p}(t, t_0) = (e_p(t, t_0))^\alpha, \tag{10}$$

and

$$1 + \mu(\alpha \odot p) = (1 + \mu p)^\alpha, \tag{11}$$

where p is a regressive function and α is a constant, see [16].

The followings are the variation of constants formulas, see Theorems 2.74 and 2.77 in [17].

The equation

$$x^\Delta = p(t)x + f(t) \tag{12}$$

is called *regressive* if $x^\Delta = p(t)x$ is regressive (i.e., p is regressive) and $f: \mathbb{T} \to \mathbb{R}$ is rd- continuous.

Theorem 0.2. *Suppose (12) is regressive. Let $t_0 \in \mathbb{T}$ and $x_0 \in \mathbb{R}$. The unique solution of the IVP*

$$x^\Delta = p(t)x + f(t), \quad x(t_0) = x_0$$

is given by

$$x(t) = e_p(t, t_0)x_0 + \int_{t_0}^t e_p(t, \sigma(\tau))f(\tau)\Delta\tau.$$

Theorem 0.3. *Suppose (12) is regressive. Let $t_0 \in \mathbb{T}$ and $x_0 \in \mathbb{R}$. The unique solution of the IVP*

$$x^\Delta = -p(t)x^\sigma + f(t), \quad x(t_0) = x_0$$

is given by

$$x(t) = e_{\ominus p}(t, t_0)x_0 + \int_{t_0}^t e_{\ominus p}(t, \tau)f(\tau)\Delta\tau.$$

First type gompertz dynamic equations

In this section, we will introduce Gompertz dynamic curves motivated by the 4-parameter Gompertz curve

$$\omega(t) = B + A \exp\left(-\exp(-K(t - t_0))\right), t \in \mathbb{R} \tag{13}$$

given in [19] for the growth curve analyses of bacterial counts. Here, K can be found as the growth rate coefficient, t_0 is the initial time, $A + B$ is the carrying capacity of the environment for the population. To explain the carrying capacity notion we can say that every environment has its own limits, therefore it is impossible for species to grow up infinitely. Thus, the number of the population should be finite.

In order to obtain the Gompertz model in the continuous case, we differentiate Eq (13) and obtain

$$\begin{aligned}\omega' &= AK \exp\{-\exp\{-K(t-t_0)\}\}\exp\{-K(t-t_0)\} \\ &= [\omega(t) - B]K\exp\{-K(t-t_0)\}.\end{aligned}$$

In addition, note that we have

$$\begin{aligned}e_{\ominus(K\odot\ominus e_{\ominus K})}(t, t_0) &= \frac{1}{e_{K\odot\ominus e_{\ominus K}}(t, t_0)} = \left(\frac{1}{e_{\ominus e_{\ominus K}}(t, t_0)}\right)^K \\ &= (e_{e_{\ominus K}}(t, t_0))^K = e_{K\odot e_{\ominus K}}(t, t_0)\end{aligned} \tag{14}$$

on $[t_0, \infty)_\mathbb{T}$, where we use Theorem 0.1 and (10). Since $e_{\ominus K}(t, t_0) = e^{-K(t-t_0)}$ for $t \in \mathbb{R}$, and (14) holds, then we obtain

$$\begin{aligned} e_{\ominus(K\odot e_{\ominus K})}(t, t_0) &= \exp\left\{\frac{\exp(-K(t-t_0))}{K} - \frac{1}{K}\right\}^{-K} \\ &= e\exp\{-\exp\{-K(t-t_0)\}\}, t \in \mathbb{R}. \end{aligned} \qquad (15)$$

Motivated by the calculation above, we have the following initial value problem modeling 4-parameter Gompertz curve on time scales.

Theorem 0.4. *The initial value problem*

$$\begin{aligned} \omega^\Delta &= -(K \odot \ominus e_{\ominus K}(t, t_0))\omega^\sigma + B(K \odot \ominus e_{\ominus K}(t, t_0)) \\ \omega(t_0) &= \omega_0 \end{aligned} \qquad (16)$$

has the solution of the form

$$\omega = B + (\omega_0 - B)e_{K \odot e_{\ominus K}}(t, t_0)$$

$t \in [t_0, \infty)_\mathbb{T}$, *where K is the growth rate and t_0 is the initial time, ω_0 is the value of the function at the initial time and B is the coefficient that has an impact on carrying capacity.*

Proof. We notice that the positivity of K implies the positivity of $e_{\ominus K}$ by Theorem 0.1. Since $1 + \mu(\ominus e_{\ominus K}) = \frac{1}{1+\mu e_{\ominus K}} > 0$, we have the positively regressivity of $\ominus e_{\ominus K}$. Since $1 + \mu(K \odot \ominus e_{\ominus K}) = (1 + \mu(\ominus e_{\ominus K}))^K > 0$ by (11), the dynamic equation in the IVP (16) is regressive. Therefore, we apply Theorem 0.3 and obtain the unique solution for $t \in [t_0, \infty)_\mathbb{T}$

$$\begin{aligned} \omega &= e_{\ominus(K\odot\ominus e_{\ominus K})}(t, t_0)\omega_0 + B\int_{t_0}^t e_{\ominus(K\odot\ominus e_{\ominus K})}(t, \tau)(K \odot \ominus e_{\ominus K}(\tau, t_0))\Delta\tau \\ &= e_{K\odot e_{\ominus K}}(t, t_0)\omega_0 \\ &\quad + Be_{K\odot\ominus e_{\ominus K}}(t_0, t)\int_{t_0}^t e_{K\odot\ominus e_{\ominus K}}(\tau, t_0)(K \odot \ominus e_{\ominus K}(\tau, t_0))\Delta\tau \\ &= e_{K\odot e_{\ominus K}}(t, t_0)\omega_0 + Be_{K\odot\ominus e_{\ominus K}}(t_0, t)\int_{t_0}^t (e_{K\odot\ominus e_{\ominus K}}(\tau, t_0))^\Delta\Delta\tau \\ &= e_{K\odot e_{\ominus K}}(t, t_0)\omega_0 + Be_{K\odot\ominus e_{\ominus K}}(t, t_0)[e_{K\odot\ominus e_{\ominus K}}(t, t_0) - 1] \\ &= B + (\omega_0 - B)e_{K\odot e_{\ominus K}}(t, t_0), \end{aligned} \qquad (17)$$

where we use (14) and Theorem 0.1.

Example 0.5. Let $\mathbb{T} = \mathbb{R}$. Then the continuous Gompertz curve

$$\omega = B + e(\omega_0 - B)\exp\{-\exp\{-K(t-t_0)\}\} \qquad (18)$$

is obtained from (17) for $t \in [t_0, \infty)_\mathbb{R}$ by using Eqs (8) and (15). This is compatible with the continuous Gompertz growth curve (13) by taking $A = e(\omega_0 - B)$ in (13).

Example 0.6. Let $\mathbb{T} = \mathbb{Z}$. Since $e_{\ominus K}(t, t_0) = (1 + K)^{-(t-t_0)}$ for $t \in [t_0, \infty)_\mathbb{Z}$ by (9), (14) yields

$$\begin{aligned}
e_{K \odot e_{\ominus K}}(t, t_0) &= [e_{e_{\ominus K}}(t, t_0)]^K \\
&= \left[\exp\left(\sum_{s=t_0}^{t-1} \ln\left(1 + \frac{1}{(1+K)^{s-t_0}}\right)\right)\right]^K \\
&= \left[\exp\left(\ln\left(\prod_{s=t_0}^{t-1} 1 + \frac{1}{(1+K)^{s-t_0}}\right)\right)\right]^K \\
&= \left[\prod_{s=t_0}^{t-1} \frac{1 + (1+K)^{s-t_0}}{(1+K)^{s-t_0}}\right]^K, \quad t \in [t_0, \infty)_\mathbb{Z}
\end{aligned}$$

and so

$$e_{\ominus(K \odot e_{\ominus K})}(t, t_0) = \left[\prod_{\tau=t_0}^{t-1} \frac{(1+K)^{\tau-t_0}}{1 + (1+K)^{\tau-t_0}}\right]^{-K} \tag{19}$$

for $t \in [t_0, \infty)_\mathbb{Z}$. Thus, the discrete Gompertz growth curve

$$\omega = B + (\omega_0 - B)\left[\prod_{\tau=t_0}^{t-1} \frac{1 + (1+K)^{\tau-t_0}}{(1+K)^{\tau-t_0}}\right]^K \tag{20}$$

again follows from (17) for $t \in [t_0, \infty)_\mathbb{Z}$.

Motivated by the first variation of constant formula, Theorem 0.2, we derive another Gompertz curve on time scales.

Theorem 0.7. *The initial value problem*

$$\begin{aligned}
\omega^\Delta &= \ominus(K \odot e_{\ominus K}(t, t_0))\omega + B(\ominus(K \odot e_{\ominus K}(t, t_0))) \\
\omega(t_0) &= \omega_0
\end{aligned} \tag{21}$$

has the solution of the form

$$\omega = (\omega_0 + B)e_{\ominus(K \odot e_{\ominus K})}(t, t_0) - B \tag{22}$$

for $t \in [t_0, \infty)_\mathbb{T}$, where K is the decay rate coefficient and regressive, t_0 is the initial time, ω_0 is the value of the function at the initial time and B is the coefficient that has an impact on carrying capacity.

Proof. Since K is regressive, $e_{\ominus K}$ is also regressive by (7). The dynamic equation in the IVP

(21) is regressive. Then, in order to obtain the unique solution (22) we apply Theorem 0.3 for $t \in [t_0, \infty)_\mathbb{T}$

$$\begin{aligned}
\omega &= e_{\ominus(K\odot e_{\ominus K})}(t,t_0)\omega_0 + B\int_{t_0}^t e_{\ominus(K\odot e_{\ominus K})}(t,\sigma(\tau))(\ominus(K\odot e_{\ominus K}(\tau,t_0)))\Delta\tau \\
&= e_{\ominus(K\odot e_{\ominus K})}(t,t_0)\omega_0 \\
&\quad - Be_{\ominus(K\odot e_{\ominus K})}(t,t_0)\int_{t_0}^t e_{K\odot e_{\ominus K}}(\tau,t_0)(K\odot e_{\ominus K}(\tau,t_0))\Delta\tau \\
&= e_{\ominus(K\odot e_{\ominus K})}(t,t_0)\omega_0 - Be_{\ominus(K\odot e_{\ominus K})}(t,t_0)\int_{t_0}^t (e_{K\odot e_{\ominus K}}(\tau,t_0))^\Delta \Delta\tau \\
&= e_{\ominus(K\odot e_{\ominus K})}(t,t_0)\omega_0 - Be_{\ominus(K\odot e_{\ominus K})}(t,t_0)[e_{K\odot e_{\ominus K}}(t,t_0) - 1] \\
&= (\omega_0 + B)e_{\ominus(K\odot e_{\ominus K})}(t,t_0) - B,
\end{aligned}$$

where we use (14) and Theorem 0.1.

Example 0.8. Let $\mathbb{T} = \mathbb{R}$. Then the alternative continuous Gompertz curve

$$\omega = \frac{1}{e}(\omega_0 + B)\exp\{\exp\{-K(t-t_0)\}\} - B \qquad (23)$$

is obtained from (22) for $t \in \mathbb{R}$. It is worth to mention that Eq (23) is a new Gompertz curve in the continuous case.

Example 0.9. Let $\mathbb{T} = \mathbb{Z}$. Then using (19), we have

$$e_{\ominus(K\odot e_{\ominus K})}(t,t_0) = \left[\prod_{\tau=t_0}^{t-1}\frac{(1+K)^{\tau-t_0}}{1+(1+K)^{\tau-t_0}}\right]^K$$

for $t \in [t_0, \infty)_\mathbb{Z}$. Since

$$e_{\ominus(K\odot e_{\ominus K})} = \frac{1}{e_{K\odot e_{\ominus K}}} = \left[\frac{1}{e_{e_{\ominus K}}}\right]^K,$$

the alternative discrete Gompertz growth curve

$$\omega = (\omega_0 + B)\left[\prod_{\tau=t_0}^{t-1}\frac{(1+K)^{\tau-t_0}}{1+(1+K)^{\tau-t_0}}\right]^K - B \qquad (24)$$

again follows from (22) for $t \in [t_0, \infty)_\mathbb{Z}$.

The Gompertz growth curve (23) is given as 4-parameter Gompertz growth curve in [19].

From this point of view, the 3-parameter Gompertz growth curve on time scales

$$\omega = \omega_0 e_{K\odot e_{\ominus K}}(t,t_0) \qquad (25)$$

is obtained from (17) when $B = 0$, and so the 3-parameter continuous and discrete Gompertz curves are

$$\omega = e\omega_0 \exp\{-\exp\{-K(t - t_0)\}\} \tag{26}$$

for $t \in [t_0, \infty)_{\mathbb{R}}$ and for $t \in [t_0, \infty)_{\mathbb{Z}}$

$$\omega = \omega_0 \left[\prod_{\tau=t_0}^{t-1} \frac{1 + (1+K)^{\tau-t_0}}{(1+K)^{\tau-t_0}}\right]^K, \tag{27}$$

respectively. From (22) when $B\ 0$, the alternative 3-parameter continuous and discrete Gompertz curves

$$\omega = \frac{1}{e}\omega_0 \exp\{\exp\{-K(t - t_0)\}\} \tag{28}$$

and

$$\omega = \omega_0 \left[\prod_{\tau=t_0}^{t-1} \frac{1 + (1+K)^{\tau-t_0}}{(1+K)^{\tau-t_0}}\right]^{-K}, \tag{29}$$

are gained to the literature, respectively.

The zwietering modification of gompertz growth curve

The Gompertz growth curve is reparameterized in order to model the bacteria growth population in food and is stated as

$$w = A \exp\{-\exp\{\frac{eK_z}{A}(T - t) + 1\}\} \tag{30}$$

in [14] for $t \in \mathbb{R}$, where K_z is the absolute growth rate at time T, so called lag time, which is interpreted as the time between when a microbial population is transferred to a new habitat recovers and when a considerable cell division occurs.

In order to find a corresponding dynamic model, we rewrite (30) as

$$w = A(\exp\{-\exp\{-\frac{eK_z}{A}t\}\})^{e^{\frac{eK_z}{A}T+1}} \tag{31}$$

so that we can use the property of circle dot (10) for the unification the continuous and dis- crete cases. Therefore, using (15) and (14) yield the dynamic Zwietering Modification Gompertz curve

$$w = Ae^{-(\frac{eK_z}{A}T+1)} \left(e_{\ominus\left(\frac{eK_z}{A}\odot\ominus e_{\ominus\frac{eK_z}{A}}\right)}(t,0)\right)^{e^{\frac{eK_z}{A}T+1}}$$

$$= Ae^{-\left(\frac{eK_z}{A}T+1\right)} e_{e^{\left(\frac{eK_z}{A}T+1\right)}\odot\ominus}\left(\frac{eK_z}{A}\odot e_{\ominus\frac{eK_z}{A}}\right)(t,0) \qquad (32)$$

$$= Ae^{-\left(\frac{eK_z}{A}T+1\right)} e_{e^{\left(\frac{eK_z}{A}T+1\right)}\odot}\left(\frac{eK_z}{A}\odot e_{\ominus\frac{eK_z}{A}}\right)(t,0).$$

Therefore, we claim that (32) is the solution of the IVP

$$\omega^\Delta = \left(e^{\left(\frac{eK_z}{A}T+1\right)}\odot\left(\frac{eK_z}{A}\odot e_{\ominus\frac{eK_z}{A}}\right)\right)w \qquad (33)$$

$$\omega(0) = Ae^{-\left(\frac{eK_z}{A}T+1\right)}.$$

Since (30) is the continuous modified Gompertz growth curve, the discrete modified Gom- pertz growth curve follows from (32).

Example 0.10. Let $\mathbb{T} = \mathbb{R}$. Then the discrete Zwietering modification of Gompertz growth curve is

$$\omega = Ae^{-\left(\frac{eK_z}{A}T+1\right)}\left[\prod_{\tau=0}^{t-1}\frac{\left(1+\frac{eK_z}{A}\right)^\tau}{1+\left(1+\frac{eK_z}{A}\right)^\tau}\right]^{-\frac{eK_z}{A}e^{\left(\frac{eK_z}{A}T+1\right)}} \qquad (34)$$

obtained from (19) for $t \in [0,\infty)_\mathbb{Z}$.

Gompertz-laird growth curve

This model is mainly used for the modeling of tumor growth. The Laird re-parameterization prevails even today as the most frequently fitted Gompertz version in cancer research, and is now also commonly fitted to growth data in other fields such as those of domestic (e.g. poultry and livestock, marine (e.g. molluscs, fish, and dolphins) animals.

The continuous Gompertz-Laird growth curve is given by

$$\omega = \omega_0 e^{-\frac{L}{K}(e^{-Kt}-1)}$$

for $t \in [0,\infty)_\mathbb{R}$ in [14], which is equivalent to

$$w = w_0 e^{\frac{L}{K}}(e^{-e^{-Kt}})^{\frac{L}{K}} \qquad (35)$$

for $t \in [0,\infty)_\mathbb{R}$, where the parameter L describes the initial specific growth rate that is not a notion that measures the relative growth or absolute growth. More precisely, we can say that the absolute growth rate at $t = 0$ is $\omega_0.L$ Thus, the term L can be described as division of the ini- tial absolute growth rate with the initial value.

Similarly, by using (15) we obtain the Gompertz-Laird growth curve on time scales as

$$w = w_0\left(e_{\ominus(K\odot e_{\ominus K})}(t,0)\right)^{\frac{L}{K}}$$

$$= w_0 e_{\frac{L}{K}\odot(\ominus(K\odot e_{\ominus K}))}(t,0) \tag{36}$$
$$= w_0 e_{\frac{L}{K}\odot(K\odot e_{\ominus K})}(t,0)$$

for $t \in [0,\infty)_\mathbb{T}$ and so (36) is the solution of the IVP

$$w^\Delta = \frac{L}{K} \odot (K \odot e_{\ominus K})w \tag{37}$$
$$\omega(0) = \omega_0.$$

Since (35) is the continuous Gompertz-Laird growth curve, the following example gives the discrete Gompertz-Laird growth curve.

Example 0.11. Let $\mathbb{T} = \mathbb{Z}$. Then we obtain

$$\omega = \omega_0 \left[\prod_{\tau=0}^{t-1} \frac{(1+K)^\tau}{1+(1+K)^\tau} \right]^{-L} \tag{38}$$

as the discrete Gompertz-Laird growth curve for $t \in [t_0,\infty)_\mathbb{Z}$, where we use again (19).

If $L = Km$ in (36), the **Zweifel and Lasker** re-parametrization dynamic equation is obtained for studying fish growth. Moreover, the continuous and discrete curves are given in (35), (38), respectively. Similarly, if $L = \ln\left(\frac{A}{\omega_0}\right)^K$ in (36), we derive the dynamic form of **Simpler** W_0 **form** of Gompertz Laird growth curve which prevails on. Moreover, the continuous and dis- crete curves are given in (35) and (38), respectively. Note that all of first type Gompertz curves in the discrete case are new.

Second type gompertz dynamic equations

It is clear that (13) is not a Gompertz model when the dependent variable is log-transformed. In this subsection, we will derive Gompertz dynamic equations involving logarithmic func- tions. This idea of the derivation of Gompertz dynamic equations is inspired from the Gom- pertz difference equation

$$\ln G(t+1) = a + b \ln G(t), \quad t \in \mathbb{Z} \tag{39}$$

where a is taken as the growth rate and b is taken as the exponential rate of growth decelera- tion, which was firstly given by Bassukas et. al. [3]. The equivalent form of (39) is given by

$$\Delta \ln G(t) = a + (b-1)\ln G(t), \quad t \in \mathbb{Z} \tag{40}$$

and so the continuous version becomes

$$[\ln G(t)]' = a + (b-1)\ln G(t), \quad t \in \mathbb{R}. \tag{41}$$

Unifying (40) and (41), we end up by

$$[\ln G(t)]^\Delta = a + (b-1)\ln G(t), \quad t \in \mathbb{T}. \tag{42}$$

By the second variation of constants formula, Theorem 0.3, we get the alternative dynamic equation

$$[\ln G(t)]^\Delta = a - (b-1)\ln G(\sigma(t)), \quad t \in \mathbb{T}. \tag{43}$$

Notice that Eq (42) turns out to be (39) when $\mathbb{T} = \mathbb{Z}$ while (43) turns out to be

$$\ln G(t+1) = \frac{a}{b} + \frac{\ln G(t)}{b}, \tag{44}$$

for a nonzero constant b and when $\mathbb{T} = \mathbb{R}$, we obtain the following Gompertz differential equation from the Gompertz dynamic Eq (43) as

$$[\ln G(t)]' = a - (b-1)\ln G(t), \tag{45}$$

which is equivalent to Gompertz differential Eq (1) when $a = 0$, $b - 1 = \beta$, and $G = x$ in (45). In [8], Gompertz differential Eq (41) with $a = 0$ is called the Mirror Gompertz differential equation, and is equivalent to (2) when $a = 0$, $b - 1 = \beta$, and $G = 1 - x$ in (41).

From now on, we take $a = \alpha$ and $b - 1 = \beta$ in (42) and (43) and assume

$$\ln G(t_0) = g_0, \tag{46}$$

where g_0 is a real number. The following theorems yield the second type Gompertz dynamic curves.

Theorem 0.12. *Suppose that β is regressive and $\alpha > 0$. Then the solution of the IVP (42)–(46) is given by*

$$G(t) = \exp\left(e_\beta(t, t_0)\left[g_0 + \alpha \int_{t_0}^t e_\beta(t_0, \sigma(\tau))\Delta\tau\right]\right) \tag{47}$$

for $t \in [t_0, \infty)_\mathbb{T}$.

Proof. If $\ln G(t)$ is taken as $u(t)$, then the IVP (42)–(46) becomes $u^\Delta(t) = \alpha + \beta u(t)$, $u(t_0) = g_0$, $t \in g_0, t \in [t_0, \infty)_\mathbb{T}$. Then by Theorems 0.1 and 0.2, we obtain

$$\begin{aligned} u(t) &= e_\beta(t, t_0)g_0 + \alpha \int_{t_0}^t e_\beta(t, \sigma(\tau))\Delta\tau \\ &= e_\beta(t, t_0)[g_0 + \alpha \int_{t_0}^t e_{\ominus\beta}(t, t_0)e_\beta(t, \sigma(\tau))\Delta\tau] \\ &= e_\beta(t, t_0)[g_0 + \alpha \int_{t_0}^t e_\beta(t_0, \sigma(\tau))\Delta\tau], \quad t \in [t_0, \infty)_\mathbb{T}. \end{aligned}$$

Since $G = e^u$, (47) is obtained as the solution of the IVP (42)–(46).

Theorem 0.13. *Suppose β is regressive and $\alpha > 0$. Then the solution of the IVP (43)–(46) is given by*

$$G(t) = \exp\left(e_{\ominus\beta}(t,t_0)\left[g_0 + \alpha \int_{t_0}^{t} e_{\ominus\beta}(t_0,\tau)\Delta\tau\right]\right) \qquad (48)$$

for $t \in [t_0, \infty)_{\mathbb{T}}$.

Proof. If $\ln G(t)$ is taken as $u(t)$, then the IVP (43)–(46) turns out to be $u^\Delta(t) = \alpha - \beta u^\sigma(t)$, $u(t_0) = g_0$, $t \in [t_0, \infty)_{\mathbb{T}}$. Again, Theorems 0.1 and 0.3 yield

$$\begin{aligned} u(t) &= e_{\ominus\beta}(t,t_0)g_0 + \alpha \int_{t_0}^{t} e_{\ominus\beta}(t,\tau)\Delta\tau \\ &= e_{\ominus\beta}(t,t_0)[g_0 + \alpha \int_{t_0}^{t} e_{\beta}(t,t_0)e_{\ominus\beta}(t,\tau)\Delta\tau] \\ &= e_{\ominus\beta}(t,t_0)[g_0 + \alpha \int_{t_0}^{t} e_{\ominus\beta}(t_0,\tau)\Delta\tau] \end{aligned}$$

for $t \in [t_0, \infty)_{\mathbb{T}}$. Since $G = u$, (48) is obtained as the solution of the IVP (43)–(46).

Example 0.14. Let $\mathbb{T} = \mathbb{R}$. Then the solutions of the IVPs (42)–(46) and (43)–(46) are

$$\begin{aligned} G(t) &= \exp\left(e^{\beta(t-t_0)}[g_0 + \alpha \int_{t_0}^{t} e^{-\beta(\tau-t_0)}d\tau]\right) \\ &= \exp\left(e^{\beta(t-t_0)}\left[g_0 + \frac{\alpha}{\beta}(1 - e^{-\beta(t-t_0)})\right]\right) \qquad (49) \\ &= \exp\left(\left(g_0 + \frac{\alpha}{\beta}\right)e^{\beta(t-t_0)} - \frac{\alpha}{\beta}\right), \end{aligned}$$

and

$$\begin{aligned} G(t) &= \exp\left(e^{-\beta(t-t_0)}[g_0 + \alpha \int_{t_0}^{t} e^{\beta(\tau-t_0)}d\tau]\right) \\ &= \exp\left(e^{-\beta(t-t_0)}\left[g_0 + \frac{\alpha}{\beta}(e^{\beta(t-t_0)} - 1)\right]\right) \qquad (50) \\ &= \exp\left(\left(g_0 - \frac{\alpha}{\beta}\right)e^{-\beta(t-t_0)} + \frac{\alpha}{\beta}\right), \end{aligned}$$

respectively for $t \in [t_0, \infty)_{\mathbb{T}}$. and here we use (8).

Example 0.15. Let $\mathbb{T} = \mathbb{Z}$. Then the solutions of the IVPs (42)–(46) and (43)–(46) are

$$\begin{aligned} G(t) &= \exp\left(e_\beta(t,t_0)[g_0 + \alpha \int_{t_0}^{t} e_\beta(t_0,\sigma(\tau))\Delta\tau]\right) \\ &= \exp\left((\beta+1)^{(t-t_0)}[g_0 + \alpha \sum_{\tau=t_0}^{t-1}(1+\beta)^{-(\tau+1-t_0)}]\right) \end{aligned}$$

$$\begin{aligned}
&= \exp\left((\beta+1)^{(t-t_0)}[g_0 + \alpha(1+\beta)^{-(t-t_0)}\sum_{\tau=t_0}^{t-1}(1+\beta)^{\tau-t_0}]\right) \qquad (51)\\
&= \exp\left((\beta+1)^{(t-t_0)}\left[g_0 + \alpha(1+\beta)^{-(t-t_0)}\frac{(1+\beta)^{t-t_0}-1}{\beta}\right]\right)\\
&= \exp\left(\left(g_0+\frac{\alpha}{\beta}\right)(\beta+1)^{t-t_0} - \frac{\alpha}{\beta}\right)
\end{aligned}$$

and

$$\begin{aligned}
G(t) &= \exp\left(e_{\ominus\beta}(t,t_0)[g_0 + \alpha\int_{t_0}^{t} e_{\ominus\beta}(t_0,\tau)\Delta\tau]\right)\\
&= \exp\left((\beta+1)^{-(t-t_0)}[g_0 + \alpha\sum_{\tau=t_0}^{t-1}(1+\beta)^{\tau-t_0}]\right)\\
&= \exp\left((\beta+1)^{-(t-t_0)}\left[g_0 + \alpha\frac{(1+\beta)^{t-t_0}-1}{\beta}\right]\right) \qquad (52)\\
&= \exp\left(\left(g_0-\frac{\alpha}{\beta}\right)(\beta+1)^{-(t-t_0)} + \frac{\alpha}{\beta}\right),
\end{aligned}$$

respectively for $t \in [t_0,\infty)_{\mathbb{Z}}$ and here we use (9).

In (51) by taking $t_0 = 0$, $\beta = b-1$ and $\alpha = a$ Equation 3.1 is obtained in [11]. Both continuous Gompertz growth curves (18) and (50) are obtained from the IVPs (13) and (43)–(46). If we let $B = 0$, $e\omega_0 = e^{\frac{\alpha}{\beta}}$, $K = \beta$ in (18), and $g_0 - \frac{\alpha}{\beta} = -1$ in (50), we observe that

$$e.e^{g_0}e^{-\frac{\alpha}{\beta}} = e.e^{-1} = 1$$

which implies $\omega_0 = e^{g_0}$. Similarly, the continuous Gompertz curves (23) and (49) are the solutions of the IVPs (21) and (42)–(46). If we let $B = 0$, $\frac{1}{e}\omega_0 = e^{-\frac{\alpha}{\beta}}$, $\beta = -K$ in (23), and $g_0 + \frac{\alpha}{\beta} = 1$ in (49), we observe that

$$e^{-1}.e^{g_0}e^{\frac{\alpha}{\beta}} = e^{-1}e = 1$$

which implies $\omega_0 = e^{g_0}$. From these observations, we conclude the 3-parameter first type continuous Gompertz curve and the second type continuous Gompertz curve are identical. However, since such an intimate relation among the discrete curves cannot be observed, considering first and second type discrete Gompertz curves contributes to the literature.

Logistic dynamic equations

In this section, we derive 4-parameter and 3-parameter Logistic continuous and discrete curves from Logistic dynamic equations.

3- Parameter logistic dynamic curves

Since there are two versions of linear equations

$$u^\Delta = p(t)u + f(t), \qquad u^\Delta = -p(t)u^\sigma + f(t),$$

there are two Logistic dynamic equations

$$L^\Delta = [\ominus(p(t) + f(t)L)]L, \qquad (53)$$

and

$$L^\Delta = [p(t) \ominus (f(t)L)]L, \qquad (54)$$

respectively, see [16]. By using the definition of circle minus (6), Logistic dynamic Eqs (53) and (54) turn out to be the typical Logistic differential Eq (3) under certain conditions on p and f when $\mathbb{T} = \mathbb{R}$. In [16], it is shown that the solutions of (53) and (54) subject to

$$L(t_0) = l_0 \neq 0 \qquad (55)$$

are given by

$$L(t) = \frac{e_{\ominus p}(t, t_0)}{\frac{1}{l_0} + \int_{t_0}^{t} e_{\ominus p}(\sigma(\tau), t_0) f(\tau) \Delta \tau}, \qquad (56)$$

and

$$L(t) = \frac{e_p(t, t_0)}{\frac{1}{l_0} + \int_{t_0}^{t} e_p(\tau, t_0) f(\tau) \Delta \tau}, \qquad (57)$$

see Theorem 4.2 in [16]. Here, we assume that p is regressive, f is a rd-continuous function. We now calculate continuous and discrete Logistic solutions in order to compare their data fit- ting. In population dynamics, one often assumes that there exists a constant $N \neq 0$ such that $p(t) = Nf(t)$ for all $t \in \mathbb{T}$.

Example 0.16. Let $\mathbb{T} = \mathbb{R}$, $t_0 = 1$, $f = \alpha$ and $p = \beta$, where α and β are constants. Then (56) and (57) turn out to be

$$L = \frac{1}{-\frac{\alpha}{\beta} + \left(\frac{1}{l_0} + \frac{\alpha}{\beta}\right) e^{\beta(t-1)}} \qquad (58)$$

and

$$L = \frac{1}{\frac{\alpha}{\beta} + \left(\frac{1}{l_0} - \frac{\alpha}{\beta}\right) e^{-\beta(t-1)}}, \qquad (59)$$

respectively.

Example 0.17. Let $\mathbb{T} = \mathbb{Z}$, $t_0 = 1$, $f = \alpha$ and $p = \beta$, where α and β are constants. (56) and (57) turn out to be

$$L = \frac{1}{-\frac{\alpha}{\beta} + \left(\frac{1}{l_0} + \frac{\alpha}{\beta}\right)(1+\beta)^{t-1}} \tag{60}$$

and

$$L = \frac{1}{\frac{\alpha}{\beta} + \left(\frac{1}{l_0} - \frac{\alpha}{\beta}\right)(1+\beta)^{-t+1}}, \tag{61}$$

respectively.

Parameter logistic dynamic curves

The 4-parameter Logistic curve

$$\omega(t) = f - \frac{1}{\frac{b}{f} + \left(\frac{1}{f-s} - \frac{b}{f}\right)e^{kt}}, \quad t \in \mathbb{R} \tag{62}$$

is introduced and discussed in [15], where k, f, and s are positive constants and b is a real number. If we let $b = 1$ in (62), then (62) and (3) are equivalent. Equivalently, we have

$$\omega(t) = f - \frac{f}{b + \left(\frac{f}{f-s} - b\right)e^{kt}}, \quad t \in \mathbb{R}. \tag{63}$$

Motivated by (63), we purpose the 4-parameter logistic dynamic curve as

$$\omega(t) = f - \frac{f}{b + \left(\frac{f}{f-s} - b\right)e_k(t,0)}, \quad t \in \mathbb{T}. \tag{64}$$

To obtain the 4-parameter logistic dynamic equation, we differentiate Eq (64) and derive

$$\begin{aligned}
\omega^\Delta &= \frac{f\left(\frac{f-s}{b} - b\right)ke_k(t,0) + kfb - kfb}{\left(b + \left(\frac{f}{f-s} - b\right)e_k(t,0)\right)\left(b + \left(\frac{f}{f-s} - b\right)e_k^\sigma(t,0)\right)} \\
&= \frac{kf\left[b + \left(\frac{f-s}{b} - b\right)e_k(t,0)\right] - kfb}{\left(b + \left(\frac{f}{f-s} - b\right)e_k(t,0)\right)\left(b + \left(\frac{f}{f-s} - b\right)e_k^\sigma(t,0)\right)} \\
&= \frac{kf}{\left(b + \left(\frac{f}{f-s} - b\right)e_k^\sigma(t,0)\right)} - \frac{kfb}{\left(b + \left(\frac{f}{f-s} - b\right)e_k(t,0)\right)\left(b + \left(\frac{f}{f-s} - b\right)e_k^\sigma(t,0)\right)} \\
&= -k\omega^\sigma + kf - \frac{kfb}{\left(b + \left(\frac{f}{f-s} - b\right)e_k(t,0)\right)\left(b + \left(\frac{f}{f-s} - b\right)e_k^\sigma(t,0)\right)}
\end{aligned} \tag{65}$$

$$+ \frac{kfb}{b+\left(\frac{f}{f-s}-b\right)e_k(t,0)} - \frac{kfb}{b+\left(\frac{f}{f-s}-b\right)e_k(t,0)}$$

$$= -k\omega^\sigma + kf + \frac{kfb}{b+\left(\frac{f}{f-s}-b\right)e_k(t,0)}\omega^\sigma - \frac{kfb}{b+\left(\frac{f}{f-s}-b\right)e_k(t,0)}$$

$$= -k\left[1 - \frac{b}{b+\left(\frac{f}{f-s}-b\right)e_k(t,0)}\right]\omega^\sigma + kf\left[1 - \frac{b}{b+\left(\frac{f}{f-s}-b\right)e_k(t,0)}\right].$$

Hence, we obtain the 4-parameter logistic dynamic equation as:

$$\omega^\Delta(t) = -p(t)\omega^\sigma(t) + fp(t), \quad t \in \mathbb{T} \tag{66}$$

where

$$p(t) = k\left[1 - \frac{b}{b+\left(\frac{f}{f-s}-b\right)e_k(t,0)}\right], \quad t \in \mathbb{T}. \tag{67}$$

Theorem 0.18. *Let k, f, s be positive constants. Consider the 4-parameter logistic dynamic Eq (66) with the initial condition*

$$\omega(0) = s. \tag{68}$$

Then,

$$\omega(t) = f + (s-f)e_{\ominus p}(t,0), \quad t \in \mathbb{T} \tag{69}$$

is the unique solution of the IVP (66)–(68) where p is defined as in (67).

Proof. To get the desired result, we use Theorem 0.3. Therefore, we have

$$\begin{aligned}\omega(t) &= se_{\ominus p}(t,0) + \int_0^t fpe_{\ominus p}(t,\tau)\Delta\tau \\ &= se_{\ominus p}(t,0) + \int_0^t fp(\tau)e_p(\tau,t)\Delta\tau \\ &= se_{\ominus p}(t,0) + f(1 - e_{\ominus p}(t,0)) \\ &= f + (s-f)e_{\ominus p}(t,0), \quad t \in \mathbb{T},\end{aligned}$$

which completes the proof.

Example 0.19. If $\mathbb{T} = \mathbb{Z}$, then the solution in (69) turns out to be

$$w(t) = f + (s-f)\frac{1}{\prod_{\tau=0}^{t-1}\left[1+k\left(-\frac{b}{b+\left(\frac{f}{f-s}-b\right)(1+k)^\tau} + 1\right)\right]}, \tag{70}$$

where we use (9).

Let $b = 1$ in (67). Note that $p = \frac{k}{f}\omega$ and this means that Eq (66) turns out to be

$$\begin{aligned}\omega^\Delta &= -\frac{k}{f}\omega\omega^\sigma + k\omega \\ &= k\omega\left(1 - \frac{\omega^\sigma}{f}\right).\end{aligned} \quad (71)$$

If $\mathbb{T} = \mathbb{R}$, then we obtain (3) from (71). One of the logistic dynamic equations is (54). By taking (54) into account and using the definition of minus circle, we get

$$L^\Delta = \left(\frac{p - f_1 L}{1 + \mu f_1 L}\right)x.$$

This implies that

$$L^\Delta + \mu L^\Delta f_1 L = (p - f_1 L)L.$$

By the simple useful formula, we have

$$L^\Delta + (L^\sigma - L)f_1 L = (p - f_1 L)L.$$

Solving the above equation for L^Δ, we get

$$\begin{aligned}L^\Delta &= (p - f_1 L^\sigma)L \\ &= pL\left(1 - \frac{f_1}{p}L^\sigma\right).\end{aligned} \quad (72)$$

If we take $L = \omega$, $p = k$ and $\frac{f_1}{p} = \frac{1}{f}$ in (72), then we obtain Eq (71).

At this point the following question arises: Is it possible to find an alternative 4-parameter logistic dynamic equation which turns out to be the equivalent form of the 3-parameter logistic dynamic equation? To find the answer of this question consider Eq (65) where e_k is replaced by e_k^σ in the last two terms of Eq (65). Then we obtain that

$$\begin{aligned}\omega^\Delta &= -k(\omega + \mu\omega^\Delta) + kf - \frac{kfb}{\left(b + \left(\frac{f}{f-s} - b\right)e_k(t,0)\right)\left(b + \left(\frac{f}{f-s} - b\right)e_k^\sigma(t,0)\right)} \\ &\quad + \frac{kfb}{b + \left(\frac{f}{f-s} - b\right)e_k^\sigma(t,0)} - \frac{kfb}{b + \left(\frac{f}{f-s} - b\right)e_k^\sigma(t,0)}.\end{aligned} \quad (73)$$

Solving the above equation for ω^Δ yields

$$\omega^\Delta(1 + k\mu) = -k\omega + kf + \frac{kb}{b + \left(\frac{f}{f-s} - b\right)e_k^\sigma(t,0)}\omega - \frac{kfb}{b + \left(\frac{f}{f-s} - b\right)e_k^\sigma(t,0)},$$

or

$$\omega^\Delta = -k\omega \left[\frac{1 - \frac{b}{b+\left(\frac{f}{f-s}-b\right)e_k^\sigma(t,0)}}{1+k\mu}\right] + \frac{kf - \frac{kfb}{b+\left(\frac{f}{f-s}-b\right)e_k^\sigma(t,0)}}{1+k\mu}$$

$$= (\ominus k)\omega\left[1 - \frac{b}{b+\left(\frac{f}{f-s}-b\right)e_k^\sigma(t,0)}\right] - f(\ominus k)\left[1 - \frac{b}{b+\left(\frac{f}{f-s}-b\right)e_k^\sigma(t,0)}\right].$$

Hence, we get the following logistic dynamic equation

$$\omega^\Delta = (\ominus k)q\omega - f(\ominus k)q, \quad t \in \mathbb{T}, \tag{74}$$

where

$$q(t) = 1 - \frac{b}{b+\left(\frac{f}{f-s}-b\right)e_k^\sigma(t,0)}, \quad t \in \mathbb{T}. \tag{75}$$

Theorem 0.20. *Let k, f, s be positive constants and q be taken as in (75). Then, Eq (74) with the initial condition (68) has the solution*

$$\omega = (s-f)e_{(\ominus k)q}(t,0) + f, \quad t \in \mathbb{T}. \tag{76}$$

Proof. By using the definition of minus circle (6), we have

$$\begin{aligned}
1 + \mu[\ominus((\ominus k)q)] &= 1 + \mu\left[\ominus\frac{-kq}{1+\mu k}\right] \\
&= 1 + \mu\frac{kq}{1+\mu k - \mu k q} \tag{77} \\
&= \frac{1+\mu k}{1+\mu k - \mu k q}.
\end{aligned}$$

By using Theorem 0.2 and the properties of exponential functions given Theorem 0.1, we arrive at the unique solution as follows:

$$\begin{aligned}
\omega(t) &= se_{(\ominus k)q}(t,0) - f\int_0^t e_{(\ominus k)q}(t,\sigma(\tau))(\ominus k)q(\tau)\Delta\tau \\
&= se_{(\ominus k)q}(t,0) - fe_{(\ominus k)q}(t,0)\int_0^t e_{\ominus((\ominus k)q)}(\sigma(\tau),0)(\ominus k)q(\tau)\Delta\tau \\
&= se_{(\ominus k)q}(t,0) \\
&\quad - fe_{(\ominus k)q}(t,0)\int_0^t e_{\ominus((\ominus k)q)}(\tau,0)[1+\mu(\tau)\ominus((\ominus k)q(\tau))]\frac{-kq(\tau)}{1+\mu(\tau)k}\Delta\tau
\end{aligned}$$

$$= se_{(\ominus k)q}(t,0)$$
$$+ fe_{(\ominus k)q}(t,0) \int_0^t e_{\ominus((\ominus k)q)}(\tau,0) \frac{1+\mu(\tau)k}{1+\mu(\tau)k - \mu(\tau)kq(\tau)} \frac{kq(\tau)}{1+\mu(\tau)k} \Delta\tau$$
$$= se_{(\ominus k)q}(t,0) + fe_{(\ominus k)q}(t,0) \int_0^t e_{\ominus((\ominus k)q)}(\tau,0) \ominus ((\ominus k)q(\tau)) \Delta\tau$$
$$= se_{(\ominus k)q}(t,0) + f(1 - e_{(\ominus k)q}(t,0))$$
$$= (s-f)e_{(\ominus k)q}(t,0) + f, \quad t \in \mathbb{T},$$

which completes the proof.

Example 0.21. If $\mathbb{T} = \mathbb{Z}$, then the solution in (76) turns out to be

$$w(t) = f + (s-f) \prod_{\tau=0}^{t-1} \left[1 + \frac{k}{k+1} \left(\frac{b}{b + \left(\frac{f}{f-s} - b\right)(1+k)^{\tau+1}} - 1 \right) \right]. \tag{78}$$

If we take $b = 1$ in (75), then we get $q = \frac{\omega^\sigma}{f}$. Then Eq (74) turns out to be

$$\omega^\Delta = \frac{\ominus k}{f} \omega^\sigma (\omega - f). \tag{79}$$

Furthermore, if $\mathbb{T} = \mathbb{R}$, (79) turns out to be the logistic differential Eq (3). Note that logistic dynamic Eq (79) is equal neither (53) nor (54). Therefore, (78) with $b = 1$ is different than (60) and (61). It means that obtaining 3-parameter logistic curves from 4-parameter logistic curves yields new 3-parameter logistic discrete curves. The following example consists of two new 3-parameter logistic discrete curves.

Example 0.22. If we let $\mathbb{T} = \mathbb{Z}$ and $b = 1$ in (70) and (78), then we obtain

$$w(t) = f + (s-f) \frac{1}{\prod_{\tau=0}^{t-1} \left[1 + k \left(-\frac{1}{1 + \left(\frac{f}{f-s} - 1\right)(1+k)^\tau} + 1 \right) \right]}, \tag{80}$$

and

$$w(t) = f + (s-f) \prod_{\tau=0}^{t-1} \left[1 + \frac{k}{k+1} \left(\frac{1}{1 + \left(\frac{f}{f-s} - 1\right)(1+k)^{\tau+1}} - 1 \right) \right], \tag{81}$$

respectively.

Goodness-of-fit test for gompertz and logistic curves and conclusion

The main aim of this study for the statistical analysis of Gompertz and Logistic curves is to determine whether their equations are able to model Pseudomonas Putita and tumor data sets given in [10] and [11]. In order to achieve our goal, *p*-values of the parameters, adjusted R- squared values and six types of errors, namely, RMSE (root mean square error), RRMSE (Rela- tive Root Mean Square Error), MAPE (Mean Absolute Percent Error), MAE (mean absolute error), U1 (Theil inequality coefficient, Theil's U1), U2 (Theil inequality coefficient, Theil's U2) for each data set calculated, where

$$RMSE = \sqrt{\frac{\sum_{t=1}^{T} e_t^2}{T}}, \quad RRMSE = \sqrt{\frac{\sum_{t=1}^{T} |\frac{e_t}{y_t}|^2}{T}}, \quad MAE = \frac{\sum_{t=1}^{T} |e_t|}{T}$$

$$MAPE = \frac{\sum_{t=1}^{T} \frac{|e_t|}{y_t}}{T}, \quad U1 = \frac{\sqrt{\frac{\sum_{t=1}^{T} e_t^2}{T}}}{\sqrt{\frac{\sum_{t=1}^{T} y_t^2}{T}} + \sqrt{\frac{\sum_{t=1}^{T} \hat{y}_t^2}{T}}}, \quad U2 = \frac{\sqrt{\frac{\sum_{t=1}^{T} e_t^2}{T}}}{\sqrt{\frac{\sum_{t=1}^{T} y_t^2}{T}}}$$

Here, e_t shows the error component, y_t the original time series values, $\hat{y}_t$ the estimated val- ues of the time series, and T the number of observations of the series. The criteria to determine an equation showing better performance in terms of goodness of fit is to have statistically sig- nif- icant coefficients; in other words, meaningful *p*-values for each coefficient, the larger adjusted R-squared value and smaller errors (see S1 and S2 Figs). The coefficient estimates of these mod- els are obtained by the Mathematica 11.0 and the Wolfram Language uses "Conjuga- teGradient", "Gradient", "LevenbergMarquardt", "Newton", "NMinimize", and "QuasiNew- ton" methods.

The curves which successfully model Pseudomonas putita data are 4-parameter first type contin- uous Gompertz curve (18), 3-parameter first type Gompertz curves (26), (27), (29), Zwietering modification of continuous Gompertz curve (31), Gompertz-Liard curves (35), (38), and 2-pa- rameter second type Gompertz curves 49α0, 50α0, 51α0 and 52α0 that are obtained from (49), (50), (51), (52) with $t_0 = 1$ and $\alpha = 0$. According to the results in S1 Fig Eq (18) has the best fit among Gompertz curves. Therefore, we observe that the performance of 4-parameter Gompertz curves for bacteria is better than 3 and 2-parameter Gompertz growth curves. In addition to these, Eqs (27), (29) and (38) are the 3- parameter discrete Gompertz growth curves are new, thus, contribute to the literature.

When we concentrate attention on the Logistic type growth curves, from S1 Fig, it is appar- ent that all of the Logistic type equations are successful in modeling the bacteria data set. In addition, growth curves (70), (78), (80), (81) are the new 4-parameter discrete Logistic curves that are obtained in this study. According the results in S1 Fig, among the Logistic type growth curves, 4-parameter Logistic growth curves are better in modeling when it is compared with 3- parameter ones.

Moreover, we can infer that the 4-parameter continuous Logistic curve (63) and (70), (78) model better than Eq (18). Thus, Pseudomonas Putita data is modeled by Logistic type equa- tions bet- ter than Gompertz type equations. In addition, Eq (63) highlighted with orange, (70) and (78) highlighted with green in S1 Fig, have the smallest errors among the other Logistic equations.

Eqs (70) and (78) have smaller errors after the eighth decimal when they are com- pared with Eq (63) highlighted yellow in S1 Fig. Therefore, new discrete growth curves (70) and (78) are the best in modeling bacteria data.

Eq (29) highlighted with yellow, (35) highlighted with green, $49\alpha0$, $50\alpha0$, $51\alpha0$ and $52\alpha0$ highlighted with orange in S2 Fig are the curves that successfully model the tumor data among the Gompertz curves. Eq (35) has the best fit in modeling based on S2 Fig. This equation is the Gompertz Liard continuous equation that was developed for tumor modeling in the literature, so our result is compatible with the one in [14]. On the other hand, 3-parameter discrete Gompertz growth curve (29) also models the tumor data set and this equation is developed in this study as well. In addition, 4-parameter continuous and discrete second type Gompertz curves and (52) are also new. 2-parameter version $51a0$ of (51) was studied in [8] as Mirror Gom- pertz curve. Nevertheless, its discrete version $52\alpha0$ is a new contribution to the literature. At this point, we declare that 3-parameter Gompertz curves are more successful in modeling tumor data than 4-parameter Gompertz curves. Although the errors of the Logistic type curves are smaller than the errors of Gompertz type curves, all of their parameters are not statistically significant. Therefore, the Gompertz-Liard curve (35) is the best curve in modeling when it is compared with all the other curves and so one can say that Gompertz type curves have better fitting than Logistic type curves for tumor data.

As a result, Logistic curves are better in modeling bacteria data whereas tumor data is mod- eled better by Gompertz curves. Increasing the number of parameters of Logistic curves give favorable results for bacteria data while decreasing the number of the parameters of Gompertz curves for tumor data turns out to be reasonable. S3 Fig gives us the curve fittings of 4-parame- ter discrete Logistic curve (70) and 3-parameter discrete Gompertz curve (29) for bacteria data set and also shows the importance of the number of parameters.

Supporting information

S1 Fig. Bacteria. Fitted parameters and statistical error analysis for bacteria. *: significant at.10 level, **: significant at.05 level and ***: significant at.01 level.
(PNG)

S2 Fig. Tumor. Fitted parameters and statistical error analysis for tumor data. *: significant at.10 level, **: significant at.05 level and ***: significant at.01 level.
(PNG)

S3 Fig. Compare. Compare with 4-parameter discrete Logistic curve and 3-parameter discrete Gompertz curve for bacteria data set.
(PDF)

S1 File. Bacteria. Data set for bacteria.
(PDF)

S2 File. Tumor. Data set for tumor.
(PDF)

Author Contributions

Formal analysis: Elvan Akın, Neslihan Nesliye Pelen.

Methodology: Elvan Akın, Neslihan Nesliye Pelen, Ismail Uğur Tiryaki, Fusun Yalcin.

Software: Ismail Uğur Tiryaki, Fusun Yalcin.

Supervision: Elvan Akın.

Validation: Elvan Akın, Neslihan Nesliye Pelen, Ismail Uğur Tiryaki, Fusun Yalcin.

Writing – original draft: Elvan Akın, Neslihan Nesliye Pelen.

Writing – review & editing: Elvan Akın.

References

1. Alverez-Arenas Arturo, Belmonte-Beitia J., and Calvo Gabriel Nonlinear waves in a simple model of hig- grade glioma. Applied Mathemaicts and Nonlinear Sciences. Volume 1: Issue 2, p.405–422 (2016).

2. Atıcı F. M., Atıcı M., Hrushesky W.J.M., Nguyen N. Modeling Tumor Volume with Basic Functions of Fractional Calculus. Progr. Fract. Differ. Appl. 1, No. 4, 229–241 (2015).

3. Bassukas I. D., Schultze B. M. The recursion formula of the Gompertz function: A simple method for the estimation and comparison of tumor growth curves. Growth Dev. Aging, Vol.52, (1988), pp.113–122.

4. Domingues Jose Sergio. Gompertz Model: Resolution and Analysis for Tumors. Journal of Mathemati- cal Modelling and Application 1, No. 7, 70–77 (2012).

5. Durbin P. W., Jeung N., Williams M. H., and Arnold J. S. Construction of a Growth Curve for Mammary Tumors of the Rat. Cancer Research. Volume 27, p.1341–1347 (1967).

6. Rojas Clara, and Belmonte-Beitia J. Optimal control problems for differential equations applied to tumor growth: state of the art. Applied Mathemaicts and Nonlinear Sciences. Volume 3: Issue 2, p.375–402 (2018).

7. Dudek Krzysztof, Kedzia Wojciech, Kedzia Emilia, and Kedzia Alicja. Mathematical modelling of the growth of human fetus anatomical structures. Anat Sci Int.

8. Skiadas C.H.Comparing the Gompertz-Type Models with a First Passage Time density Model. Advances in Data Analysis. Springer/Birkhauser (2010), pp. 203–209.

9. Kelley W., and Peterson A. The Theory of Differential Equations: Classical and Qualitative. Springer, Second Edition.

10. Annadurai G., Rajesh Babu S., Srinivasamoorthy V. R. Development of mathematical models (Logistic, Gompertz and Richards models) describing the growth pattern of Pseudomonas putida(NICM 2174). Bioprocess Engineering, Vol.23, (2000), pp.607–612.

11. Şengü"l, S. Discrete fractional calculus and its applications to tumor growth. Master thesis, Paper 161. http://digitalcommons.wku.edu/theses/161, 2010.

12. Espinosa-Urgel M., Salido A., Ramos J. Genetic Analysis of Functions Involved in Adhesion of Pseudo- monas putida to Seeds. Journal of Bacteriology. Volume 182, p.2363–2369 (2000).

13. Perz J., Craig A. S., Stratton C. W., Bodner S. J., Philipps W. E. Jr., and Schaffner W. Pseudomonas putida Septicemia in a Special Care Nursery Due to Contaminated Flush Solutions Prepared in a Hospi- tal Pharmacy. Journal of Clinical Microbiology, volume 43, p.5316–5318 (2005).

14. Tjorve K.M.C., Tjorve E. The use of Gompertz models in growth analyses, and new Gompertz-model approach: An addition to the Unified-Richards family. PLOS ONE https://doi.org/10.1371/journal.pone. 0178691. (2017). PMID: 28582419

15. Wan X., Wang M., Wang G., Zhong W. A new 4 parameter generalized logistic equation and its applica- tions to mammalian somatic growth. Acta Theriologica 45(2):145–153,2000.

16. Akın-Bohner E. and Bohner M. Miscellaneous Dynamic Equations. Methods and Applications of Analy- sis. Vol 10, No. 1, pp. 011–030, (2003).

17. Bohner M. and Peterson A. C. Dynamic Equations on Time Scales: An Introduction with Applications. Birkhauser (2001).

18. Bohner M. and Peterson A. C. Advances in Dynamic Equations on Time Scales. Birkhauser (2003).

19. Gibson A.M, Bratchell N., Roberts T.A. Predicting microbial growth: growth responses of salmonellae in a laboratory medium as affected by pH, sodium chloride and storage temperature. Int. J. Food Micro- biol. 1988; 6:155–78. https://doi.org/10.1016/0168-1605(88)90051-7 PMID: 3275296

Mean almost periodicity and moment exponential stability of semi-discrete random cellular neural networks with fuzzy operations

Sufang Han[1], Guoxin Liu[1], Tianwei Zhang[2]*

1 School of Mathematics and Statistics, Central South University, Changsha, China, **2** City College, Kunming University of Science and Technology, Kunming, China

These authors contributed equally to this work.

* zhang@kmust.edu.cn

Editor: Jun Ma, Lanzhou University of Technology, CHINA

Competing interests: The authors have declared that no competing interests exist.

Abstract

By using the semi-discretization technique of differential equations, the discrete analogue of a kind of cellular neural networks with stochastic perturbations and fuzzy operations is for- mulated, which gives a more accurate characterization for continuous-time models than that by Euler scheme. Firstly, the existence of at least one p-th mean almost periodic sequence solution of the semi-discrete stochastic models with almost periodic coefficients is investi- gated by using Minkowski inequality, Hö lder inequality and Krasnoselskii's fixed point theo- rem. Secondly, the p-th moment global exponential stability of the semi-discrete stochastic models is also studied by using some analytical skills and the proof of contradiction. Finally, a problem of stochastic stabilization for discrete cellular neural networks is studied.

Introduction

Cellular neural networks (CNNs) [1] have been widely applied in psychophysics, parallel computing, perception, robotics associative memory, image processing pattern recognition and combinatorial optimization. Most of these applications heavily depend on the (almost) periodicity and global exponential stability. Specifically, there are many scholars focusing on the study of the equilibrium points, (almost) periodic solutions and global exponential stability of CNNs with time delays in literatures [2–7]. For instance, Xu [7] considered the following CNNs with time delays:

$$\frac{dx_i(t)}{dt} = -a_i(t)x_i(t) + \sum_{j=1}^{n} b_{ij}(t)f_j(x_j(t)) + \sum_{j=1}^{n} c_{ij}(t)g_j(x_j(t - \tau_{ij}(t))) + I_i(t), \quad (1)$$

where n denotes the number of units in a neural network, $x_i(t)$ corresponds to the state of the ith unit at time t, $a_i > 0$ represents the passive decay rates at time t, f_j and g_j are the neuronal output signal functions, $b_{ij}(t)$ and $c_{ij}(t)$ denote the strength of the jth unit on the ith unit at time t, $I_i(t)$ denotes the external inputs at time t, the continuous function $\tau_{ij}(t)$ corresponds to the transmission delay at time t, $i, j = 1, 2, \ldots, n$. In [7], the author studied the existence and exponential stability of anti-periodic solutions of system (1).

In real world applications, most of the problems are uncertain. They should be described by uncertain models and studied by using the research techniques for uncertain models. Stochas- tic and fuzzy theories are the most general and practical techniques for the research of uncer- tain models. On one hand, in the actual situations, uncertainties have a consequence on the performance of neural networks. The connection weights of the neurons depend on certain resistance and capacitance values that include modeling errors or uncertainties during the parameter identification process. Therefore, many neural network models described by sto- chastic differential equations [8, 9] have been widely studied over the last two decades, see [10–17]. On the other hand, fuzzy theory was conceived in the 1960s by L.A. Zadeh, it took about 20 years until the broader use of this theory in practice. Fuzzy technology joined forces with artificial neural networks and genetic algorithms under the title of computational intelli- gence or soft computing. In recent years, the research on the dynamical behaviours of fuzzy neural networks has attracted much attention, see [18–22]. To summarize, we consider the fol- lowing CNNs with stochastic perturbations and fuzzy operations:

$$dx_i(t) = \left[-a_i(t)x_i(t) \right.$$

$$+ \sum_{j=1}^{n} b_{ij}(t)f_j(x_j(t)) + \sum_{j=1}^{n} c_{ij}(t)g_j(x_j(t - \tau_{ij}(t))) + \bigwedge_{j=1}^{n} \alpha_{ij}g_j(x_j(t - \tau_{ij}(t)))$$

$$+ \bigwedge_{j=1}^{n} T_{ij}\mu_j + \bigvee_{j=1}^{n} \beta_{ij}g_j(x_j(t - \tau_{ij}(t))) + \bigvee_{j=1}^{n} S_{ij}\mu_j + I_i(t) \right] dt \quad (2)$$

$$+ \sum_{j=1}^{n} d_{ij}(t)\sigma_j(x_j(t - \eta_{ij}(t)))dw_j(t),$$

where α_{ij}, β_{ij}, T_{ij} and S_{ij} are elements of fuzzy feedback MIN, MAX template, fuzzy feed for- ward MIN and MAX template, respectively; V and W denote the fuzzy AND and fuzzy OR operation, respectively; d_{ij}, η_{ij} and σ_j are similarly specified as that in system (1), w_j is the stan- dard Brownian motion defined on a complete probability space, $i, j = 1, 2, \ldots, n$.

Periodicity often appears in implicit ways in various natural phenomena. Though one can deliberately periodically fluctuate environmental parameters in laboratory experiments, fluctuations in nature are hardly periodic. Almost periodicity is more likely to accurately describe natural fluctuations [23–30]. The concept of mean almost periodicity is important in probability especially for investigations on stochastic processes. In particular, mean almost periodicity enables us to understand the impact of the noise or stochastic perturbation on the corresponding recurrent motions, is of great concern in the study of stochastic differential equations and random dynamical systems. The notion of almost periodic stochastic process was proposed in the 1980s and since then almost periodic solutions to stochastic differential equations driven have been studied by many authors. On the other hand, the problem of stability analysis of dynamic systems has a rich, long history of literature [31–35]. All the applications of such stochastic dynamical systems depend on qualitative behavior such as stability, existence and uniqueness, convergence and so on. In particular, exponential stability is a significant one in the design and applications of neural networks. Therefore, the mean almost periodicity and moment exponential stability of various kinds of stochastic neural networks has been reported in [36–41].

The discrete-time neural networks become more important than the continuous-time counterparts when implementing the neural networks in a digital way. In order to investigate the dynamical characteristics with respect to digital signal transmission, it is essential to formulate the discrete analog of neural networks. A large number of literatures have been obtained for the dynamics of discrete-time neural networks formulated by Euler scheme [42–46]. Mohamad and Gopalsamy [47, 48] proposed a novel method (i.e., semi-discretization technique) in formulating a discrete-time analogue of the continuous-time neural networks, which faithfully preserved the characteristics of their continuous-time counterparts. In [47], the authors employed computer simulations to show that semi-discrete models give a more accurate characterization for the corresponding continuous-time models than that by Euler scheme. With the help of the semi-discretization technique [47], many scholars obtained the semi-discrete analogue of the continuous-time neural networks and some meaningful results were gained for the dynamic behaviours of the semi-discrete neural networks, such as periodic solutions, almost periodic solutions and global exponential stability, see [49–55]. For instance, Huang et al. [52] discussed the following semi-discrete cellular neural networks:

$$x_i(k+1) = e^{-a_i(k)}x_i(k) + \frac{1-e^{-a_i(k)}}{a_i(k)}\left[\sum_{j=1}^{n}b_{ij}(k)f_j(x_j(k)) + I_i(k)\right], \qquad (3)$$

where $k \in \mathbb{Z}$, $\mathbb{Z}$ denotes the set of integers, $i = 1, 2, \ldots, n$. In [52], sufficient conditions were obtained for the existence of a unique stable almost periodic sequence solution of system (3) under assumption of almost periodicity of coefficients of system (3). Similarly, Ji [55] considered a kind of semi-discrete Cohen-Grossberg neural networks with delays and the same problems as that in [52] were studied. In 2014, by using semi-discretization technique [47], Huang et al. [53] obtained the following semi-discrete models for a class of general neural networks:

$$x_i(k+1) = e^{-a_i(k)}x_i(k) + \frac{1-e^{-a_i(k)}}{a_i(k)}\left[\sum_{l=1}^{m}\sum_{j=1}^{n}b_{ijl}(k)f_j(x_j(k-\tau_{ijl})) + I_i(k)\right], \qquad (4)$$

where $k \in \mathbb{Z}$, $i = 1, 2, \ldots, n$. The authors [53] derived the existence of locally exponentially convergent $2N$ almost periodic sequence solutions of system (4). Kong and Fang [50] in 2018 investigated a class of semi-discrete neutral-type neural networks with delays and some results are acquired for the existence of a unique pseudo almost periodic sequence solution which is globally attractive and globally exponentially stable.

However, the disquisitive models in literatures [49–55] are deterministic. Stimulated by this point, we should consider random factors in the studied models, such as system (2). By using the semi-discretization technique [47], Krasnoselskii's fixed point theorem and stochastic theory, the main aim of this paper is to establish some decision theorems for the existence of p-th mean almost periodic sequence solutions and p-th moment global exponential stability for the semi-discrete analogue of uncertain system (2). The work of this paper is a continuation of that in [52–55] and the results in this paper complement the corresponding results in [52–55]. The main contributions of this paper are summed up as: **(1)** The semi-discrete analogue is established for stochastic fuzzy CNNs (2); **(2)** A Volterra additive equation is derived for the solution of the semi-discrete stochastic fuzzy CNNs; **(3)** The existence of p-th mean almost periodic sequence solutions is obtained; **(4)** A decision theorem is acquired for the p-th moment global exponential stability; **(5)** A problem of stochastic stabilization for discrete CNNs is proposed and researched.

Throughout this paper, we use the following notations. Let $\mathbb{R}$ denote the set of real numbers. $\mathbb{R}^n$ denotes the n-dimensional real vector space. Let $(\Omega, \mathcal{F}, P)$ be a complete probability space. Denote by $BC(\mathbb{Z}, L^p(\Omega, \mathbb{R}^n))$ the vector space of all bounded continuous functions from $\mathbb{Z}$ to $L^p(\Omega, \mathbb{R}^n)$, where $L^p(\Omega, \mathbb{R}^n)$ denotes the collection of all p-th integrable $\mathbb{R}^n$-valued random variables. Then $BC(\mathbb{Z}, L^p(\Omega, \mathbb{R}^n))$ is a Banach space with the norm $\|X\|_p = \sup_{k \in \mathbb{Z}} |X|_p$, $|X|_p = \max_{1 \le i \le n} (E|x_i(k)|^p)^{\frac{1}{p}}$, $\forall X = \{x_i\} := (x_1, x_2, \ldots, x_n)^T \in BC(\mathbb{Z}, L^p(\Omega, \mathbb{R}^n))$, where $p > 1$ and $E(\cdot)$ stands for the expectation operator with respect to the given probability measure P. Set $\overline{f} = \sup_{k \in \mathbb{Z}} |f(k)|$ and $\underline{f} = \inf_{k \in \mathbb{Z}} |f(k)|$ for bounded real function f defined on $\mathbb{Z}$.

$$[a,b]_{\mathbb{Z}} = [a,b] \cap \mathbb{Z}, \forall a, b \in \mathbb{R}.$$

Discrete analogue and preliminaries

The semi-discretization model

For the sake of gaining the discrete analogue of system (2) with the semi-discretization technique [47], the following uncertain CNNs with piecewise constant arguments corresponding to system (2) have been taken into account:

$$\begin{aligned}
dx_i(t) = &\left[-a_i([t])x_i(t) + \sum_{j=1}^n b_{ij}([t])f_j(x_j([t])) + \sum_{j=1}^n c_{ij}([t])g_j(x_j([t]-\tau_{ij}([t])))\right.\\
&+ \bigwedge_{j=1}^n \alpha_{ij}g_j(x_j([t]-\tau_{ij}([t]))) + \bigwedge_{j=1}^n T_{ij}\mu_j + \bigvee_{j=1}^n \beta_{ij}g_j(x_j([t]-\tau_{ij}([t])))\\
&\left.+ \bigvee_{j=1}^n S_{ij}\mu_j + \sum_{j=1}^n d_{ij}([t])\sigma_j(x_j([t]-\eta_{ij}([t])))\Delta w_j([t]) + I_i([t]) \right] dt,
\end{aligned}$$

where [t] denotes the integer part of t, $i = 1, 2, \ldots, n$. Here the discrete analogue of the stochastic parts of system (2) is obtained by Euler scheme, i.e., $dw_j(t) = \Delta w_j([t])dt = [w_j([t] + 1) - w_j([t])]dt$, $j = 1, 2, \ldots, n$. For each t, there exists an integer k such that $k \le t < k + 1$. Then the above equation becomes

$$\begin{aligned}
dx_i(t) &= \Bigg[-a_i(k)x_i(t) + \sum_{j=1}^n b_{ij}(k)f_j(x_j(k)) + \sum_{j=1}^n c_{ij}(k)g_j(x_j(k - \tau_{ij}(k))) \\
&+ \bigwedge_{j=1}^n \alpha_{ij} g_j(x_j(k - \tau_{ij}(k))) + \bigwedge_{j=1}^n T_{ij}\mu_j + \bigvee_{j=1}^n \beta_{ij} g_j(x_j(k - \tau_{ij}(k))) \\
&+ \bigvee_{j=1}^n S_{ij}\mu_j + \sum_{j=1}^n d_{ij}(k)\sigma_j(x_j(k - \eta_{ij}(k)))\Delta w_j(k) + I_i(k) \Bigg] dt,
\end{aligned}$$

where $i = 1, 2, \ldots, n$. Integrating the above equation from k to t and letting $t \to k + 1$, we achieve the discrete analogue of system (2) as follows:

$$\begin{aligned}
x_i(k+1) &= e^{-a_i(k)} x_i(k) \\
&+ \frac{1 - e^{-a_i(k)}}{a_i(k)} \Bigg[\sum_{j=1}^n b_{ij}(k)f_j(x_j(k)) + \sum_{j=1}^n c_{ij}(k)g_j(x_j(k - \tau_{ij}(k))) \\
&+ \bigwedge_{j=1}^n \alpha_{ij} g_j(x_j(k - \tau_{ij}(k))) + \bigwedge_{j=1}^n T_{ij}\mu_j + \bigvee_{j=1}^n \beta_{ij} g_j(x_j(k - \tau_{ij}(k))) \\
&+ \bigvee_{j=1}^n S_{ij}\mu_j + \sum_{j=1}^n d_{ij}(k)\sigma_j(x_j(k - \eta_{ij}(k)))\Delta w_j(k) + I_i(k) \Bigg],
\end{aligned} \quad (5)$$

where $k \in \mathbb{Z}$, $i = 1, 2, \ldots, n$.

Volterra additive equation for the solution of system (5)

Lemma 1. $X = \{x_i\}$ is a solution of system (5) if and only if

$$\begin{aligned}
x_i(k) &= \prod_{s=k_0}^{k-1} e^{-a_i(s)} x_i(k_0) + \sum_{v=k_0}^{k-1} \prod_{s=v+1}^{k-1} \frac{e^{-a_i(s)}[1 - e^{-a_i(v)}]}{a_i(v)} \Bigg[\sum_{j=1}^n b_{ij}(v)f_j(x_j(v)) \\
&+ \sum_{j=1}^n c_{ij}(v)g_j(x_j(v - \tau_{ij}(v))) + \bigwedge_{j=1}^n \alpha_{ij} g_j(x_j(v - \tau_{ij}(v))) \\
&+ \bigwedge_{j=1}^n T_{ij}\mu_j + \bigvee_{j=1}^n \beta_{ij} g_j(x_j(v - \tau_{ij}(v))) + \bigvee_{j=1}^n S_{ij}\mu_j \\
&+ \sum_{j=1}^n d_{ij}(v)\sigma_j(x_j(v - \eta_{ij}(v)))\Delta w_j(v) + I_i(v) \Bigg],
\end{aligned} \quad (6)$$

where $k_0 \in \mathbb{Z}$, $k \in [k_0 + 1, +\infty)_{\mathbb{Z}}$, $i = 1, 2, \ldots, n$.

Proof. Let

$$F_i(k, x) := \sum_{j=1}^n b_{ij}(k) f_j(x_j(k)) + \sum_{j=1}^n c_{ij}(k) g_j(x_j(k - \tau_{ij}(k)))$$

$$+ \bigwedge_{j=1}^n \alpha_{ij} g_j(x_j(k - \tau_{ij}(k))) + \bigwedge_{j=1}^n T_{ij} \mu_j + \bigvee_{j=1}^n \beta_{ij} g_j(x_j(k - \tau_{ij}(k))) + \bigvee_{j=1}^n S_{ij} \mu_j$$

$$+ \sum_{j=1}^n d_{ij}(k) \sigma_j(x_j(k - \eta_{ij}(k))) \Delta w_j(k) + I_i(k), \quad k \in \mathbb{Z}, \ i = 1, 2, \ldots, n.$$

Assume that $X = \{x_i\}$ is a solution of system (5). By $\Delta[u(k)v(k)] = [\Delta u(k)]v(k) + u(k+1)[\Delta v(k)]$ and system (5), it gets

$$\Delta \left[\prod_{s=0}^{k-1} e^{a_i(s)} x_i(k) \right] = \prod_{s=0}^{k} \frac{e^{a_i(s)}[1 - e^{-a_i(k)}]}{a_i(k)} F_i(k, x), \quad k \in \mathbb{Z}, \ i = 1, 2, \ldots, n.$$

So

$$\sum_{v=k_0}^{k-1} \Delta \left[\prod_{s=0}^{v-1} e^{a_i(s)} x_i(v) \right] = \sum_{v=k_0}^{k-1} \prod_{s=0}^{v} \frac{e^{a_i(s)}[1 - e^{-a_i(v)}]}{a_i(v)} F_i(v, x)$$

is equivalent to

$$\prod_{s=0}^{k-1} e^{a_i(s)} x_i(k) = \prod_{s=0}^{k_0-1} e^{a_i(s)} x_i(k_0) + \sum_{v=k_0}^{k-1} \prod_{s=0}^{v} \frac{e^{a_i(s)}[1 - e^{-a_i(v)}]}{a_i(v)} F_i(v, x),$$

where $i = 1, 2, \ldots, n$, $k \in \mathbb{Z}$. By the above equations, we can easily derive (6).

If $X = \{x_i\}$ satisfies (6), then

$$x_i(k) = \prod_{s=k_0}^{k-1} e^{-a_i(s)} x_i(0) + \sum_{v=k_0}^{k-1} \prod_{s=v+1}^{k-1} \frac{e^{a_i(s)}[1 - e^{-a_i(v)}]}{a_i(v)} F_i(v, x),$$

which implies that

$$x_i(k+1) = \prod_{s=k_0}^{k} e^{-a_i(s)} x_i(0) + \sum_{v=k_0}^{k} \prod_{s=v+1}^{k} \frac{e^{a_i(s)}[1 - e^{-a_i(v)}]}{a_i(v)} F_i(v, x)$$

$$= e^{-a_i(k)} \left[\prod_{s=k_0}^{k-1} e^{-a_i(s)} x_i(0) \right.$$

$$+ \sum_{v=k_0}^{k-1} \prod_{s=v+1}^{k-1} \frac{e^{a_i(s)}[1-e^{-a_i(v)}]}{a_i(v)} F_i(v,x) \Bigg] + \frac{1-e^{-a_i(k)}}{a_i(k)} F_i(k,x)$$

$$= e^{-a_i(k)} x_i(k) + \frac{1-e^{-a_i(k)}}{a_i(k)} F_i(k,x),$$

where $i = 1, 2, \ldots, n$, $k \in \mathbb{Z}$. Therefore, $X = \{x_i\}$ is a solution of system (5). This completes the proof.

Some lemmas

Lemma 2. ([56]) (Minkowski inequality) Assume that $p \geq 1$, $E|\xi|^p < \infty$, $E|\eta|^p < \infty$, then

$$(E|\xi + \eta|^p)^{1/p} \leq (E|\xi|^p)^{1/p} + (E|\eta|^p)^{1/p}.$$

Lemma 3. ([56]) (Hölder inequality) Assume that $p > 1$, then

$$\sum_k |a_k b_k| \leq \left[\sum_k |a_k|\right]^{1-1/p} \left[\sum_k |a_k||b_k|^p\right]^{1/p}.$$

If $p = 1$, then $\sum_k |a_k b_k| \leq (\sum_k |a_k|)(\sup_k |b_k|)$.

Lemma 4. ([9]) Suppose that $g \in L^2([a,b], \mathbb{R})$, then

$$E\left[\sup_{t \in [a,b]} \left|\int_a^t g(s)d\omega(s)\right|^p\right] \leq C_p E\left[\int_a^b |g(t)|^2 dt\right]^{\frac{p}{2}},$$

where

$$C_p = \begin{cases} (32/p)^{p/2}, & 0 < p < 2, \\ 4, & p = 2, \\ \left[\frac{p^{p+1}}{2(p-1)^{(p-1)}}\right]^{\frac{p}{2}}, & p > 2. \end{cases}$$

Lemma 5. Assume that $\{x(k) : k \in \mathbb{Z}\}$ s real-valued stochastic process and $w(k)$ is the standard Brownian motion, then

$$E|x(k)\Delta w(k)|^p \leq C_p E|x(k)|^p, \quad k \in \mathbb{Z},$$

where C_p is defined as that in Lemma 4, $p > 0$.

Proof. By Lemma 4, it follows that

$$E|x(k)\Delta w(k)|^p = E\left|\int_k^{k+1} x(k)\, dw(s)\right|^p \leq C_p E\left|\int_k^{k+1} x^2(k)\, ds\right|^{\frac{p}{2}} \leq C_p E|x(k)|^p,$$

where $k \in \mathbb{Z}$. This completes the proof.

Lemma 6. ([57]) Suppose $X = \{x_j\}$ and $Y = \{y_j\}$ are two states of system (5), then we have

$$\left|\bigwedge_{j=1}^{n} \alpha_{ij} f_j(x_j) - \bigwedge_{j=1}^{n} \alpha_{ij} f_j(y_j)\right| \leq \sum_{j=1}^{n} |\alpha_{ij}||f_j(x_j) - f_j(y_j)|$$

and

$$\left|\bigvee_{j=1}^{n} \beta_{ij} f_j(x_j) - \bigvee_{j=1}^{n} \beta_{ij} f_j(y_j)\right| \leq \sum_{j=1}^{n} |\beta_{ij}||f_j(x_j) - f_j(y_j)|, \quad i = 1, 2, \ldots, n.$$

p-th mean almost periodic sequence solution

Definition 1. ([8]) A stochastic process $X \in BC(\mathbb{Z}; L^p(\Omega; \mathbb{R}^n))$ is said to be p-th mean almost periodic sequence if for each $\epsilon > 0$, there exists an integer $l(\epsilon) > 0$ such that each interval of length $l(\epsilon)$ contains an integer ω for which

$$|X(k+\omega) - X(k)|_p = \max_{1 \leq i \leq n} (E|x_i(k+\omega) - x_i(k)|^p)^{\frac{1}{p}} < \epsilon, \quad \forall k \in \mathbb{Z}.$$

A stochastic process X, which is 2-nd mean almost periodic sequence will be called square-mean almost periodic sequence. Like for classical almost periodic functions, the number ω will be called an -translation of X.

Lemma 7. ([58]) Assume that Λ is a closed convex nonempty subset of a Banach space $\mathbb{X}$. Suppose further that $\mathcal{B}$ and $\mathcal{C}$ map Λ into $\mathbb{X}$ such that

1. $\mathcal{B}$ is continuous and $\mathcal{B}\Lambda$ is contained in a compact set,

2. $x, y \in \Lambda$ implies that $\mathcal{B}x + \mathcal{C}y \in L$,

3. $\mathcal{C}$ is a contraction mapping.

Then there exists a $z \in \Lambda$ such that $z = \mathcal{B}z + \mathcal{C}z$.

Throughout this paper, we always assume that the following conditions are satisfied:

(H_1) $a_i > 0, i = 1, 2, \ldots, n.$

(H_2) There are several positive constants L_j^f, L_j^g and L_j^σ such that

$$|f_j(u) - f_j(v)| \leq L_j^f |u - v|, \tag{7}$$

$$|g_j(u) - g_j(v)| \leq L_j^g |u - v|, \tag{8}$$

$$|\sigma_j(u) - \sigma_j(v)| \leq L_j^\sigma |u - v|, \tag{9}$$

$\forall u, v \in \mathbb{R}$, where $j = 1, 2, \ldots, n$.

Define

$$\bar{a} := \max_{1 \leq i \leq n} \bar{a}_i, \quad \underline{a} := \min_{1 \leq i \leq n} \underline{a}_i, \quad D^* := \max_{1 \leq i \leq n} \sum_{j=1}^{n} \{\bar{b}_{ij} L_j^f + (|\alpha_{ij}| + |\beta_{ij}| + \bar{c}_{ij}) L_j^g\},$$

$$K^* := \max_{1 \leq i \leq n} \sum_{j=1}^{n} \bar{d}_{ij} L_j^\sigma, \quad r_p := \frac{(1 - e^{-\bar{a}})}{\underline{a}(1 - e^{-\underline{a}})} \left\{ D^* + K^* C_p^{\frac{1}{p}} \right\}, \quad \beta_p := \frac{\alpha_p}{1 - r_p},$$

$$\alpha_p := \frac{(1 - e^{-\bar{a}})}{\underline{a}(1 - e^{-\underline{a}})} \max_{1 \leq i \leq n} \left[\sum_{j=1}^{n} \left(\bar{b}_{ij} |f_j(0)| + \bar{c}_{ij} |g_j(0)| \right) \right.$$

$$\left. + \sum_{j=1}^{n} \left(|\alpha_{ij}| + |\beta_{ij}| \right) |g_j(0)| + \sum_{j=1}^{n} \left(|T_{ij}| + |S_{ij}| \right) |\mu_j| + \bar{I}_i + \sum_{j=1}^{n} \bar{d}_{ij} \sigma_j(0) C_p^{\frac{1}{p}} \right].$$

Theorem 1. *Assume that all coefficients in system (5) excluding the Brownian motions are almost periodic sequences, (H_1)-(H_2) hold and the following condition is satisfied:*

(H_3) $r_p < 1$, where $p > 1$.

Then there exists a p-th mean almost periodic sequence solution X of system (5) with $\|X\|_p \leq \beta_p$.

Proof. Let $\Lambda \subseteq BC(\mathbb{Z}; L^p(\Omega; \mathbb{R}^n))$ be the collection of all p-th mean almost periodic sequences $X = \{x_i\}$ satisfying $\|X\|_p \leq \beta_p$.

Firstly, $X = \{x_i\}$ is described by

$$x_i(k) = \sum_{v=-\infty}^{k-1} \prod_{s=v+1}^{k-1} \frac{e^{-a_i(s)}[1 - e^{-a_i(v)}]}{a_i(v)} \left[\sum_{j=1}^{n} b_{ij}(v) f_j(x_j(v)) \right.$$

$$+ \sum_{j=1}^{n} c_{ij}(v) g_j(x_j(v - \tau_{ij}(v))) + \bigwedge_{j=1}^{n} \alpha_{ij} g_j(x_j(v - \tau_{ij}(v))) + \bigwedge_{j=1}^{n} T_{ij} \mu_j$$

$$+ \bigvee_{j=1}^{n} \beta_{ij} g_j(x_j(v - \tau_{ij}(v))) + \bigvee_{j=1}^{n} S_{ij} \mu_j$$

$$\left. + \sum_{j=1}^{n} d_{ij}(v) \sigma_j(x_j(v - \eta_{ij}(v))) \Delta w_j(v) + I_i(v) \right], \tag{10}$$

where $i = 1, 2, \ldots, n$, $\in \mathbb{Z}$. Obviously, (10) is well defined and satisfies (6). So we define $\Phi X(k) = \mathcal{B} X(k) + \mathcal{C} X(k)$, where

$$\Phi X(k) = ((\Phi X)_1(k), (\Phi X)_2(k), \ldots, (\Phi X)_n(k))^T,$$

$$(\Phi X)_i(k) = (\mathcal{B}X)_i(k) + (\mathcal{C}X)_i(k), \tag{11}$$

$$\begin{aligned}(\mathcal{B}X)_i(k) = \sum_{v=-\infty}^{k-1} \prod_{s=v+1}^{k-1} \frac{e^{-a_i(s)}[1-e^{-a_i(v)}]}{a_i(v)} &\left[\sum_{j=1}^{n} b_{ij}(v)f_j(x_j(v))\right.\\
&+\sum_{j=1}^{n} c_{ij}(v)g_j(x_j(v-\tau_{ij}(v))) + \bigwedge_{j=1}^{n}\alpha_{ij}g_j(x_j(v-\tau_{ij}(v)))\\
&+\bigwedge_{j=1}^{n} T_{ij}\mu_j + \bigvee_{j=1}^{n}\beta_{ij}g_j(x_j(v-\tau_{ij}(v)))\\
&\left.+\bigvee_{j=1}^{n} S_{ij}\mu_j + I_i(v)\right],\end{aligned} \tag{12}$$

$$(\mathcal{C}X)_i(k) = \sum_{v=-\infty}^{k-1} \prod_{s=v+1}^{k-1} \frac{e^{-a_i(s)}[1-e^{-a_i(v)}]}{a_i(v)} \sum_{j=1}^{n} d_{ij}(v)\sigma_j(x_j(v-\eta_{ij}(v)))\Delta w_j(v), \tag{13}$$

where $i = 1, 2, \ldots, n, \epsilon \mathbb{Z}$.

Let $X^0 = \{x_i^0\}$ be defined as

$$\begin{aligned}x_i^0(k) = \sum_{v=-\infty}^{k-1} \prod_{s=v+1}^{k-1} \frac{e^{-a_i(s)}[1-e^{-a_i(v)}]}{a_i(v)} &\left[\sum_{j=1}^{n} b_{ij}(v)f_j(0)\right.\\
&+\sum_{j=1}^{n} c_{ij}(v)g_j(0) + \bigwedge_{j=1}^{n}\alpha_{ij}g_j(0) + \bigwedge_{j=1}^{n} T_{ij}\mu_j\\
&\left.+\bigvee_{j=1}^{n}\beta_{ij}g_j(0) + \bigvee_{j=1}^{n} S_{ij}\mu_j + \sum_{j=1}^{n} d_{ij}(v)\sigma_j(0)\Delta w_j(v) + I_i(v)\right],\end{aligned}$$

where $i = 1, 2, \ldots, n, \epsilon \mathbb{Z}$. By Minkoswki inequality in Lemma 2, we have

$$\begin{aligned}\|X^0\|_p &\\
\leq \max_{1\leq i\leq n}\sup_{k\in\mathbb{Z}}\Bigg\{&\left[E\left|\sum_{v=-\infty}^{k-1}\prod_{s=v+1}^{k-1}\frac{e^{-a_i(s)}[1-e^{-a_i(v)}]}{a_i(v)}\sum_{j=1}^{n}\Big(b_{ij}(v)f_j(0)+c_{ij}(v)g_j(0)\Big)\right|^p\right]^{\frac{1}{p}}\\
&+\left[E\left|\sum_{v=-\infty}^{k-1}\prod_{s=v+1}^{k-1}\frac{e^{-a_i(s)}[1-e^{-a_i(v)}]}{a_i(v)}\left(\bigwedge_{j=1}^{n}\alpha_{ij}+\bigvee_{j=1}^{n}\beta_{ij}\right)g_j(0)\right|^p\right]^{\frac{1}{p}}\\
&+\left[E\left|\sum_{v=-\infty}^{k-1}\prod_{s=v+1}^{k-1}\frac{e^{-a_i(s)}[1-e^{-a_i(v)}]}{a_i(v)}\left(\bigwedge_{j=1}^{n} T_{ij}+\bigvee_{j=1}^{n} S_{ij}\right)\mu_j\right|^p\right]^{\frac{1}{p}}\\
&+\left[E\left|\sum_{v=-\infty}^{k-1}\prod_{s=v+1}^{k-1}\frac{e^{-a_i(s)}[1-e^{-a_i(v)}]}{a_i(v)}\sum_{j=1}^{n}d_{ij}(v)\sigma_j(0)\Delta w_j(v)\right|^p\right]^{\frac{1}{p}}\end{aligned}$$

$$+\left[E\left|\sum_{v=-\infty}^{k-1}\prod_{s=v+1}^{k-1}\frac{e^{-a_i(s)}[1-e^{-a_i(v)}]}{a_i(v)}I_i(v)\right|^p\right]^{\frac{1}{p}}\right\}.$$

From Lemma 6 and Ho¨lder inequality in Lemma 3, it gets from the above inequality that

$$\begin{aligned}\|X^0\|_p &\leq \max_{1\leq i\leq n}\sup_{k\in\mathbb{Z}}\left\{\frac{(1-e^{-\bar{a}})}{\underline{a}(1-e^{-\underline{a}})}\left[\sum_{j=1}^n\left(\bar{b}_{ij}|f_j(0)|+\bar{c}_{ij}|g_j(0)|\right)\right.\right.\\
&\quad\left.+\sum_{j=1}^n(|\alpha_{ij}|+|\beta_{ij}|)|g_j(0)|+\sum_{j=1}^n(\sum_{j=1}^n|T_{ij}|+|S_{ij}|)|\mu_j|+\bar{I}_i\right]\\
&\quad+\sum_{j=1}^n\bar{d}_{ij}\sigma_j(0)\left[\sum_{v=-\infty}^{k-1}\prod_{s=v+1}^{k-1}\frac{e^{-a_i(s)}[1-e^{-a_i(v)}]}{a_i(v)}\right]^{1-\frac{1}{p}}\\
&\quad\left.\times\left[\sum_{v=-\infty}^{k-1}\prod_{s=v+1}^{k-1}\frac{e^{-a_i(s)}[1-e^{-a_i(v)}]}{a_i(v)}E|\Delta w_j(v)|^p\right]^{\frac{1}{p}}\right\}\\
&\leq\frac{(1-e^{-\bar{a}})}{\underline{a}(1-e^{-\underline{a}})}\max_{1\leq i\leq n}\left[\sum_{j=1}^n\left(\bar{b}_{ij}|f_j(0)|+\bar{c}_{ij}|g_j(0)|\right)+\sum_{j=1}^n\left(|\alpha_{ij}|+|\beta_{ij}|\right)|g_j(0)|\right.\\
&\quad\left.+\sum_{j=1}^n\left(|T_{ij}|+|S_{ij}|\right)|\mu_j|+\bar{I}_i+\sum_{j=1}^n\bar{d}_{ij}\sigma_j(0)C_p^{\frac{1}{p}}\right]:=\alpha_p.\end{aligned}\qquad(14)$$

It follows from (11), (12) and (13) that

$$\begin{aligned}&\|\Phi X-X^0\|_p\\
&\leq \max_{1\leq i\leq n}\sup_{k\in\mathbb{Z}}\sum_{j=1}^n\bar{b}_{ij}L_j^f\left\{E\left[\sum_{v=-\infty}^{k-1}\prod_{s=v+1}^{k-1}\frac{e^{-a_i(s)}[1-e^{-a_i(v)}]}{a_i(v)}|x_j(v)|\right]^p\right\}^{\frac{1}{p}}\\
&\quad+\max_{1\leq i,j\leq n}\sup_{k\in\mathbb{Z}}D_i^{**}\left\{E\left[\sum_{v=-\infty}^{k-1}\prod_{s=v+1}^{k-1}\frac{e^{-a_i(s)}[1-e^{-a_i(v)}]}{a_i(v)}|x_j(v-\tau_{ij}(v))|\right]^p\right\}^{\frac{1}{p}}\\
&\quad+\max_{1\leq i,j\leq n}\sup_{k\in\mathbb{Z}}K^*\left\{E\left[\sum_{v=-\infty}^{k-1}\prod_{s=v+1}^{k-1}\frac{e^{-a_i(s)}[1-e^{-a_i(v)}]}{a_i(v)}|x_j(v-\eta_{ij}(v))\Delta w_j(v)|\right]^p\right\}^{\frac{1}{p}},\end{aligned}$$

which yields from Lemma 3 that

$$\begin{aligned}&\|\Phi X-X^0\|_p\\
&\leq \max_{1\leq i\leq n}\sup_{k\in\mathbb{Z}}\sum_{j=1}^n\bar{b}_{ij}L_j^f\left\{\left[\sum_{v=-\infty}^{k-1}\prod_{s=v+1}^{k-1}\frac{e^{-a_i(s)}[1-e^{-a_i(v)}]}{a_i(v)}\right]^{p-1}\right.\\
&\quad\left.\times\sum_{v=-\infty}^{k-1}\prod_{s=v+1}^{k-1}\frac{e^{-a_i(s)}[1-e^{-a_i(v)}]}{a_i(v)}E|x_j(v)|^p\right\}^{\frac{1}{p}}\\
&\quad+\max_{1\leq i,j\leq n}\sup_{k\in\mathbb{Z}}D_i^{**}\left\{\left[\sum_{v=-\infty}^{k-1}\prod_{s=v+1}^{k-1}\frac{e^{-a_i(s)}[1-e^{-a_i(v)}]}{a_i(v)}\right]^{p-1}\right.\end{aligned}\qquad(15)$$

$$\times \sum_{v=-\infty}^{k-1} \prod_{s=v+1}^{k-1} \frac{e^{-a_i(s)}[1-e^{-a_i(v)}]}{a_i(v)} E|x_j(v-\tau_{ij}(v))|^p \bigg\}^{\frac{1}{p}}$$

$$+ \max_{1 \leq i,j \leq n} \sup_{k \in \mathbb{Z}} K^* \left\{ \left[\sum_{v=-\infty}^{k-1} \prod_{s=v+1}^{k-1} \frac{e^{-a_i(s)}[1-e^{-a_i(v)}]}{a_i(v)} \right]^{p-1} \right.$$

$$\left. \times \sum_{v=-\infty}^{k-1} \prod_{s=v+1}^{k-1} \frac{e^{-a_i(s)}[1-e^{-a_i(v)}]}{a_i(v)} E|x_j(v-\eta_{ij}(v))\Delta w_j(v)|^p \right\}^{\frac{1}{p}},$$

where $D_i^{**} = D^* - \sum_{j=1}^{n} \bar{b}_{ij} L_j^f$, $i = 1, 2, \ldots, n$. Applying Lemma 5 to the above inequality, it derives

$$\|\Phi X - X^0\|_p \leq \frac{(1-e^{-\bar{a}})}{\underline{a}(1-e^{-\underline{a}})} \left\{ D^* + K^* C_p^{\frac{1}{p}} \right\} \|X\|_p = r_p \|X\|_p \leq \frac{r_p \alpha_p}{1-r_p}. \tag{16}$$

Hence, $\forall X = \{x_i\} \in \Lambda$, it leads from (14) and (16) to

$$\|\Phi X\|_p \leq \|X^0\|_p + \|\Phi X - X^0\|_p \leq \alpha_p + \frac{r_p \alpha_p}{1-r_p} = \frac{\alpha_p}{1-r_p} := \beta_p. \tag{17}$$

Similar to the argument as that in (17), it is easy to verify that $\mathcal{B}\Lambda$ is uniformly bounded and continuous. Together with the continuity of $\mathcal{B}$, for any bounded sequence $\{\varphi_n\}$ in Λ, we know that there exists a subsequence $\{\varphi_{n_k}\}$ in Λ such that $\{\mathcal{B}(\varphi_{n_k})\}$ is convergent in $\mathcal{B}(\Lambda)$. Therefore, $\mathcal{B}$ is compact on Λ. Then condition (1) of Lemma 7 is satisfied.

The next step is proving condition (2) of Lemma 7. Now, we consist in proving the p-th mean almost periodicity of $\mathcal{B}X(\cdot)$ and $\mathcal{C}X(\cdot)$. Since $X(\cdot)$ is a p-th mean almost periodic sequence and all coefficients in system (5) are almost periodic sequences, for any $\epsilon > 0$ there exists $l_\epsilon > 0$ such that every interval of length $l_\epsilon > 0$ contains a ω with the prop- erty that

$$[E|x_i(k+\omega) - x_i(k)|^p]^{\frac{1}{p}} < \epsilon, \quad |a_i(k+\omega) - a_i(k)| < \epsilon,$$

$$|b_{ij}(k+\omega) - b_{ij}(k)| < \epsilon, \quad |c_{ij}(k+\omega) - c_{ij}(k)| < \epsilon, \quad |d_{ij}(k+\omega) - d_{ij}(k)| < \epsilon,$$

$$|\tau_{ij}(k+\omega) - \tau_{ij}(k)| < \epsilon, \quad |\eta_{ij}(k+\omega) - \eta_{ij}(k)| < \epsilon, \quad |I_i(k+\omega) - I_i(k)| < \epsilon,$$

where $i, j = 1, 2, \ldots, n$, $\epsilon \in \mathbb{Z}$. By (12), (13) and (H_2), we could easily find a positive constant M such that

$$[E|(\mathcal{B}X)_i(k+\omega) - (\mathcal{B}X)_i(k)|^p]^{\frac{1}{p}} \leq M \max_{1 \leq i \leq n} \sup_{k \in \mathbb{Z}} [E|x_i(k+\omega) - x_i(k)|^p]^{\frac{1}{p}} < M\epsilon, \tag{18}$$

$$[E|(\mathcal{C}X)_i(k+\omega) - (\mathcal{C}X)_i(k)|^p]^{\frac{1}{p}} \leq M \max_{1 \leq i \leq n} \sup_{k \in \mathbb{Z}} [E|x_i(k+\omega) - x_i(k)|^p]^{\frac{1}{p}} < M\epsilon, \tag{19}$$

where $i = 1, 2, \ldots, n, \in \mathbb{Z}$. From (18) and (19), $\mathcal{B}X(\cdot)$ and $\mathcal{C}X(\cdot)$ are p-th mean almost peri-odic processes. Further, by (17), it is easy to obtain that $\mathcal{B}X + \mathcal{C}Y \in \Lambda, \forall X, Y \in \Lambda$. Then condition (2) of Lemma 7 holds.

Finally, $\forall X = \{x_i\}, Y = \{y_i\} \in \Lambda$, from (13), it yields

$$\begin{aligned}
\|\mathcal{C}X - \mathcal{C}Y\|_p &\leq \frac{[1-e^{-\bar{a}}]}{\underline{a}} \max_{1 \leq i \leq n} \sup_{k \in \mathbb{Z}} \Bigg\{ E \Bigg[\sum_{v=-\infty}^{k-1} \prod_{s=v+1}^{k-1} e^{-\underline{a}} \\
&\quad \times \sum_{j=1}^{n} d_{ij}(v)(\sigma_j(x_j(v-\eta_{ij}(v))) - \sigma_j(y_j(v-\eta_{ij}(v))))\Delta w_j(v) \Bigg]^p \Bigg\}^{\frac{1}{p}} \\
&\leq \frac{[1-e^{-\bar{a}}]}{\underline{a}} \max_{1 \leq i,j \leq n} \sup_{k \in \mathbb{Z}} K^* \Bigg\{ \Bigg[\sum_{v=-\infty}^{k-1} \prod_{s=v+1}^{k-1} e^{-\underline{a}} \Bigg]^{p-1} \\
&\quad \times \sum_{v=-\infty}^{k-1} \prod_{s=v+1}^{k-1} e^{-\underline{a}} E|[x_j(v-\eta_{ij}(v)) - y_j(v-\eta_{ij}(v))]\Delta w_j(v)|^p \Bigg\}^{\frac{1}{p}} \\
&\leq \frac{K^* C_p^{\frac{1}{p}}(1-e^{-\bar{a}})}{\underline{a}(1-e^{-\underline{a}})}\|X-Y\|_p \\
&\leq r_p \|X-Y\|_p.
\end{aligned} \quad (20)$$

In view of (H_3), $\mathcal{C}$ is a contraction mapping. Hence condition (3) of Lemma 7 is satisfied. Therefore, all conditions in Lemma 7 hold. By Lemma 7, system (5) has a p-th mean almost periodic sequence solution. This completes the proof.

p-th moment global exponential stability

Suppose that $X = \{x_i\}$ with initial value $\varphi = \{\varphi_i\}$ and $X^* = \{x_i^*\}$ with initial value $\varphi^* = \{\varphi_i^*\}$ are i i arbitrary two solutions of system (5). For convenience, let

$$\gamma_p = \max_{1 \leq i \leq n} \sup_{s \in [-\mu_0, 0]_\mathbb{Z}} \{(E|\varphi_i(s) - \varphi_i^*(s)|^p)^{\frac{1}{p}}\}, \mu_0 = \max_{1 \leq i,j \leq n}\{\bar{\tau}_{ij}, \bar{\eta}_{ij}\}.$$

Definition 2. ([9]) System (5) is said to be p-th moment global exponential stability if there are positive constants k_0, M and λ such that

$$|X(k) - X^*(k)|_p = \max_{1 \leq i \leq n}(E|x_i(k) - x_i^*(k)|^p)^{\frac{1}{p}} < M\gamma_p e^{-\lambda k}, \quad \forall k > k_0, \; k \in \mathbb{Z}.$$

The 2-nd moment global exponential stability will be called square-mean global exponential stability.

Theorem 2. *Assume that (H_1)-(H_3) hold, then system (5) is p-th moment globally exponentially stable, $p > 1$.*

Proof. By Lemma 1, it follows that

$$|x_i(k) - x_i^*(k)|$$

$$
\begin{aligned}
\leq\ & \prod_{s=0}^{k-1} e^{-a_i(s)} |\varphi_i(0) - \varphi_i^*(0)| + \frac{(1-e^{-\bar{a}})}{\underline{a}} \sum_{v=0}^{k-1} \prod_{s=v+1}^{k-1} e^{-a_i(s)} \sum_{j=1}^{n} \Big\{ \bar{b}_{ij} L_j^f |x_j(v) - x_j^*(v)| \\
& + (\bar{c}_{ij} + |\alpha_{ij}| + |\beta_{ij}|) L_j^g |x_j(v - \tau_{ij}(v)) - x_j^*(v - \tau_{ij}(v))| \\
& + \bar{d}_{ij} L_j^\sigma |x_j(v - \eta_{ij}(v)) - x_j^*(v - \eta_{ij}(v))| |\Delta w_j(v)| \Big\} \\
\leq\ & e^{-\underline{a}k} |\varphi_i(0) - \varphi_i^*(0)| + \frac{(1-e^{-\bar{a}})}{\underline{a}} \sum_{v=0}^{k-1} e^{-\underline{a}(k-v-1)} \sum_{j=1}^{n} \Big\{ \bar{b}_{ij} L_j^f |x_j(v) - x_j^*(v)| \\
& + (\bar{c}_{ij} + |\alpha_{ij}| + |\beta_{ij}|) L_j^g |x_j(v - \tau_{ij}(v)) - x_j^*(v - \tau_{ij}(v))| \\
& + \bar{d}_{ij} L_j^\sigma |x_j(v - \eta_{ij}(v)) - x_j^*(v - \eta_{ij}(v))| |\Delta w_j(v)| \Big\},
\end{aligned}
\tag{21}
$$

where $i = 1, 2, \ldots, n, k \in [1, +\infty)_{\mathbb{Z}}$. For convenience, let $a_0 = \frac{1-e^{-\bar{a}}}{\underline{a}}$ and $Z(k) = \{z_i(k)\}$, $z_i(k) = x_i(k) - x_i^*(k)$, $i = 1, 2, \ldots, n, k \in \mathbb{Z}$. By Lemmas 2 and 3, it gets from (21) that

$$
\begin{aligned}
|Z(k)|_p =\ & |X(k) - X^*(k)|_p \\
\leq\ & e^{-\underline{a}k} \gamma_p \\
& + \max_{1 \leq i \leq n} \sum_{j=1}^{n} a_0 \bar{b}_{ij} L_j^f \Bigg\{ \bigg[\sum_{s=0}^{k-1} e^{-\underline{a}(k-s-1)} \bigg]^{p-1} \\
& \sum_{s=0}^{k-1} e^{-\underline{a}(k-s-1)} E|x_j(s) - x_j^*(s)|^p \Bigg\}^{\frac{1}{p}} \\
& + \max_{1 \leq i \leq n} \sum_{j=1}^{n} a_0 (\bar{c}_{ij} + |\alpha_{ij}| + |\beta_{ij}|) L_j^g \Bigg\{ \bigg[\sum_{s=0}^{k-1} e^{-\underline{a}(k-s-1)} \bigg]^{p-1} \\
& \times \sum_{s=0}^{k-1} e^{-\underline{a}(k-s-1)} E|x_j(s - \tau_{ij}(s)) - x_j^*(s - \tau_{ij}(s))|^p \Bigg\}^{\frac{1}{p}} \\
& + \max_{1 \leq i \leq n} \sum_{j=1}^{n} a_0 \bar{d}_{ij} L_j^\sigma \Bigg\{ \bigg[\sum_{s=0}^{k-1} e^{-\underline{a}(k-s-1)} \bigg]^{p-1} \\
& \times \sum_{s=0}^{k-1} e^{-\underline{a}(k-s-1)} E|[x_j(s - \eta_{ij}(s)) - x_j^*(s - \eta_{ij}(s))] \Delta w_j(s)|^p \Bigg\}^{\frac{1}{p}} \\
\leq\ & e^{-\underline{a}k} \gamma_p + \max_{1 \leq i \leq n} \sum_{j=1}^{n} a_0 \bar{b}_{ij} L_j^f \Bigg\{ \bigg[\sum_{s=0}^{k-1} e^{-\underline{a}(k-s-1)} \bigg]^{p-1} \sum_{s=0}^{k-1} e^{-\underline{a}(k-s-1)} |Z(s)|_p^p \Bigg\}^{\frac{1}{p}} \\
& + \max_{1 \leq i \leq n} \sum_{j=1}^{n} a_0 (\bar{c}_{ij} + |\alpha_{ij}| + |\beta_{ij}|) L_j^g \Bigg\{ \bigg[\sum_{s=0}^{k-1} e^{-\underline{a}(k-s-1)} \bigg]^{p-1} \\
& \sum_{s=0}^{k-1} e^{-\underline{a}(k-s-1)} |Z(s - \tau_{ij}(s))|_p^p \Bigg\}^{\frac{1}{p}}
\end{aligned}
\tag{22}
$$

$$+ \max_{1\le i\le n}\sum_{j=1}^{n} a_0 C_p^{\frac{1}{p}} \bar{d}_{ij} L_j^{\sigma} \left\{ \left[\sum_{s=0}^{k-1} e^{-\underline{a}(k-s-1)}\right]^{p-1} \right.$$
$$\left. \sum_{s=0}^{k-1} e^{-\underline{a}(k-s-1)} |Z(s-\eta_{ij}(s))|_p^p \right\}^{\frac{1}{p}}.$$

Be aware of (H_3) in Theorem 1, there exists a constant $\lambda > 0$ small enough such that

$$\max_{1\le i\le n}\sum_{j=1}^{n} \frac{e^{\lambda} a_0}{1 - e^{-(\underline{a}-2p\lambda)}} \left[\bar{b}_{ij} L_j^f + e^{\mu_0\lambda}(\bar{c}_{ij} + |\alpha_{ij}| + |\beta_{ij}|)L_j^g + e^{\mu_0\lambda} C_p^{\frac{1}{p}} \bar{d}_{ij} L_j^{\sigma}\right] \stackrel{\text{def}}{=} \rho \le 1.$$

Next, we claim that there exists a constant $M_0 > 1$ such that

$$|Z(k)|_p \le M_0 \gamma_p e^{-\lambda k}, \quad \forall k \in [-\mu_0, +\infty)_{\mathbb{Z}}. \tag{23}$$

If (23) is invalid, then there must exist $k_0 \in (0, +\infty)_{\mathbb{Z}}$ such that

$$|Z(k_0)|_p > M_0 \gamma_p e^{-\lambda k_0} \tag{24}$$

and

$$|Z(k)|_p \le M_0 \gamma_p e^{-\lambda k}, \quad \forall k \in [-\mu_0, k_0)_{\mathbb{Z}}. \tag{25}$$

In view of (22), it follows from (25) that

$$|Z(k_0)|_p \le e^{-\underline{a}k_0}\gamma_p$$
$$+\max_{1\le i\le n}\sum_{j=1}^{n} a_0 \bar{b}_{ij} L_j^f M_0 \gamma_p \left\{\left[\sum_{s=0}^{k_0-1} e^{-\underline{a}(k_0-s-1)}\right]^{p-1} \sum_{s=0}^{k_0-1} e^{-\underline{a}(k_0-s-1)} e^{-p\lambda s}\right\}^{\frac{1}{p}}$$
$$+\max_{1\le i\le n}\sum_{j=1}^{n} a_0 M_0 \gamma_p \left[(\bar{c}_{ij}+|\alpha_{ij}|+|\beta_{ij}|)L_j^g + C_p^{\frac{1}{p}}\bar{d}_{ij}L_j^{\sigma}\right]$$
$$\times \left\{\left[\sum_{s=0}^{k_0-1} e^{-\underline{a}(k_0-s-1)}\right]^{p-1} \sum_{s=0}^{k_0-1} e^{-\underline{a}(k_0-s-1)} e^{-p\lambda(s-\mu_0)}\right\}^{\frac{1}{p}}$$
$$\le e^{-\underline{a}k_0}\gamma_p$$
$$+\max_{1\le i\le n}\sum_{j=1}^{n} a_0 \bar{b}_{ij} L_j^f M_0 \gamma_p e^{-\lambda k_0} e^{\lambda} \left[\frac{1-e^{-\underline{a}k_0}}{1-e^{-\underline{a}}}\right]^{1-\frac{1}{p}} \left[\sum_{s=0}^{k_0-1} e^{-(\underline{a}-p\lambda)(k_0-s-1)}\right]^{\frac{1}{p}}$$
$$+\max_{1\le i\le n}\sum_{j=1}^{n} a_0 M_0 \gamma_p \left[(\bar{c}_{ij}+|\alpha_{ij}|+|\beta_{ij}|)L_j^g + C_p^{\frac{1}{p}}\bar{d}_{ij}L_j^{\sigma}\right]$$
$$\times e^{-\lambda k_0} e^{(\mu_0+1)\lambda} \left[\frac{1-e^{-\underline{a}k_0}}{1-e^{-\underline{a}}}\right]^{1-\frac{1}{p}} \left[\sum_{s=0}^{k_0-1} e^{-(\underline{a}-p\lambda)(k_0-s-1)}\right]^{\frac{1}{p}} \tag{26}$$

$$\leq e^{-\underline{a}k_0}\gamma_p + \max_{1\leq i\leq n}\sum_{j=1}^{n} a_0 M_0 \gamma_p e^{-\lambda k_0}\left[\bar{b}_{ij}L_j^f + e^{\mu_0\lambda}(\bar{c}_{ij} + |\alpha_{ij}| + |\beta_{ij}|)L_j^g\right.$$

$$\left. + e^{\mu_0\lambda}\bar{C}_p^{\frac{1}{p}}\bar{d}_{ij}L_j^\sigma\right]e^{\lambda}\left[\frac{1-e^{-\underline{a}k_0}}{1-e^{-\underline{a}}}\right]^{1-\frac{1}{p}}\left[\frac{1-e^{-(\underline{a}-p\lambda)k_0}}{1-e^{-(\underline{a}-p\lambda)}}\right]^{\frac{1}{p}}$$

$$\leq M_0\gamma_p e^{-\lambda k_0}\left\{\frac{1}{M_0}e^{-(\underline{a}-\lambda)k_0} + \max_{1\leq i\leq n}\sum_{j=1}^{n} a_0\left[\bar{b}_{ij}L_j^f\right.\right.$$

$$\left.\left. + e^{\mu_0\lambda}(\bar{c}_{ij} + |\alpha_{ij}| + |\beta_{ij}|)L_j^g + e^{\mu_0\lambda}\bar{C}_p^{\frac{1}{p}}\bar{d}_{ij}L_j^\sigma\right]\frac{e^{\lambda}[1-e^{-(\underline{a}-\lambda)k_0}]}{1-e^{-(\underline{a}-p\lambda)}}\right\}$$

$$\leq M_0\gamma_p e^{-\lambda k_0}\left\{e^{-(\underline{a}-\lambda)k_0} + \rho[1-e^{-(\underline{a}-\lambda)k_0}]\right\}$$

$$\leq M_0\gamma_p e^{-\lambda k_0}.$$

In the fourth inequality from the bottom of (26), we use the fact $[1-e^{-\underline{a}k_0}]^{1-\frac{1}{p}}[1-e^{-(\underline{a}-p\lambda)k_0}]^{\frac{1}{p}} \leq 1-e^{-(\underline{a}-\lambda)k_0}$ and $[1-e^{-\underline{a}}]^{\frac{1}{p}} \geq [1-e^{-(\underline{a}-p\lambda)}]^{\frac{1}{p}}$. p. (26) contradicts (24). Hence, (23) is satisfied.

Therefore, system (5) is p-th moment globally exponentially stable. This completes the proof. Together with Theorem 1, we have

Theorem 3. *Assume that all conditions in Theorem 1 hold, then system (5) admits a p-th mean almost periodic sequence solution, which is p-th moment globally exponentially stable. Fur- ther, if all coefficients in system (5) are periodic sequences, then system (5) admits at least one p- th mean periodic sequence solution, which is globally exponentially stable.*

Proof. The result can be easily obtained by Theorem 2, so we omit it. This completes the proof.

In system (5), if we remove the effects of uncertain factors, then the following deterministic model is obtained:

$$\begin{aligned}x_i(k+1) &= e^{-a_i(k)}x_i(k) + \frac{1-e^{-a_i(k)}}{a_i(k)}\left[\sum_{j=1}^{n}b_{ij}(k)f_j(x_j(k))\right.\\&\left. + \sum_{j=1}^{n}c_{ij}(k)g_j(x_j(k-\tau_{ij}(k))) + I_i(k)\right],\end{aligned} \quad (27)$$

where $k \in \mathbb{Z}$, $i = 1, 2, \ldots, n$.

Define

$$\hat{r} := \max_{1\leq i\leq n}\frac{(1-e^{-\bar{a}_i})}{\underline{a}_i(1-e^{-\underline{a}_i})}\sum_{j=1}^{n}(\bar{b}_{ij}L_j^f + \bar{c}_{ij}L_j^g).$$

Corollary 1. *Assume that* (H_1) *and* (7) *and* (8) *in* (H_2) *hold. Suppose further that all of coeffi- cients of model* (27) *are almost periodic sequences, and* $\hat{r} < 1$, *then model* (27) *admits at least one almost periodic sequence solution, which is globally exponentially stable. Moreover, if all of coefficients of model* (27) *are periodic sequences, then model* (27) *admits at least one periodic solution, which is globally exponentially stable.*

Remark 1. In literature [52], Huang et al. studied model (27) with $c_{ij} \equiv 0 (i, j = 1, 2, \ldots, n)$ and obtained some sufficient conditions for the existence of a unique almost periodic sequence solution which is globally attractive. In [53], they considered system (4) and studied the dynamics of 2N almost periodic sequence solutions. But neither of them considered the uncertain factors. Therefore, the work in this paper complements the corresponding results in [52, 53].

Remark 2. Assume that $X(k) = (x_1(k), x_2(k), \ldots, x_n(k))$ is a solution of (27), the length of $X(k)$ is usually measured by $\|X\|_\infty = \sup_{k \in \mathbb{R}} \max_{1 \le i \le n} |x_i(k)|$. However, if $X(k)$ is a solution of stochastic system (5), its length should not be measured by $\|X\|_\infty$ because $X(k)$ is a random1variable. In this paper, we use norm $\|X\|_p = \max_{1 \le i \le n} \sup_{k \in \mathbb{Z}} (E|x_i(k)|^p)^{\frac{1}{p}} (p > 1)$ for random variable $X(k)$. Owing to the expectation E and order p in $\|X\|_p$, the computing processes of this paper are more complicated than that in literatures [49–55]. It is worth mentioning that Min- koswki inequality in Lemma 2 and Hö̈lder inequality in Lemma 3 are crucial to the computing processes. The facts above are obvious from the computations of (14), (15), (22) and (26) in Theorems 1 and 2. Further, the stochastic term $d_{ij}\sigma_j \Delta w_j (i, j = 1, 2, \ldots, n)$ in system (5) also increases the complexity of computing. This point is also clear from the computations of (20) and (22) in Theorems 1 and 2.

Stochastic stabilization

In this section, we consider the following stochastic cellular neural networks:

$$dx_i(t) = \left[-a_i(t)x_i(t) + \sum_{j=1}^{n} b_{ij}(t)f_j(x_j(t)) + I_i(t) \right] dt + \kappa x_i(t) dw(t), \tag{28}$$

where $w(t)$ is a standard Brownian motion, $t \in \mathbb{R}$, $i = 1, 2, \ldots, n$.

Let $\kappa = 0$ in system (28), the following deterministic cellular neural networks is derived:

$$\frac{dx_i(t)}{dt} = -a_i(t)x_i(t) + \sum_{j=1}^{n} b_{ij}(t)f_j(x_j(t)) + I_i(t), \tag{29}$$

where $t \in \mathbb{R}$, $i = 1, 2, \ldots, n$. Noting that the unique distinction between (28) and (29) is the stochastic disturbance.

The semi-discretization models of systems (28) and (29)

Regarding the following stochastic differential equations (SDEs):

$$du(t) = -a(t)u(t)dt + F(t, u(t))dt + \kappa u(t)dw(t), \quad t \in \mathbb{R},$$

which yields the following SDEs with piecewise constant arguments:

$$du(t) = -a([t])u(t)dt + F([t], u([t]))dt + \kappa u(t)dw(t),$$

where $t \in \mathbb{R}$, $[t]$ denotes the integer part of t. For each $t \in \mathbb{R}$, there exists an integer $k \in \mathbb{Z}$, such that $k \leq t < k+1$. Then the above equation becomes

$$du(t) = -a(k)u(t)dt + F(k, u(k))dt + \kappa u(t)dw(t), \quad t \in \mathbb{R}, k \in \mathbb{Z}. \tag{30}$$

Let $z_k(t) = a(k)t + 0.5\kappa^2 t - \kappa w(t)$, $\forall t \in \mathbb{R}, k \in \mathbb{Z}$. By using Itô formula and formula of integration by parts in stochastic theory, it obtains from (30) that

$$\begin{aligned} d(e^{z_k(t)}u(t)) &= u(t)de^{z_k(t)} + e^{z_k(t)}du(t) + (de^{z_k(t)}) \cdot (du(t)) \\ &= (a(k) + 0.5\kappa^2)u(t)e^{z_k(t)}dt - \kappa u(t)e^{z_k(t)}dw(t) \\ &\quad + 0.5\kappa^2 u(t)e^{z_k(t)}dt + e^{z_k(t)}du(t) - \kappa^2 u(t)e^{z_k(t)}dt \\ &= e^{z_k(t)}F(k, u(k))dt, \quad t \in \mathbb{R}, k \in \mathbb{Z}. \end{aligned}$$

Integrating the above equation from k to t and letting $t \to k+1$, the following equation is obtained:

$$\begin{aligned} u(k+1) &= e^{p(k)}u(k) + e^{-z_k(k+1)}F(k, u(k))\int_k^{k+1} e^{z_k(s)}ds \\ &\approx e^{p(k)}u(k) + \frac{(1 - e^{-a(k) - 0.5\kappa^2})e^{\kappa \Delta w(k)}}{a(k) + 0.5\kappa^2}F(k, u(k)), \end{aligned} \tag{31}$$

where $p(k) = -a(k) - 0.5\kappa^2 + \kappa \Delta w(k)$, $\Delta w(k) = w(k+1) - w(k), k \in \mathbb{Z}$. In (31), we use the fact $\int_k^{k+1} e^{z_k(s)}ds \approx e^{-\kappa w(k)}\int_k^{k+1} e^{a(k)s + 0.5\kappa^2 s}ds, k \in \mathbb{Z}$.

By a similar discussion as that in system (31), we gets the semi-discrete analogue for system (28) as follows:

$$\begin{aligned} x_i(k+1) &= e^{p_i(k)}x_i(k) \\ &\quad + \frac{(1 - e^{-a_i(k) - 0.5\kappa^2})e^{\kappa \Delta w(k)}}{a_i(k) + 0.5\kappa^2}\left[\sum_{j=1}^n b_{ij}(t)f_j(x_j(t)) + I_i(k)\right], \end{aligned} \tag{SM}$$

where $p_i(k) = -a_i(k) - 0.5\kappa^2 + \kappa \Delta w(k)$, $\Delta w(k) = w(k+1) - w(k), k \in \mathbb{Z}, i = 1, 2, \ldots, n$.

Let $\kappa = 0$ in system (SM), the semi-discrete analogue for system (29) is obtained as follows:

$$x_i(k+1) = e^{-a_i(k)}x_i(k) + \frac{1 - e^{-a_i(k)}}{a_i(k)}\left[\sum_{j=1}^n b_{ij}(t)f_j(x_j(t)) + I_i(k)\right], \tag{DM}$$

where $k \in \mathbb{Z}, i = 1, 2, \ldots, n$. Also, the unique difference between (SM) and (DM) is the stochastic disturbance.

Stability analysis of systems (SM) and (DM)

Assume that $X = \{x_i\}$ with initial value $X = \{x_i\}$ with initial value $X_0 = \{x_{i0}\} \in \mathbb{R}^n$ and $X^* = \{x_i^*\}$ with initial value $X_0^* = \{x_{i0}^*\} \in \mathbb{R}^n$ are arbitrary two solutions of system (SM) or (DM).

Definition 3. ([9]) System (SM) or (DM) is said to be exponential stability if

$$\lim_{k \to +\infty} \frac{\ln\left[\sum_{i=1}^n |x_i(k) - x_i^*(k)|\right]}{k} < 0, \quad \forall X_0, X_0^* \in \mathbb{R}^n.$$

System (SM) or (DM) is said to be exponential instability if

$$\lim_{k \to +\infty} \frac{\ln\left[\sum_{i=1}^n |x_i(k) - x_i^*(k)|\right]}{k} > 0, \quad \forall X_0, X_0^*, X_0 - X_0^* \in \mathbb{R}^n \setminus \{0\}.$$

Lemma 8. ([9]) Assume that w is a standard Brownian motion, then $w(0) = 0$ and $\lim_{t \to \infty} \frac{w(t)}{t} = 0$, a.s..

Theorem 4. *Assume that (H_2) holds. Suppose further that*

$$(H_4) \; \Theta = \max_{1 \leq i \leq n}\left[e^{-a_i^- - 0.5\kappa^2} + \frac{1}{\underline{a}_i + 0.5\kappa^2} \sum_{j=1}^n \bar{b}_{ij} L_j^f\right] < 1, \text{ where } a_i^- = \min_{k \in \mathbb{Z}} a_i(k), \; i = 1, 2, \ldots, n.$$

Then system (SM) is exponentially stable.

Proof. From (SM), it gets

$$\begin{aligned}
|x_i(k+1) - x_i^*(k+1)| &\leq e^{p_i(k)} |x_i(k) - x_i^*(k)| + \frac{(1 - e^{-a_i(k) - 0.5\kappa^2})e^{\kappa \Delta w(k)}}{a_i(k) + 0.5\kappa^2} \sum_{j=1}^n \bar{b}_{ij} L_j^f |x_j(k) - x_j^*(k)| \\
&\leq \Theta e^{\kappa \Delta w(k)} \max_{1 \leq i \leq n} |x_i(k) - x_i^*(k)|, \quad i = 1, 2, \ldots, n,
\end{aligned}$$

which derives

$$\max_{1 \leq i \leq n} |x_i(k) - x_i^*(k)| \leq \Theta^k e^{\kappa w(k)} \max_{1 \leq i \leq n} |x_i(0) - x_i^*(0)|, \quad k \in \mathbb{Z},$$

which implies

$$\frac{\ln\left[\max_{1 \leq i \leq n}|x_i(k) - x_i^*(k)|\right]}{k} \leq \ln \Theta + \frac{|\kappa w(k)|}{k} + \frac{\ln \gamma_0}{k}, \quad k \in [1, +\infty)_{\mathbb{Z}}.$$

From Lemma 8, it leads to

$$\lim_{k \to +\infty} \frac{\ln\left[\max_{1 \leq i \leq n}|x_i(k) - x_i^*(k)|\right]}{k} \leq \ln \Theta < 0.$$

Then system (SM) is exponential stability. This completes the proof.

Let $\kappa = 0$ in Theorem 4, it has

Theorem 5. *Assume that* (H_2) *holds. Suppose further that*

$$(H_5) \max_{1 \le i \le n} \left[e^{-a_i^-} + \frac{1}{\underline{a}_i} \sum_{j=1}^{n} \overline{b}_{ij} L_j^f \right] < 1.$$

Then system (DM) is exponentially stable.

Similar to the argument as that in Theorem 4, the exponential instability of system (DM) is easily derived as follows:

Theorem 6. *Assume that* (H_2) *holds. Suppose further that*

$$(H_6) \min_{1 \le i \le n} \left[e^{-a_i^+} - \frac{1}{\underline{a}_i} \sum_{j=1}^{n} \overline{b}_{ij} L_j^f \right] > 1, \text{ where } a_i^+ = \max_{k \in \mathbb{Z}} a_i(k), i = 1, 2, \ldots, n.$$

Then system (DM) is exponentially instable.

Definition 4. ([9]) Assume that system (DM) is exponential instability and there exists a suitable stochastic disturbance coefficient κ ensuring that system (SM) is exponential stable, then system (SM) is a stochastic stabilization system of system (DM). Together with Theorems 4 and 6, it gains.

Theorem 7. *Assume that* (H_2), (H_4) *and* (H_6) *are satisfied. Then system* (SM) *is a stochastic stabilization system of system* (DM).

Remark 3. If (H_6) is valid, (DM) is exponentially instable. Meanwhile, (H_5) is invalid. By viewing (H_4), one could select a suitable stochastic disturbance coefficient κ ensuring that (H_4) is satisfied, which yields system (SM) is exponentially stable. Therefore, stochastic disturbance could be a useful method, which brings unstable system to be stable. More details could be observed in Example 2.

Examples and computer simulations

Example 1. Consider the following continuous-time uncertain cellular neural networks with random perturbations and fuzzy operations:

$$\begin{cases} dx_1(t) = \Big[-x_1(t) + 0.01 \sin(\sqrt{5}t) \sin(x_1(t)) + 0.05 \sin(\sqrt{7}t) \cos(x_2(t-1)) \\ \qquad\qquad + \bigwedge_{j=1}^{2} 0.1 x_j(t-1) + \bigvee_{j=1}^{2} 0.02 x_j(t-1) + 0.01 \cos^2(\sqrt{17}t) \Big] dt \\ \qquad\qquad + 0.01 \cos(\sqrt{3}t) dw(t), \\ dx_2(t) = \Big[-0.2 x_2(t) + 0.02 \cos(\sqrt{5}t) \cos(x_2(t)) + 0.03 \cos(\sqrt{2}t) \sin(x_1(t-1)) \\ \qquad\qquad + \bigwedge_{j=1}^{2} 0.04 x_j(t-1) + \bigvee_{j=1}^{2} 0.2 x_j(t-1) - 0.02|\sin(\sqrt{33}t)| \Big] dt \\ \qquad\qquad + 0.01 \sin(\sqrt{2}t) dw(t), \quad \forall t \in \mathbb{R}. \end{cases} \qquad (32)$$

(1) Semi-discrete model: base on model (32), we obtain the following semi-discrete model by using the semi-discretization technique:

$$\begin{cases} x_1(k+1) = e^{-1}x_1(k) + (1-e^{-1})\Big[0.01\sin(\sqrt{5}k)\sin(x_1(k)) \\ \qquad\qquad +0.05\sin(\sqrt{7}k)\cos(x_2(k-1)) + \bigwedge_{j=1}^{2}0.1x_j(k-1) \\ \qquad\qquad +\bigvee_{j=1}^{2}0.02x_j(k-1) + 0.01\cos(\sqrt{3}k)\Delta w(k) + 0.01\cos^2(\sqrt{17}k)\Big], \\ x_2(k+1) = e^{-0.2}x_2(k) + \dfrac{1-e^{-0.2}}{0.2}\Big[0.02\cos(\sqrt{5}k)\cos(x_2(k)) \\ \qquad\qquad +0.03\cos(\sqrt{2}k)\sin(x_1(k-1)) + \bigwedge_{j=1}^{2}0.04x_j(k-1) \\ \qquad\qquad +\bigvee_{j=1}^{2}0.2x_j(k-1) + 0.01\sin(\sqrt{2}k)\Delta w(k) - 0.02|\sin(\sqrt{33}k)|\Big], \end{cases} \quad (33)$$

where $k \in \mathbb{Z}$.

(2) Discrete model formulated by Euler scheme: base on model (32), we obtain the following discrete-time model by using Euler method:

$$\begin{cases} x_1(k+1) = 0.01\sin(\sqrt{5}k)\sin(x_1(k)) + 0.05\sin(\sqrt{7}k)\cos(x_2(k-1)) \\ \qquad\qquad + \bigwedge_{j=1}^{2}0.1x_j(k-1) + \bigvee_{j=1}^{2}0.02x_j(k-1) \\ \qquad\qquad + 0.01\cos(\sqrt{3}k)\Delta w(k) + 0.01\cos^2(\sqrt{17}k), \\ x_2(k+1) = 0.8x_2(k) + 0.02\cos(\sqrt{5}k)\cos(x_2(k)) + 0.03\cos(\sqrt{2}k)\sin(x_1(k-1)) \\ \qquad\qquad + \bigwedge_{j=1}^{2}0.04x_j(k-1) + \bigvee_{j=1}^{2}0.2x_j(k-1) \\ \qquad\qquad + 0.01\sin(\sqrt{2}k)\Delta w(k) - 0.02|\sin(\sqrt{33}k)|, \end{cases} \quad (34)$$

where $k \in \mathbb{Z}$.

In Figs 1 and 2, we give two plots of numerical solutions which are produced by continuous-time model (32), semi-discrete model (33) and Euler-discretization model (34), respectively. Compared with Euler-discretization model (34), semi-discrete model (33) gives a more accurate characterization for continuous-time model (32).

Remark 4. In literature [43, 44], the authors discussed the dynamics of periodic solutions of discrete-time neural networks formulated by Euler scheme. From the above discussion, semi-discrete stochastic system (5) gives a more accurate and realistic formulation for studying the dynamics of discrete-time neural networks. In a way, the work of this paper complements and improves some corresponding results in [43, 44].

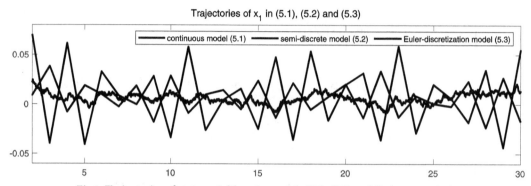

Fig 1. Trajectories of state variable x_1 in models (32), (33) and (34), respectively.

Corresponding to system (5), we have $\underline{a} = 1, \bar{a} = 2, L_i^f = L_i^g = L_i^\sigma = 1, \bar{b}_{ij} = 0.02, \bar{c}_{ij} = 0.05, \alpha_{11} = \alpha_{12} = 0.1, \beta_{11} = \beta_{12} = 0.02, \alpha_{21} = \alpha_{22} = 0.04, \beta_{21} = \beta_{22} = 0.2, \bar{d}_{ij} = 0.01, i, j = 1, 2$.

Taking $p = 2$, by simple calculation,

$$C_2^{1/2} = 2, \quad D^* \approx 0.74, \quad K^* \approx 0.02, \quad r_4 \approx 0.85 < 1.$$

According to Theorems 1 and 2, system (32) admits a square-mean almost periodic sequence solution, which is square-mean globally exponentially stable.

Fig 3 depicts a numerical solution (x_1, x_2) of semi-discrete stochastic model (33). Observe that the trajectories of (x_1, x_2) demonstrate almost periodic oscillations. Figs 4 and 5 display three numerical solutions of semi-discrete stochastic model (33) at different initial values (1.5, 1.5), (0.5, 2.5) and (0.1, 0.2), respectively. They are shown that semi-discrete stochastic model (33) is square-mean globally exponentially stable.

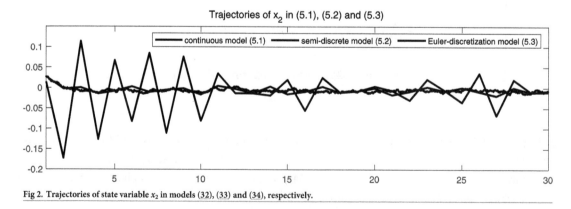

Fig 2. Trajectories of state variable x_2 in models (32), (33) and (34), respectively.

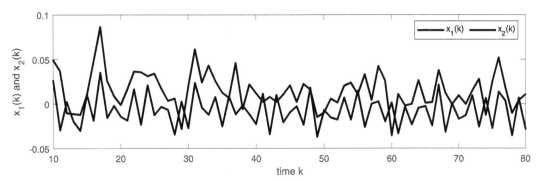

Fig 3. Square-mean almost periodicity of state variables $(x_1, x_2)^T$ in model (33).

Example 2. Considering the following deterministic cellular neural networks:

$$\begin{cases} \dot{x}_1(t) = 0.2x_1(t) + 0.01\cos t|x_1(t)| + 0.01\sin t \sin x_2(t) + \sin(0.1t), \\ \dot{x}_2(t) = 0.3x_2(t) + 0.02\sin t|x_1(t)| + 0.01\cos(\sqrt{5}t)\sin x_2(t) + \cos t, \end{cases} \quad (35)$$

where $t \in \mathbb{R}$. The following semi-discrete model for system (35) is obtained:

$$\begin{cases} x_1(k+1) = e^{0.2}x_1(k) \\ \qquad - \dfrac{1-e^{0.2}}{0.2}\Big[0.01\cos k|x_1(k)| + 0.01\sin k \sin x_2(k) + \sin(0.1k)\Big], \\ x_2(k+1) = e^{0.3}x_2(k) \\ \qquad - \dfrac{1-e^{0.3}}{0.3}\Big[0.02\sin k|x_1(k)| + 0.01\cos(\sqrt{5}k)\sin x_2(k) + \cos k\Big], \end{cases} \quad (36)$$

where $k \in \mathbb{Z}$. Obviously, system (36) satisfies (H_6) in Theorem 6. So system (36) is exponen- tially instable, see Figs 6 and 7.

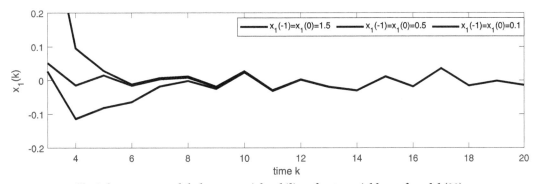

Fig 4. Square-mean global exponential stability of state variable x_1 of model (33).

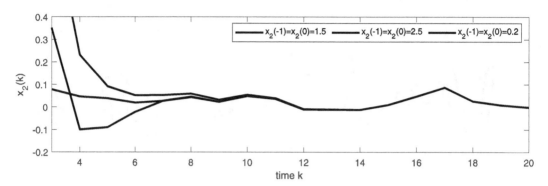

Fig 5. Square-mean global exponential stability of state variable x_2 of model (33).

Similar to the discussion as that in (SM), we consider a random perturbation in system (35) and a semi-discrete model with random perturbation is achieved as follows:

$$\begin{cases} x_1(k+1) = e^{p_1(k)}x_1(k) + \dfrac{(1-e^{0.2-0.5\kappa^2})e^{\kappa\Delta w(k)}}{-0.2+0.5\kappa^2}\bigg[0.01\cos k|x_1(k)| \\ \qquad\qquad\qquad +0.01\sin k \sin x_2(k) + \sin(0.1k)\bigg], \\ x_2(k+1) = e^{p_2(k)}x_2(k) + \dfrac{(1-e^{0.3-0.5\kappa^2})e^{\kappa\Delta w(k)}}{-0.3+0.5\kappa^2}\bigg[0.02\sin k|x_1(k)| \\ \qquad\qquad\qquad +0.01\cos(\sqrt{5}k)\sin x_2(k) + \cos k\bigg], \end{cases} \quad (37)$$

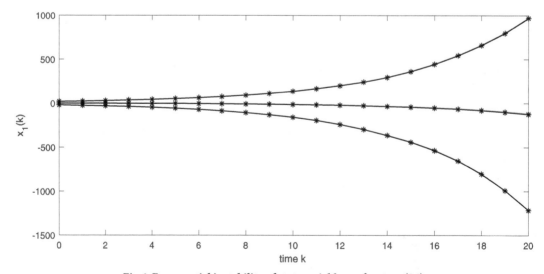

Fig 6. Exponential instability of state variable x_1 of system (36).

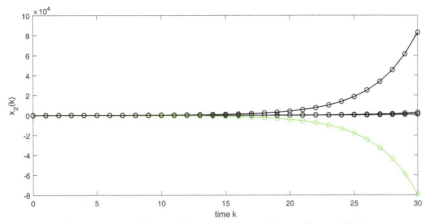

Fig 7. Exponential instability of state variable x_2 of system (36).

where $p_1(k) = 0.2 - 0.5\kappa^2 + \kappa\Delta w(k)$, $p_2(k) = 0.3 - 0.5\kappa^2 + \kappa\Delta w(k)$, $\Delta w(k) = w(k+1) - w(k)$, $k \in \mathbb{Z}$. Here we choose stochastic disturbance coefficient $\kappa = 1$. It easily calculate (H_4) in Theorem 4 is satisfied. Then system (37) is exponentially stable, see Figs 8 and 9. By Theorem 7, system (37) is a stochastic stabilization system of system (36).

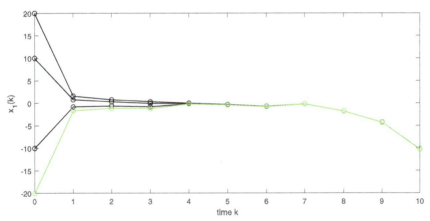

Fig 8. Exponential stability of state variable x_1 of system (37).

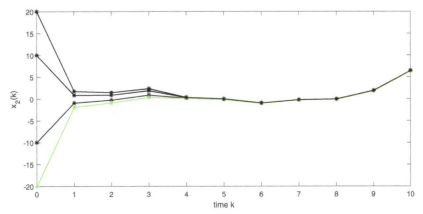

Fig 9. Exponential stability of state variable x_2 of system (37).

Conclusions and future works

In this paper, we formulate a discrete analogue of cellular neural networks with stochastic perturbations and fuzzy operations by using semi-discretization technique. The existence of p-th mean almost periodic sequence solutions and p-th moment global exponential stability for the above models are investigated with the help of Krasnoselskii's fixed point theorem and stochastic theory. The main results obtained in this paper are completely new and the methods used in this paper provide a possible technique to study p-th mean almost periodic sequence solution and p-th moment global exponential stability of semi-discrete models with stochastic perturbations and fuzzy operations.

With a careful observation of Theorems 1 and 2, it is not difficult to discover that

1. $p > 1$ is crucial to the p-th mean almost periodicity and moment global exponential stability of system (5).

2. From Example 2, stochastic disturbance could be a useful method, which brings unstable system to be stable.

3. The time delays have no effect on the existence of p-th mean almost periodicity and p-th moment global exponential stability of system (5).

In the future, the following aspects can be explored further:

1. The methods used in this paper can be applied to study other types of neural networks, such as impulsive neural networks, high-order neural networks, neural networks on time scales, etc.

2. Other types of fuzzy neural networks could be investigated, such as Takagi-Sugeno fuzzy neural networks.

3. Other dynamic behaviours of system (5) should be further discussed.

4. The case of $p \in (0, 1]$ could be further explored.

Acknowledgments

The authors would like to extend their thanks to the referees for their careful reading of the manuscript and insightful comments.

Author Contributions

Conceptualization: Sufang Han, Guoxin Liu. **Formal analysis:** Tianwei Zhang.

Investigation: Sufang Han, Guoxin Liu, Tianwei Zhang.

Writing – original draft: Sufang Han.

Writing – review & editing: Guoxin Liu, Tianwei Zhang.

References

1. Chua LO, Yang L. Cellular neural networks: theory and applications. IEEE Transactions on Circuits and Systems. 1988; 35:1257–1290. https://doi.org/10.1109/31.7601

2. Zhou LQ. Global asymptotic stability of cellular neural networks with proportional delays. Nonlinear Dynamics 2014; 77:41–47. https://doi.org/10.1007/s11071-014-1271-y

3. Feng ZG, Lam J. Stability and dissipativity analysis of distributed delay cellular neural networks. Applied Mathematics and Computation. 2004; 154:683–695.

4. Huang ZK, Mohamad S, Feng CH, New results on exponential attractivity of multiple almost periodic solutions of cellular neural networks with time-varying delays. Math Comput Modell. 2010; 52:1521– 1531. https://doi.org/10.1016/j.mcm.2010.06.013

5. Li YK, Zhao L, Chen XR. Existence of periodic solutions for neutral type cellular neural networks with delays. Appl Math Modell. 2012; 36:1173–1183. https://doi.org/10.1016/j.apm.2011.07.090

6. Chen AP, Cao JD. Existence and attractivity of almost periodic solutions for cellular neural networks with distributed delays and variable coefficients. Appl Math Comput. 2003; 134:125–140.

7. Xu CJ. Existence and exponential stability of anti-periodic solutions in cellular neural networks with time-varying delays and impulsive effects. Electronic Journal of Differential Equations. 2016; 2016:1– 14.

8. Bezandry PH, Diagana T. Almost Periodic Stochastic Processes. Springer, New York. 2010.

9. Hu SG, Huang CM, Wu FK. Stochastic Differential Equations. Science Press, Beijing, China. 2008.

10. Wu T, Xiong LL, Cao JD, Zhang HY. Stochastic stability and extended dissipativity analysis for uncer- tain neutral systems with semi-Markovian jumping parameters via novel free matrix-based integral inequality. International Journal of Robust and Nonlinear Control. 2019; 29:2525–2545. https://doi.org/ 10.1002/rnc.4510

11. Xiong LL, Zhang HY, Li YK, Liu ZX. Improved stability and H infinity performance for neutral systems with uncertain Markovian jump. Nonlinear Analysis-Hybrid Systems. 2016; 19:13–25.

12. Ren Y, He Q, Gu YF, Sakthivel R. Mean-square stability of delayed stochastic neural networks with impulsive effects driven by G-Brownian motion. Statistics and Probability Letters. 2018; 143:56–66. https://doi.org/10.1016/j.spl.2018.07.024

13. Selvaraj P, Sakthivel R, Kwon OM. Finite-time synchronization of stochastic coupled neural networks subject to Markovian switching and input saturation. Neural Networks. 2018; 105:154–165. https://doi. org/10.1016/j.neunet.2018.05.004 PMID: 29886328

14. Arunkumar A, Sakthivel R, Mathiyalagan K, Park JH. Robust stochastic stability of discrete-time fuzzy Markovian jump neural networks. ISA Transactions. 2014; 53:1006–1014. https://doi.org/10.1016/j. isatra.2014.05.002 PMID: 24933353

15. Hu MF, Cao JD, Hua AH. Mean square exponential stability for discrete-time stochastic switched static neural networks with randomly occurring nonlinearities and stochastic delay. Neurocomputing. 2014; 129:476–481. https://doi.org/10.1016/j.neucom.2013.09.011

16. Liu D, Zhu S, Chang WT. Global exponential stability of stochastic memristor-based complex-valued neural networks with time delays. Nonlinear Dynamics. 2017; 90:915–934. https://doi.org/10.1007/ s11071-017-3702-z

17. Ramasamy S, Nagamani G, Zhu QX. Robust dissipativity and passivity analysis for discrete-time sto- chastic T-S fuzzy Cohen-Grossberg Markovian jump neural networks with mixed time delays. Nonlinear Dynamics. 2016; 85:2777–2799. https://doi.org/10.1007/s11071-016-2862-6

18. Hsu CF, Chang CW. Intelligent dynamic sliding-mode neural control using recurrent perturbation fuzzy neural networks. Neurocomputing. 2016; 173, 734–743. https://doi.org/10.1016/j.neucom.2015.08.024

19. Kim EH, Oh SK, Pedrycz W. Reinforced hybrid interval fuzzy neural networks architecture: Design and analysis. Neurocomputing. 2018; 303:20–36. https://doi.org/10.1016/j.neucom.2018.04.003

20. Meng FR, Li KL, Song QK, Liu YR, Alsaadi FE. Periodicity of Cohen-Grossberg-type fuzzy neural net- works with impulses and time-varying delays. Neurocomputing. 2019; 325:254–259. https://doi.org/10. 1016/j.neucom.2018.10.038

21. Li YK, Wang C. Existence and global exponential stability of equilibrium for discrete-time fuzzy BAM neural networks with variable delays and impulses. Fuzzy Sets and Systems. 2013; 217:62–79. https:// doi.org/10.1016/j.fss.2012.11.009

22. Xu CJ, Li PL. pth moment exponential stability of stochastic fuzzy Cohen-Grossberg neural networks with discrete and distributed delays. Nonlinear Analysis: Modelling and Control. 2017; 22:531–544. https://doi.org/10.15388/NA.2017.4.8

23. Wang C, Sakthivel R. Double almost periodicity for high-order Hopfield neural networks with slight vibra- tion in time variables. Neurocomputing. 2018; 282:1–15.

24. Diagana T, Elaydi S, Yakubu AZ. Population models in almost periodic environments. Journal of Differ- ence Equations and Applications. 2007; 13:239–260. https://doi.org/10.1080/10236190601079035

25. Xu CJ, Li PL, Pang YC. Exponential stability of almost periodic solutions for memristor-based neural networks with distributed leakage delays. Neural Computation. 2016; 28:2726–2756. https://doi.org/10. 1162/NECO_a_00895 PMID: 27626965

26. Han SF, Li YQ, Liu GX, Xiong LL, Zhang TW. Dynamics of two-species delayed competitive stage- structured model described by differential-difference equations. Open Mathematics. 2019; 17:385– 401. https://doi.org/10.1515/math-2019-0030

27. Zhang TW, Yang L, Xu LJ. Stage-structured control on a class of predator-prey system in almost periodic environment. International Journal of Control. 2018. https://doi.org/10.1080/00207179.2018.1513165

28. Zhang TW, Liao YZ. Existence and global attractivity of positive almost periodic solutions for a kind of fishing model with pure-delay. Kybernetika. 2017; 53:612–629.

29. Zhang TW, Li YK, Ye Y. On the existence and stability of a unique almost periodic solution of Scho- ener's competition model with pure-delays and impulsive effects. Communications in Nonlinear Science and Numerical Simulation. 2012; 17:1408–1422. https://doi.org/10.1016/j.cnsns.2011.08.008

30. Zhang TW, Gan XR. Almost periodic solutions for a discrete fishing model with feedback control and time delays. Communications in Nonlinear Science and Numerical Simulation. 2014; 19:150–163. https://doi.org/10.1016/j.cnsns.2013.06.019

31. Zhang XM, Han QL, Zeng ZG. Hierarchical type stability criteria for delayed neural networks via canoni- cal bessel-legendre inequalities. IEEE Transactions on Cybernetics. 2017; 48:1660–1671. https://doi. org/10.1109/TCYB.2017.2776283

32. Xiong LL, Cheng J, Cao JD, Liu ZX. Novel inequality with application to improve the stability criterion for dynamical systems with two additive time-varying delays. Applied Mathematics and Computation. 2018; 321:672–688. https://doi.org/10.1016/j.amc.2017.11.020

33. Zhang HY, Qiu ZP, Xiong LL. Stochastic stability criterion of neutral-type neural networks with additive time-varying delay and uncertain semi-Markov jump. Neurocomputing. 2019; 333:395–406. https://doi. org/10.1016/j.neucom.2018.12.028

34. Zhang HY, Qiu ZP, Xiong LL, Jiang GH. Stochastic stability analysis for neutral-type Markov jump neu- ral networks with additive time-varying delays via a new reciprocally convex combination inequality. International Journal of Systems Science. 2019; 50:970–988. https://doi.org/10.1080/00207721.2019. 1586005

35. Xiong LL, Tian JK, Liu XZ. Stability analysis for neutral Markovian jump systems with partially unknown transition probabilities. Journal of the Franklin Institute-Engineering and Applied Mathematics. 2012; 349:2193–2214. https://doi.org/10.1016/j.jfranklin.2012.04.003

36. Wang P, Li B, Li YK. Square-mean almost periodic solutions for impulsive stochastic shunting inhibitory cellular neural networks with delays. Neurocomputing. 2015; 167:76–82. https://doi.org/10.1016/j. neucom.2015.04.089

37. Li YK, Yang L, Wu WQ. Square-mean almost periodic solution for stochastic Hopfield neural networks with time-varying delays on time scales. Neural Comput Appl. 2015; 26:1073–1084. https://doi.org/10. 1007/s00521-014-1784-9

38. Wang C. Existence and exponential stability of piecewise mean-square almost periodic solutions for impulsive stochastic Nicholson's blowflies model on time scales. Applied Mathematics and Computa- tion. 2014; 248:101–112. https://doi.org/10.1016/j.amc.2014.09.046

39. Arnold L, Tudor C. Stationary and almost periodic solutions of almost periodic affine stochastic differen- tial equations. Stochastics and Stochastics Reports. 1998; 64:177–193. https://doi.org/10.1080/ 17442509808834163

40. Swift RJ. Almost periodic harmonizable processes. Georgian Mathematical Journal. 1996; 3:275–292. https://doi.org/10.1007/BF02280009

41. Tudor C. Almost periodic solutions of affine stochastic evolutions equations. Stochastics and Sto- chas- tics Reports. 1992; 38:251–266. https://doi.org/10.1080/17442509208833758

42. Chen WH, Lu X, Liang DY. Global exponential stability for discrete time neural networks with variable delays. Physics Letters A. 2006; 358:186–198. https://doi.org/10.1016/j.physleta.2006.05.014

43. Yang XS, Li F, Long Y, Cui XZ. Existence of periodic solution for discrete-time cellular neural networks with complex deviating arguments and impulses. Journal of the Franklin Institute. 2010; 347:559–566. https://doi.org/10.1016/j.jfranklin.2009.12.004

44. Gao S, Shen R, Chen TR. Periodic solutions for discrete-time Cohen-Grossberg neural networks with delays. Physics Letters A. 2019; 383:414–420. https://doi.org/10.1016/j.physleta.2018.11.016

45. Sun KY, Zhang AC, Qiu JL, Chen XY, Yang CD, Chen X. Dynamic analysis of periodic solution for high- order discrete-time Cohen-Grossberg neural networks with time delays. Neural Networks. 2015; 61:68–74. https://doi.org/10.1016/j.neunet.2014.10.002 PMID: 25462635

46. Wang JL, Jiang HJ, Hu C. Existence and stability of periodic solutions of discrete-time Co- hen-Gross- berg neural networks with delays and impulses. Neurocomputing. 2014; 142:542–550. https://doi.org/ 10.1016/j.neucom.2014.02.056

47. Mohamad S, Gopalsamy K. Dynamics of a class of discrete-time neural networks and their continuous- time counterparts. Mathematics and Computers in Simulation. 2000; 53:1–39. https://doi.org/10.1016/ S0378-4754(00)00168-3

48. Mohamad S, Gopalsamy K. Exponential stability of continuous-time and discrete-time cellular neural networks with delays. Applied Mathematics and Computation. 2013; 135:17–38. https://doi.org/10. 1016/S0096-3003(01)00299-5

49. Zhao HY, Sun L, Wang GL. Periodic oscillation of discrete-time bidirectional associative memory neural networks. Neurocomputing. 2007; 70:2924–2930. https://doi.org/10.1016/j.neucom.2006.11.010

50. Kong FC, Fang XW. Pseudo almost periodic solutions of discrete-time neutral-type neural networks with delays. Applied Intelligence. 2018; 48:3332–3345. https://doi.org/10.1007/s10489-018-1146-x

51. Zhang ZQ, Zhou DM. Existence and global exponential stability of a periodic solution for a discrete-time interval general BAM neural networks. Journal of the Franklin Institute. 2010; 347:763–780. https://doi. org/10.1016/j.jfranklin.2010.02.007

52. Huang ZK, Wang XH, Gao F. The existence and global attractivity of almost periodic sequence solution of discrete-time neural networks. Physics Letters A. 2006; 350:182–191. https://doi.org/10.1016/j. physleta.2005.10.022

53. Huang ZK, Mohamad S, Gao F. Multi-almost periodicity in semi-discretizations of a general class of neural networks. Mathematics and Computers in Simulation. 2014; 101:43–60. https://doi.org/10.1016/ j.matcom.2013.05.017

54. Huang ZK, Xia YH, Wang XH. The existence and exponential attractivity of κ-almost periodic sequence solution of discrete time neural networks. Nonlinear Dynamics. 2007; 50:13–26. https://doi.org/10. 1007/s11071-006-9139-4

55. Ji Y. Global attractivity of almost periodic sequence solutions of delayed discrete-time neural networks. Arabian Journal for Science and Engineering. 2011; 36:1447–1459. https://doi.org/10.1007/s13369- 011-0109-x

56. Kuang JC. Applied Inequalities. Shandong Science and Technology Press, Shandong, China. 2012.

57. Yang T, Yang LB. The global stability of fuzzy cellular neural networks. IEEE Transactions on Circuits and Systems I Fundamental Theory and Applications. 1996; 43:880–883. https://doi.org/10.1109/81. 538999

58. Smart DR. Fixed Point Theorems. Cambridge University Press, Cambridge, UK. 1980.

Local Riemannian geometry of model manifolds and its implications for practical parameter identifiability

Daniel Lill [1*], *Jens Timmer*[1,2,] *Daniel Kaschek*[1]

1 Institute of Physics, University of Freiburg, Freiburg, Germany, **2** BIOSS Centre For Biological Signalling Studies, University of Freiburg, Freiburg, Germany

Editor: Timon Idema, Delft University of Technology, NETHERLANDS

Funding: This work was supported by Bundesministerium fü̈r Bildung und Forschung (Grantee: DL [Daniel Lill], Grant number: BMBF 031L0048, URL: lisym.org).

Competing interests: The authors have declared that no competing interests exist.

* daniel.lill@physik.uni-freiburg.de

Abstract

When non-linear models are fitted to experimental data, parameter estimates can be poorly constrained albeit being identifiable in principle. This means that along certain paths in parameter space, the log-likelihood does not exceed a given statistical threshold but remains bounded. This situation, denoted as practical non-identifiability, can be detected by Monte Carlo sampling or by systematic scanning using the profile likelihood method. In con- trast, any method based on a Taylor expansion of the log-likelihood around the optimum, e.g., parameter uncertainty estimation by the Fisher Information Matrix, reveals no informa- tion about the boundedness at all. In this work, we present a geometric approach, approxi- mating the original log-likelihood by geodesic coordinates of the model manifold. The Christoffel Symbols in the geodesic equation are fixed to those obtained from second order model sensitivities at the optimum. Based on three exemplary non-linear models we show that the information about the log-likelihood bounds and flat parameter directions can already be contained in this local information. Whereas the unbounded case represented by the Fisher Information Matrix is embedded in the geometric framework as vanishing Chris- toffel Symbols, non-vanishing constant Christoffel Symbols prove to define prototype non- linear models featuring boundedness and flat parameter directions of the log-likelihood. Finally, we investigate if those models could allow to approximate and replace computationally expensive objective functions originating from non-linear models by a surrogate objec- tive function in parameter estimation problems.

Introduction

Parameter estimation by the maximum-likelihood method has numerous applications in different fields of physics, engineering, and other quantitative sciences. In systems biology, e.g., ordinary differential equation (ODE) models are used to describe cell-biological processes [1, 2]. Parameter estimation in these non-linear models can easily become time-consuming. Solv- ing the ODEs and computing model sensitivities for numerical optimization is computation- ally demanding. The difficulty is further increased if many experimental conditions contribute to the evaluation of the likelihood function because the model ODE needs to be solved inde- pendently for each condition.

Upon successful parameter estimation, thorough investigation of the log-likelihood around the optimum frequently reveals that some parameters, although having a unique optimum, cannot be constrained to finite confidence intervals. This situation is denoted as practical non-identifiability [3]. The reason for practical non-identifiability is the non-linear relationship between model parameters and model predictions. The non-linearity culmi- nates in the boundedness of model predictions for all possible combinations of parameters and, consequently, in upper limits of the negative log-likelihood that are not exceeded along certain paths. Based on the likelihood-ratio test statistics, log-likelihood thresholds relative to the value at the optimum can be derived [4] that, when being exceeded by the choice of parameters, allow to reject the model specification. Conversely, if the derived thresholds are above the upper limit of the negative log-likelihood, the model cannot be rejected over an infinite range of parameter values.

In this work we discuss that for certain models, practical non-identifiability can already be detected from local information, i.e., *second order model sensitivities* at the optimum. The approach even allows to construct an approximated log-likelihood function uniting both, the local shape around the optimum and the asymptotic shape in the limit of arbitrarily large/ small parameter values. The construction is based on a differential geometric point of view on least squares estimation as laid out in [5, 6]. The geometry of least squares estimation has already previously been discussed, e.g., in [7]. Also the usage of second order model sensitivi- ties to derive equations for parameter transformations providing the log-likelihood with a more quadratic shape around the optimum has been suggested in earlier statistical works, see [8, 9]. However, these previous attempts have been too general to be either solved analytically or be feasible numerically.

In contrast, by sticking to a local approximation of Christoffel Symbols, i.e., the connec- tion coefficients of the Levi-Civita connection on the model manifold, [10], we can show that these are sufficient to construct a globally defined parameter transformation with bounded co-domain that, in the best case, turns the original log-likelihood into a purely quadratic function of the new coordinates. The boundary value problem underlying the parameter transformation can be solved efficiently by numerical methods [11]. The result is that despite being based on purely local second order sensitivity information, the log-likelihood function constructed in this way reflects a fundamental property of the original log-likelihood: its boundedness.

It is thereby possible to capture not only the parameter estimates but also their correlation structure locally as well as in the limit of practical non-identifiability.

Methods

The statistical point of view

Given a mathematical model to describe a set of M data points, one is interested in the N parameters such that the model fits the data best. In this work, a data point y_D taken at the value t_m (usually time) is assumed to be described by a model $y(t, \theta)$ evaluated at the parameter $\theta = \theta_{true} \in \mathbb{P} \subseteq \mathbb{R}^N$. Additionally, data points are affected by Gaussian noise $\epsilon \sim N(0, \sigma^2)$:

$$y_{D,m} = y(t_m, \theta_{true}) + \epsilon_m. \tag{1}$$

In the case of Gaussian noise and known variance σ^2, the Maximum Likelihood Estimate for the parameters is given by

$$\hat{\theta} = \underset{\theta}{\mathrm{argmin}} \underbrace{\sum_{m=1}^{M} \left(\frac{y_{D,m} - y(t_m, \theta)}{\sigma_m} \right)^2}_{\chi^2(\theta)}. \tag{2}$$

χ^2 is a non-linear function of the parameters, and therefore can be bounded in certain directions of the parameter space, in which case we call this model a bounded model. This boundedness has implications for parameter estimation and confidence interval determination [3].

The confidence interval $I_{\theta_i} \subseteq \mathbb{R}$ of the ith parameter is defined as

$$I_{\theta_i} = \{\theta_i | \min_{\theta_j} \chi^2(\theta_i, \{\theta_j\}_{j \neq i}) - \chi^2(\hat{\theta}) < T_{1-\alpha}\}, \tag{3}$$

where $T_{1-\alpha}$ is the threshold to be exceeded to guarantee a confidence level of $1 - \alpha$. These confidence intervals can be either smaller or larger than those derived from the Fisher Information Matrix, or equal, in case of a purely quadratic shape of χ^2. Eq (3) implies that the boundedness of χ^2 eventually leads to infinite confidence intervals if the confidence level is chosen large enough.

The geometric point of view

A different perspective on the boundedness of χ^2 can be obtained by regarding it as a function of the normalized residuals r^m. All residuals can be combined into a vector r which is an ele- ment of the M-dimensional *data space* $\mathbb{D}$. In $\mathbb{D}$, each residual contributes one dimension and the χ^2 function is simply a quadratic function of r:

$$\chi^2 = \sum_m \left(\frac{y_{D,m} - y(t_m, \theta)}{\sigma_m} \right)^2 = \sum_m (r^m(\theta))^2 = \|r(\theta)\|^2 \tag{4}$$

The residual vector is restricted to the N-dimensional model manifold $\mathcal{M}$, which is the set of all residual vectors that can be reached by the model:

$$\mathcal{M} = \{r | r = r(\theta), \theta \in \mathbb{P}\}. \tag{5}$$

With the parameters $\theta \in \mathbb{P}$, $\mathcal{M}$ is readily equipped with a coordinate system. Fig 1A shows an example of an extrinsically flat, one-dimensional model manifold in a two-dimensional data space. Fig 1B shows χ^2 in the parameter space. This simple manifold illustrates that the bound- edness of the χ^2 function in $\mathbb{P}$ is reflected in $\mathbb{D}$ via boundaries of the model manifold. Even tun- ing the parameter θ to infinity results in a residual vector of finite length.

At the optimum $\hat{r}$, any tangent vector of the model manifold is perpendicular to the residual vector itself, as emphasized in Fig 1A, and the squared distance between a point r^* on the flat model manifold and $\hat{r}$ can be directly related to a change in χ^2:

$$\Delta \chi^2 = \chi^2(r^*) - \chi^2(\hat{r}) = \|\Delta r\|^2. \tag{6}$$

We now construct a coordinate system for $\mathcal{M}$ to take advantage of this geometric property. In the Euclidean data space, the length of the vector connecting $\hat{r}$ and r^* coincides with the arc length s of the geodesic between these points. We parameterize this geodesic $r(\tau)$, solution to the equation $\ddot{r} = 0$ by

$$r(\tau) = \hat{r} + \frac{r^* - \hat{r}}{\Delta \tau}\tau = \hat{r} + v_{\hat{r}}\tau \tag{7}$$

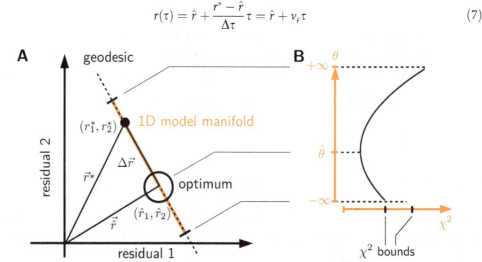

Fig 1. Model manifold with boundary. (A) The tangent at the optimum is perpendicular to its residual vector $\sim\!\hat{}r$. Boundaries of the model manifold, shown in orange, are marked by black segments. In the parameterization by the model parameter θ the boundaries are reached in the limit $\theta \,!\, \pm 1$. (B) The values of χ^2 on the interval $[-1, 1]$ are shown as graph, illustrating the boundedness of the function.

with $\Delta \tau = 1$. The squared arc length s^2 can therefore be expressed simply in terms of the veloc- ity $v_{\hat{r}} = \Delta r / \Delta \tau$ of the geodesic:

$$s^2 = \|\Delta r\|^2 = \int_0^1 \delta_{ij} \dot{r}^i \dot{r}^j d\tau = \delta_{ij} v_{\hat{r}}^i v_{\hat{r}}^j. \tag{8}$$

Here, δ_{ij} is the Euclidean metric of $\mathbb{D}$, and we make use of Einstein's summation convention. Working in the coordinate system of model parameters θ the geodesic equation needs to be solved for the metric of $\mathbb{P}$. Compared to the Euclidean data space, these expressions take the following form:

$$\delta_{ij} \rightsquigarrow g_{\mu\nu} = \frac{\partial r^m}{\partial \theta^\mu} \frac{\partial r^n}{\partial \theta^\nu} \delta_{mn}$$
$$\ddot{r} = 0 \rightsquigarrow \ddot{\theta}^\mu = -\Gamma^\mu_{\alpha\beta} \dot{\theta}^\alpha \dot{\theta}^\beta, \qquad (9)$$

with non-vanishing Christoffel Symbols of the second kind, referred to as "Christoffel Symbols" in this work:

$$\Gamma^\mu_{\alpha\beta} = \sum_m g^{\mu\nu} \frac{\partial r^m}{\partial \theta^\nu} \frac{\partial^2 r^m}{\partial \theta^\alpha \partial \theta^\beta}, \qquad (10)$$

where $g^{\mu\nu} = (g_{\mu\nu})^{-1}$ [6].

As a defining property of geodesics, the absolute value of the velocity stays constant along a geodesic. Thus, initial velocities of the geodesic at $\hat{\theta}$ suffice to express s^2 in the coordinate system of model parameters

$$s^2 = \int_0^1 g_{\mu\nu} \dot{\theta}^\mu \dot{\theta}^\nu d\tau = g_{\mu\nu}\big|_{\hat{\theta}} v^\mu_{\hat{\theta}} v^\nu_{\hat{\theta}} \qquad (11)$$

with $v_{\hat{\theta}}$ indicating that the initial velocity is now expressed in θ-coordinates. The initial velocities are in fact Riemann Normal Coordinates (RNC) for $\mathcal{M}$ [10]. For bounded model manifolds, the RNC are bounded in their domain, since the boundary can be reached by a geodesic with finite initial velocity.

Combining Eqs (6) and (11), χ^2 transforms from a non-linear into a quadratic function:

$$\chi^2(\theta) \rightsquigarrow \chi^2(v_{\hat{\theta}}) \approx \chi^2\big|_{\hat{\theta}} + g_{\mu\nu}\big|_{\hat{\theta}} v^\mu_{\hat{\theta}} v^\nu_{\hat{\theta}}. \qquad (12)$$

The boundedness of this expression is now achieved by the finite domain of the coordinates $v_{\hat{\theta}}$ rather than through the model's non-linearity.

We emphasize that Eq (12) is exact if and only if the model manifold is extrinsically flat. For non-linear models, the extrinsic curvature generally is non-zero and Eq (12) only holds locally, since in this case the assumption $\Delta r \perp \hat{r}$ in Eq (6) is violated when moving further away from the optimum and the geodesic is not a straight path in $\mathbb{D}$. In this work, we do not account for this deviation. It has been noted in [6] that extrinsic curvature of model manifolds can often be neglected.

Results

The methods described above can be used to approximate χ^2 in a new way to allow for regions in $\mathbb{P}$ where χ^2 is bounded.

To perform the coordinate change from the original parameters to the RNC, the geodesic equation has to be solved as a two-point boundary value problem. The first point is the point around which the RNC are constructed, in our case $\hat{\theta}$. The second one is the point where $\chi^2(\theta)$ is to be approximated. Since the geodesic equation is a non-linear ordinary differential equa- tion, in most cases a closed-form solution does not exist and approximations are made to solve the geodesic equation. A popular approach in literature (e.g. [6]) is to Taylor expand all objects in the geodesic equation in terms of the curve parameter τ and requiring that, locally, the geo- desic has the form of a straight line. The coordinates $v(\theta)$ obtained this way are polynomials of θ. As a polynomial, the approximated expression for $\chi^2(\theta)$ cannot be bounded and hence, the accuracy of the approximation becomes insufficient for asymptotically bounded χ^2. On the other hand, solving the geodesic equation numerically comes with high computational costs because at each integration step, the model's derivatives up to second order must be computed to evaluate the Christoffel Symbols.

Our approach approximates only the Christoffel Symbols by their values at $\hat{\theta}$ and inserts them in the otherwise unmodified geodesic equation:

$$\ddot{\theta}^\mu = -\Gamma^\mu_{\alpha\beta}\Big|_{\theta=\hat{\theta}} \dot{\theta}^\alpha \dot{\theta}^\beta. \tag{13}$$

This approximated geodesic equation can be numerically solved without repeated model evaluation. The parameter transformation between original parameters θ and RNC v is given by

$$\phi : \theta^\mu \mapsto v^\mu = \dot{\theta}^\mu(0)$$
$$\text{s.t.} \quad \theta^\mu(0) = \hat{\theta}^\mu, \quad \theta^\mu(1) = \theta^\mu, \tag{14}$$

where $\theta^\mu(\cdot)$ denotes the solution of Eq (13).

By construction, the resulting curves approximate the true geodesics in a neighborhood of $\hat{\theta}$. The approximated RNC v are inserted in Eq (12) with the metric $g_{\mu\nu}|_{\theta=\hat{\theta}}$ to obtain an approximation of χ^2: Through the Christoffel Symbols, this approximation depends on second order model sensitivities at $\hat{\theta}$, but unlike a Taylor expansion of χ^2 of order two, it allows for areas in the parameter space, in which it is bounded. This can be understood from the fact that the solution of a quadratic ordinary differential equation can diverge in finite time. In other words, infinite values for the original parameters θ can be obtained from finite values of the new parameters v. Furthermore, local skewness of χ^2 around the optimum can be better captured than by the Hessian matrix.

Summarising, the approximated objective function can be implemented by the following steps:

1. Optimize the original log-likelihood $\chi^2(\theta)$ to obtain the optimal parameter vector $\hat{\theta}$.

2. Compute the first order and second order sensitivities of the residuals at the optimum. Note that derivatives of the residuals are required, as opposed to derivatives of the objective Function $\frac{\partial r^m}{\partial \theta^\mu}\Big|_{\theta=\hat{\theta}}$ and $\frac{\partial^2 r^m}{\partial \theta^\mu \partial \theta^\nu}\Big|_{\theta=\hat{\theta}}$.

3. With these sensitivities, compute the values of the metric $\hat{g}_{\mu\nu} = g_{\mu\nu}|_{\hat{\theta}}$ and the Christoffel Symbols $\hat{\Gamma}^{\mu}_{\alpha\beta} = \Gamma^{\mu}_{\alpha\beta}|_{\hat{\theta}}$ according to Eqs (9) and (10).

4. With the Christoffel Symbols $\hat{\Gamma}$, solve the geodesic equation as boundary value problem with the constraints specified by Eq (14). To this end, it usually is helpful to reformulate the geodesic equation as a first order ODE with auxiliary variables ξ:

$$\dot{\theta}^{\mu} = \xi^{\mu}$$
$$\dot{\xi}^{\mu} = -\hat{\Gamma}^{\mu}_{\alpha\beta}\xi^{\alpha}\xi^{\beta} \tag{15}$$

5. From the solution, use the initial velocities $v = \dot{\theta}_{\tau=0}$ as new RNC to obtain the approximated $\tilde{\chi}^2$:

$$\tilde{\chi}^2(v(\theta)) \approx \chi^2|_{\hat{\theta}} + \hat{g}_{\mu\nu}v^{\mu}(\theta)v^{\nu}(\theta) \tag{16}$$

For applications relying on derivative information of the objective function, we provide formulas to obtain the gradient and an approximated Hessian of $\tilde{\chi}^2$ in other section.

Model 1: Exponential growth/decay with fixed initial amount

We discuss various features of the approximation by means of three models with increasing complexity. The first model is an example for which the approximation is exact. Consider the model

$$y = e^{-k_1 t}, \tag{17}$$

with $k_1 \in \mathbb{R}$. The sign of the parameter k_1 determines whether the model describes exponential decay or exponential growth.

If we measure only one data point $y_D = 1$ at $t = 1$ with standard deviation $\sigma = 1$, the objective function is given by $\chi^2 = (1 - e^{-k_1})^2$, the model manifold is one-dimensional, extrinsically flat and its Christoffel Symbol is given by $\Gamma^1_{11} = -1$. Clearly, the model manifold is bounded in one direction: In a one-dimensional data space, it covers the positive real numbers shifted negatively by the value of the data point. In this rare case, the geodesic equation can be solved exactly:

$$\theta(\tau) = -\log(C_1\tau + C_0) \tag{18}$$

The integration constants are chosen as

$$C_0 = 1, \tag{19}$$

$$C_1 = -v \tag{20}$$

This way, the boundary conditions at $\tau = 0$ and $\tau = 1$ are fulfilled and the initial velocity is given by $\dot{\theta}|_{\tau=0} = v$. The coordinate change can now be performed by substitution of the solution at $\tau = 1$ into the original χ^2—function:

$$\chi^2(v) = (e^{-k_1} - y_D)^2 = (e^{-(-\log(-v+1))} - y_D)^2 \stackrel{?}{=} (v)^2 \qquad (21)$$

Since the Christoffel Symbols are constant and the model manifold is extrinsically flat, this amounts to the same expression as if Eq (16) is used directly, with $g_{\mu\nu}|_{\hat{\theta}} = 1$:

$$\tilde{\chi}^2(v) = \chi^2|_{\hat{\theta}} + g_{\mu\nu}|_{\hat{\theta}} v^\mu_{\hat{\theta}} v^\nu_{\hat{\theta}} = 0 + 1 \cdot v^2 \qquad (22)$$

As previously derived, the χ^2—function is transformed back to a quadratic function by the Riemann Normal Coordinates, but the domain of v is restricted to values smaller than 1. The source of the boundedness of the χ^2—function is clearly the restricted co-domain of the model itself. We therefore conclude that boundedness of a model along a parameter axis relates to a restricted co-domain of the model: The amount of the decaying substance can never drop below zero, regardless of the rate constant. This boundedness is represented appro- priately by a model with constant Christoffel Symbol.

Model 2: Exponential decay with variable initial amount

We now modify Model 1 at two stages: On the one hand, we introduce a parameter A_0 for the initial amount:

$$\dot{A} = -k_1 A \quad \Leftrightarrow \quad A(t) = A_0 e^{-k_1 t}. \qquad (23)$$

On the other hand, we restrict both the parameter k_1 and A_0 to values greater or equal to zero. Fig 2 visualizes an exemplary χ^2 landscape for three data points the values of which are pre- sented , the contours of the original objective function are shown as solid line.

The restricted model behaviour as in Model 1 can again be observed when considering one-dimensional cross sections of the χ^2 landscape for a fixed value of the initial amount A_0. Furthermore, and somewhat trivially, the χ^2 values are bounded as soon as the parameters approach their respective boundaries. However, also the third source of boundedness can be observed, the coupling of parameters such that a flat canyon is formed which means that a change of one parameter is compensated by the other parameter.

The dashed contour lines represent the approximated objective function $\tilde{\chi}^2$. In comparison to the original objective function, the asymptotic behaviour for cross sections at constant A_0 does not appear to be bounded, but for cross sections at constant k_1. However, the parameter coupling between A_0 and k_1 is matched very well, a behavior which an approximation by a Taylor expansion could never exhibit. It is noticeable that the path of the parameter coupling is straight as opposed to the curved path of the original objective function. However, we note that usually the paths of parameter coupling tend to straighten out asymptotically.

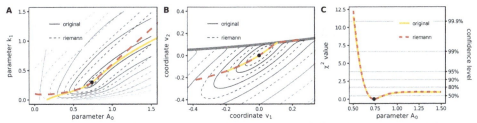

Fig 2. Landscape and profile of χ^2. (A) The shape of the landscape is visualized by solid and dashed contour lines for the original and Riemann approximated χ^2, respectively. The colored lines represent paths that are optimal with respect to the parameter k_1 for any given value of parameter A_0. (B) The non-quadratic χ^2 turns into a quadratic function in Riemannian Normal Coordinates. The paths computed for (A) are shown in the new coordinates as colored lines. (C) The χ^2 values along the exact and the approximated parameter paths agree well indicating that confidence intervals derived from either objective function coincide. Thresholds for different confidence levels are depicted in gray.

The red and yellow lines, referring to χ^2 and $\tilde{\chi}^2$, respectively, indicate the paths that for given value of A_0 minimize χ^2 with respect to k_1, the so-called profile likelihood path for parameter A_0. They each follow the paths of the parameter coupling, which, though different in parameter space, appear to have very similar objective function values along their path, which is shown in Fig 2C.

The same paths and contours are shown in Fig 2B in the new coordinates v. By construction, the dashed contour-lines of the approximated χ^2 are exactly elliptic, but with a boundary as indicated by the fat gray line. Also the original χ^2 appears more quadratic in the new coordinates. The approximated χ^2 purely based on local information around the optimum correctly describes the non-linear phenomenon of parameter coupling. The identification of parameter coupling in the limit of infinitely large/small parameters is a key element of model reduction as demonstrated in [12] and [13].

Model 3: Enzyme kinetics

Next, the approximation is tested on an enzymatic reaction modeled by mass-action kinetics. In this model, an enzyme E and its substrate S first form a complex C which can either dissociate back into E and S, or form a product P, in which case P and E are released. The corresponding ODEs are given by

$$[\dot{S}] = -k_1[S][E] + k_2[C]$$

$$[\dot{E}] = -k_1[S][E] + (k_2 + k_3)[C] = -[\dot{C}] \quad (24)$$

$$[\dot{P}] = k_3[C]$$

The enzyme model typically exhibits two time-scales: the binding and dissociation of E and S are usually much faster than the product formation, i.e., $k_1, k_2 \gg k_3$. This can lead to non-identifiable parameters k_1 and k_2. For large values of k_1 and k_2, the complex quickly reaches a quasi-equilibrium in which only the ratio of k_1 and k_2 can be determined.

Because the visualization of higher-dimensional parameter spaces by contour lines is not feasible, we have evaluated the original and approximated objective function along profile- like-lihood paths for different parameters. To identify coupled parameters, we again compare the profile likelihoods for χ^2 and $\tilde{\chi}^2$. The production rate k_3 and initial amount of substrate S are identifiable from the data as shown in Fig 3. For the parameters S, both the original and the approximated profiles are perfectly qua- dratic and coincide. The original profile of k_3 is skewed. This skewness is visible in the approximation but it is not strong enough to reproduce the exact profile. As expected, the parameters k_1 and k_2 are practically non-identifiable. Again, the effect of practical non-identifiability is well captured by the approximating $\tilde{\chi}^2$, although not as strikingly as for the simpler model.

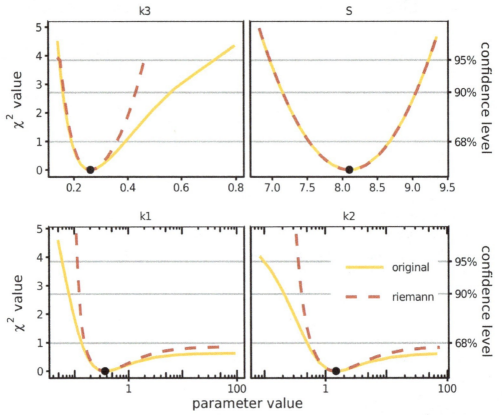

Fig 3. The profile likelihood for the parameter k_3 and initial amount S indicates finite confidence intervals according to both, the original (yellow) and Riemann-approximated (red) χ^2. The estimated parameter values are depicted as black dots. Practically non-identifiable parameters k_1 and k_2 do not even exceed the 68% confidence thresholds towards large values, correctly recognized by the approximation.

Discussion

We established a connection between the Christoffel Symbols of a model and its boundedness. While models with globally vanishing Christoffel Symbols, i.e. linear models, are representatives of unbound models, models with constant Christoffel Symbols are bound in certain parameter directions and can therefore be considered as simple representatives of bounded models. We explored a possible application of this class of models as an approximation to non-linear least-squares, for which information about boundedness can be of importance in parameter uncertainty assessment. We now discuss various additional remarks concerning the quality, applicability and use of the approximation.

First, it should be noted that Christoffel Symbols depend on the coordinates from which they are calculated. The quality of the approximation by constant Christoffel Symbols therefore depends on the original coordinate system from which the coordinate change to the approxi- mated Riemann Normal Coordinates is performed. If for example Model 1 was parameterized by $y = \frac{1}{\theta}, \theta > 0$, the Christoffel Symbols would be given by $\Gamma^1_{11} = -\frac{2}{\theta}$. Thus, for different parameterizations of the same model manifold, the approximation of the Christoffel Symbols by constants yields different qualities of the respective approximations. For more complex models, in which the Christoffel Symbols cannot be written in closed form, the introduced errors may become difficult to quantify or to predict.

Furthermore, the parameterization of the model must not be redundant, such that the change of one parameter can exactly be compensated by another parameter. This setting is called "structural non-identifiability" [3] and in this case, the model manifold has a singular metric, which cannot be inverted. In this case, the structural non-identifiability has to be elimi- nated first, e.g. by methods described in [14].

Another issue is that the approximation by constant Christoffel Symbols could erroneously predict model manifold boundaries where there are none. This is exemplarily highlighted by the fact, that an unbound, one-dimensional polynomial model $y = \theta^n$, n being an uneven integer, has Christoffel Symbol $\Gamma^1_{11} = \frac{n-1}{\theta}$, a form which appears in the previous paragraph, associated with bounded manifolds. Therefore, the approach might be best suited to problems with riori known restricted model behaviors as in the presented examples. In this case, log-likeli- hood bounds will occur in certain directions of the parameter space. A question open to fur- ther research is if Christoffel Symbols obtained by second order sensitvities at the optimum are best suited to reproduce the models bounds for extreme parameter values or if the bounds could be reproduced better by other Christoffel Symbols. These could be obtained e.g. by sam- pling χ^2 in the parameter space and might improve the distant approximation at the cost of reducing the quality of the local approximation close to the optimum.

On the statistical side regarding parameter identifiability, we explicitly note that while a model manifold or its approximation might be bounded in a certain region of parameter space, this does not necessarily imply that a parameter is practically non-identifiable, if the boundary is distant enough to the point of best fit such that the $\Delta\chi^2$ threshold corresponding to a specific confidence level is crossed.

A possible application of the approximation is its use in parameter uncertainty assessment as done for Model 3. There are scenarios in which the approximation might be computation- ally cheaper than the original objective function. The original objective function might be much arbitrarily complex, whereas the structure of the approximated model is fixed to one evaluation of second order sensitivities of χ^2 at $\hat{\theta}$ and, subsequently, to a two-point boundary value problem with N states. In practice, obtaining second order sensitivities can be challenging. For models formulated as ODEs, the numerically most stable way to obtain parameter derivatives is to integrate the sensitivity equations alongside the original ODEs [15]. Since the number of equations for second order sensitivity equations is $\mathcal{O}(n^3)$, already the algebraic derivation of these equation quickly becomes infeasible. Therefore, obtaining second order derivatives by finite differences might prove more viable in practice. Furthermore, the integration of the geodesic equation as a boundary value problem can be challenging and limits the approach to few parameters.

There are settings, though, in which the surrogate objective function $\tilde{\chi}^2$ could be evaluated faster than the original χ^2. An example is given by an ODE model with many states, but few parameters. Such systems are frequent in rule-based biochemical models [16]. In this case, a complicated ODE with very many states but relatively few parameters could be replaced by a much simpler ODE with far fewer states. In a second scenario where the approximation might be beneficial, the data to be modeled consists of many experimental conditions, as is the case for dose-response experiments. This setting involves few unknown parameters as the stimulat- ing concentrations in dose response experiments are usually fixed, but the ODE has to be eval- uated many times with slightly different parameter values.

Runtimes and profile likelihood paths for a simulated dose-response experiment of Model 3 with 51 different enzyme concentrations. In this case, computation time could be saved com- pared to the original objective function.

Conclusion

Parameter estimation in non-linear models is based on the optimization of an objective function, here denoted as χ^2, which is non-quadratic, possibly non-convex, and features certain directions in which its values are bounded. This means that the position of the optimum in parameter space or the Hessian matrix around the optimum represent only a fraction of the information necessary to determine confidence intervals and relationships between model parameters in the case of practical non-identifiability.

In this work we have presented an approach based on differential geometry that, although using only second-order derivatives of the model at the optimum, provides an approximate χ^2 that preserves the essential property of boundedness. Thereby, it allows to approximate χ^2 of models with practically non-identifiable parameters surprisingly well and correctly predicts parameter coupling in the limit of infinitely large parameter values.

Despite this intriguing result, the local approximation of Christoffel Symbols bears also possible shortcomings. In our observation, the quality of the approximation decreases with increasing model size. Also the numerical solution of second-order sensitivity equations is limited by the mere number of equations. This raises the question if other, properly selected points in parameter space could be used to derive constant Christoffel Symbols.

In conclusion, the idea to capture the entire objective function of a non-linear model in a single matrix is tempting both from a conceptual and computational point of view. In the limit of many informative data points this idea is already realized by the quadratic form defined by the Hessian matrix around the optimum. In case of insufficient data we have shown that the geodesic equation with constant Christoffel Symbols can produce objective functions that approximate the original χ^2 not only locally but also globally. Furthermore, for complex non-linear models with few parameters but high computational costs, the approximated objective function could be used to save computation time.

Supporting information

S1 Text. Details on methods and simulated data.

(PDF)

S1 File. Christoffel Symbols of Model 2 and 3, runtimes of the simulation experiments.

(ZIP)

S2 File. Computer code for simulation experiments.

(ZIP)

Author Contributions

Conceptualization: Daniel Kaschek. **Formal analysis:** Daniel Lill.

Funding acquisition: Jens Timmer.

Investigation: Daniel Lill.

Methodology: Daniel Kaschek.

Supervision: Jens Timmer.

Visualization: Daniel Lill, Daniel Kaschek.

Writing – original draft: Daniel Lill, Daniel Kaschek.

Writing – review & editing: Daniel Lill, Jens Timmer, Daniel Kaschek.

References

1. Kholodenko B. N., Demin O. V., Moehren G., and Hoek J. B. (1999). Quantification of short term signal- ing by the epidermal growth factor receptor. Journal of Biological Chemistry, 274(42), 30169–30181. https://doi.org/10.1074/jbc.274.42.30169 PMID: 10514507
2. Schoeberl B., Eichler-Jonsson C., Gilles E. D., and Mu¨ ller G. (2002). Computational modeling of the dynamics of the MAP kinase cascade activated by surface and internalized EGF receptors. Nature Bio- technology, 20(4), 370–375. https://doi.org/10.1038/nbt0402-370 PMID: 11923843
3. Raue A., Kreutz C., Maiwald T., Bachmann J., Schilling M., Klingmu¨ ller U., et al. (2009). Structural and practical identifiability analysis of partially observed dynamical models by exploiting the profile likeli- hood. Bioinformatics, 25(15), 1923–1929. https://doi.org/10.1093/bioinformatics/btp358 PMID: 19505944

4. Wilks S. S. (1938). The large-sample distribution of the likelihood ratio for testing composite hypothe- ses. The Annals of Mathematical Statistics, 9(1), 60–62. https://doi.org/10.1214/aoms/1177732360

5. Transtrum M. K., Machta B. B., and Sethna J. P. (2010). Why are nonlinear fits to data so challenging? Physical Review Letters, 104(6), 060201. https://doi.org/10.1103/PhysRevLett.104.060201 PMID: 20366807

6. Transtrum M. K., Machta B. B., and Sethna J. P. (2011). Geometry of nonlinear least squares with appli- cations to sloppy models and optimization. Physical Review E, 83(3), 036701. https://doi.org/10.1103/ PhysRevE.83.036701

7. Kass R. E. (1989). The geometry of asymptotic inference. Statist Sci., 188–219. https://doi.org/10.1214/ ss/1177012480

8. Bates D. M. and Watts D. G. (1981). Parameter transformations for improved approximate confidence regions in nonlinear least squares. The Annals of Statistics, pages 1152–1167. https://doi.org/10.1214/ aos/1176345633

9. Harris I. R. and Pauler D. K. (1992). Locally quadratic log likelihood and data-based transformations. Communications in Statistics-Theory and Methods, 21(3), 637–646. https://doi.org/10.1080/03610929208830804

10. Michor P. W. (2008). Topics in Differential Geometry, volume 93. American Mathematical Soc.

11. Cash J. R. and Mazzia F. (2011). Efficient global methods for the numerical solution of nonlinear sys- tems of two point boundary value problems. In Simos T.E., editor, Recent Advances in Computational and Applied Mathematics, pages 23–39. Springer Netherlands, Dordrecht.

12. Maiwald T., Hass H., Steiert B., Vanlier J., Engesser R., Raue A., et al. (2016). Driving the model to its limit: Profile likelihood based model reduction. PloS One, 11(9), e0162366. https://doi.org/10.1371/ journal.pone.0162366 PMID: 27588423

13. Transtrum M. K. and Qiu P. (2014). Model reduction by manifold boundaries. Physical Review Letters, 113(9), 098701. https://doi.org/10.1103/PhysRevLett.113.098701 PMID: 25216014

14. Merkt B, Timmer J, Kaschek D. Higher-order Lie symmetries in identifiability and predictability analysis of dynamic models Phys Rev E. 28; 92(1):012920 https://doi.org/10.1103/PhysRevE.92.012920

15. Stumpf M, Balding DJ, Girolami M. Handbook of Statistical Systems Biology. John Wiley & Sons; 2011

16. Faeder JR, Blinov ML, Goldstein B, Hlavacek WS. Rule-based modeling of biochemical networks. Com- plexity. 2005 Mar; 10(4):22–41.

Investigation of singular ordinary differential equations by a neuroevolutionary approach

Waseem Waseem[1], Muhammad Sulaiman[1*], Poom Kumam[2,3,4*], Muhamad Shoaib[7], Muhammad Asif Zahoor Raja[5,6,] Saeed Islam[1,8,9]

1 Department of Mathematics, Abdul Wali Khan University Mardan, KP, Pakistan, **2** KMUTTFixed Point Research Laboratory, Department of Mathematics, Faculty of Science, King Mongkut's University of Technology Thonburi (KMUTT), Bangkok, Thailand, **3** KMUTT-Fixed Point Theory and Applications Research Group, Theoretical and Computational Science Center (TaCS), Faculty of Science, King Mongkut's University of Technology Thonburi (KMUTT), Bangkok, Thailand, **4** Department of Medical Research, China Medical University Hospital, China Medical University, Taichung, Taiwan, **5** Future Technology Research Center, National Yunlin University of Science and Technology, Yunlin, Taiwan, R.O.C., **6** Department of Electrical and Computer Engineering, COMSATS University Islamabad, Attock, Pakistan, **7** Department of Mathematics, COMSATS University Islamabad, Attock, Pakistan, **8** Informetrics Research Group, Ton Duc Thang University, Ho Chi Minh City, Vietnam, **9** Faculty of Mathematics & Statistics, Ton Duc Thang University, Ho Chi Minh City, Vietnam.

Editor: Hector Vazquez-Leal, Universidad Veracruzana, MEXICO

Funding: The authors acknowledge the financial support provided by the Center of Excellence in Theoretical and Computational Science (TaCS- CoE), King Mongkut's University of Technology Thonburi (KMUTT). Funder role: Professor Poom Kumam had a role in the preparation of the manuscript, and financial support in article processing charges.

Competing interests: The authors have declared that no competing interests exist.

* msulaiman@awkum.edu.pk (MS); poom.kum@kmutt.ac.th (PK)

Abstract

In this research, we have investigated doubly singular ordinary differential equations and a real application problem of studying the temperature profile in a porous fin model. We have suggested a novel soft computing strategy for the training of unknown weights involved in the feed-forward artificial neural networks (ANNs). Our neuroevolutionary approach is used to suggest approximate solutions to a highly nonlinear doubly singular type of differential equations. We have considered a real application from thermodynamics, which analyses the temperature profile in porous fins. For this purpose, we have used the optimizer, namely, the fractional-order particle swarm optimization technique (FO-DPSO), to minimize errors

in solutions through fitness functions. ANNs are used to design the approximate series of solutions to problems considered in this paper. We find the values of unknown weights such that the approximate solutions to these problems have a minimum residual error. For global search in the domain, we have initialized FO-DPSO with random solutions, and it collects best so far solutions in each generation/ iteration. In the second phase, we have fine-tuned our algorithm by initializing FO-DPSO with the collection of best so far solutions. It is graphi- cally illustrated that this strategy is very efficient in terms of convergence and minimum mean squared error in its best solutions. We can use this strategy for the higher-order sys- tem of differential equations modeling different important real applications.

1 Introduction

Real-world problems which are modeled as a singular boundary value problem (BVP) of ordinary differential equations are often hard to solve. Such systems frequently arise in physics, or specifically astrophysics, thermodynamics, physical chemistry, nuclear technology, atomic energy, and all studies involving non-linear conic systems [1–4]. Singular BVPs are been tack- led numerically and analytically by different researchers using techniques like the monotonic iterative method of Bessel functions [5], an improved iterative technique [6], homotopy per- turbation method (HPM) [7], finite difference method with uniform mesh [8], monotonic iterative technique involving expansion of eigenfunction [9], modified adomian decomposi- tion method (MADM) [10], Borel–Laplace transformation technique [11], and approximate power series solution method [12].

A review of all these numerical methods shows that they are deterministic and require prior information about the problem. Which is a disadvantage in case we do not have any information about a problem under consideration [13–16]. One example of such problem is the class of doubly non-linear singular differential equations. Meta-heuristic techniques are better alternatives for a variety of singular differential equations like doubly non-linear singu- lar problems.

In this study, a soft computing approach based on hybridization of feed-forward neural networks and fractional-order darwinian PSO is suggested. To investigate the capability of our approach, we have solved three singular non-linear differential equation known as differential equations with doubly singularities. To further analyse our approach, we have solved nine sub-cases with different combinations of parameters. A real application is also considered in problem 4 to further highlight the effectiveness of our approach. We give a general representation of this system in Eq (1).

$$(p(x)y'(x))' = q(x)f(x, y(x)), \quad 0 < x \leq 1, \tag{1}$$

this differential equation is subject to the Dirichlet type of boundary conditions, which are given below.

$$y(0) = a_1, \quad y(1) = c_1, \tag{2}$$

some problems are also bounded by mixed boundary conditions as
$$y(0) = 0, \quad ay(1) + by'(1) = c, \tag{3}$$

here a, a_1, c_1 are non-zero positive real numbers and $b \geq 0$. On the other hand, c can be any real number. When the value of $p(0)$ is zero, the system becomes a singular differential equa- tion. If $q(x)$ is treated as a discontinuous function over the y-axis, then the problem stated in Eqs (1)–(3) becomes a doubly-singular type of differential equation.

From the above discussion, we have understood the singular doubly boundary value problem and based on this understanding we have developed our proposed soft computing approach to get better numerical solutions of these problems.

In the recent couple of years, alternate approaches based on artificial neural networks com- bined with heuristics and meta-heuristic are extensively developed to solve non-linear differ- ential equations. Some important problems which are worth mentioning, include conduction problem in electrical engineering [17, 18], thermodynamics [19, 20], non-linear pantograph differential equations [21], models of atom known as Thomas-Fermi equations [22], Fuzzy logic based problems [23], Navier-stokes equations [23], Volterra differential equations [24, 25], problems in nanofluids [26], Fredholm integro-differential equations [27], non-linear Flierl-Petviashvili differential systems [28], problems in fractional control theory [29], bilinear programming differential systems [30], flow studies of non-linear differential system of Jeff- ery-Hamel problems [28], Bratu differential systems [31], differential systems in electromagnetism [32].

In [33], two techniques namely GA and SQP are combined to tackle the doubly non-linear singular differential equations. However, this combined algorithm takes more time and is computationally expensive. Also, these techniques are local search routines which stuck in a local minimum. By viewing all these contributions, it has led us to design an easy-to-use approach based on soft computing, which can produce better solutions with less computational time in solving these problems which are already handled by classical techniques. The main disadvantage of those classical techniques was their requirement of prior information about the problem in hand. In this paper, our proposed approach is used to solve a second- order non-linear differential equation with double singularities and complex boundary conditions.

Our approach aims to train the unknown weights in ANN by minimizing the error function through a well-balanced single meta-heuristic algorithm known as FO-DPSO [34, 35]. We have considered different case studies of non-linear doubly singular BVPs to check the capabil- ities of our approach. To examine the robustness of our approach, we have performed multiple simulations to get the best values of unknown weights. A real application is considered in problem 4 to further illustrate the effectiveness of our algorithm.

Key contributions in this paper are given below:

A theoretical and graphical model that explains why our soft computing approach works, and it is novel, is presented in other section.

Series solutions based on artificial neural networks are designed with the help of fractional- order particle swarm optimizer (FO-DPSO). Our novel soft computing approach is used to solve non-linear doubly singular differential equations, see Fig 1.

- A real application problem is considered to further elaborate on the competitiveness of our approach. In this problem, we have analyzed the temperature profile in a porous fin model, see Fig 8.
- We have compared our results with GA and a variant GA-SQP algorithm.
- The statistical analysis is presented in terms of absolute errors, global mean absolute error (*GMAE*), mean absolute error (MAE), and mean value of fitness (M_{fit}).
- Computational times, maximum iterations took to solve our problems by ANN-based FO-DPSO are presented.
- Frequency plots of performance indicators fitted with normal distribution are graphically illustrated.

The rest of this paper is organized as, in section, mathematical modeling of approximate solution based on ANNs is illustrated. Fitness functions and FO-DPSO is briefly explained. Sections contain problems description, numerical results for different case studies. Sec- tion, comprises the statistical analysis based on different performance indicators. Conclu- sions and future work are given in other section.

The hybrid ANN and FO-DPSO approach

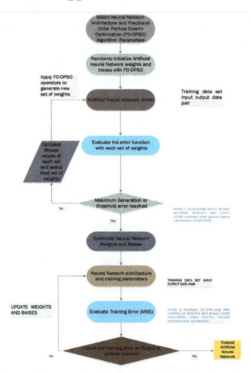

Fig 1. Graphical illustration of our soft computing procedure for doubly singular non-linear ODEs and Porous fin model.

In this section, we have presented our novel approach. This approach constructs ANN-based approximate solutions with unknown weights for the doubly singular BVPs. The unknown weights are determined such that the approximate solution satisfies the problem with a less residual error. We have presented a detailed graphical abstract of the novel procedure in Fig 1.

```
Start of FO-DPSO
    Step 1: Initialization: Randomly generate the initial
    swarm. Assign values to the parameters of "FO-DPSO"

    Step 2: Fitness Evaluation: Scrutinize the "fitness value"
    of each particle using equation (6)
    Step 3: Ranking: Rank each particle of the minimum values
    of the "fitness function".
    Step 4: Stopping Criteria: Stop if
        • Level of "fitness" achieved
        • Selected "flights/cycles" executed
    If meet the "stopping" criteria, then move to Step 5

    Step 5: Renewal: Call the "position" and "velocity", using
    equations (12) and (13).
    Step 6: Improvement: Repeat the steps 2 to 6 up to the
    whole flights are achieved.
    Step 7: Storage: Store "fitness values", which are best
    achieved values and signify as "best global particle"
End FO-DPSO
```

Fig 2. Pseudo-code of our soft computing technique.

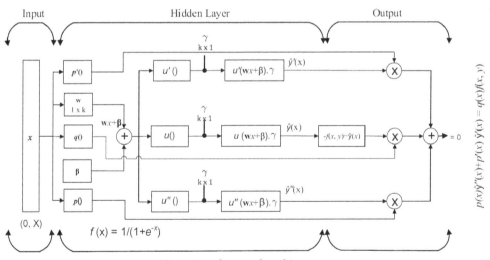

Fig 3. Neural network architecture.

ANN based approximation

Most of the real-world problems are mathematically modeled in the form of differential equations containing the derivative of integer and/ or fractional order. ANNs are frequently used to suggest approximate solutions for such problems [36–41].

In [42] the author presented a generalized method of neural networks to tackle both ODE's as well as PDE's. This method is based on approximation of a function and the capability of feed-forward artificial neural network is to construct a solution of the differential equation which is differentiable and in a closed analytic form. Evolutionary optimization algorithm has been applied to train the weights and biases of the network, which results in minimum mean squared error and turn a good approximate solution of the problem.

The mathematical form of an approximate solution for BVPs with doubly singularities is suggested by feed-forward ANN as given in Eq (4),

$$\hat{y}(x) = \sum_{i=1}^{k} \gamma_i u(w_i x + \beta_i), \tag{4}$$

$$\frac{d^n \hat{y}(x)}{dx^n} = \sum_{i=1}^{k} \gamma_i \frac{d^n}{dx^n} u(w_i x + \beta_i), \tag{5}$$

$\hat{y}(x)$ represents the approximate solution, x is the independent variable, γ_i, β_i and w_i are the unknown weights, and Eq (5) expresses the nth derivative of this approximate solution. A detailed discussion of neural networks is given in [36, 37, 41]. An activation function $u(x) = \frac{1}{1+e^{-x}}$ also known as Log-sigmoid mapping function is used to train the unknown weights. We substitute the approximate solutions in Eqs (1)–(3) to design an ANN-based solution for doubly singular BVPs. We show a detailed architecture of interaction among input-output and hidden stages of ANN in Fig 3. The objective function includes the mini- mization of mean squared errors in the approximate solutions.

$$\text{Minimize} \quad e = e_1 + e_2, \tag{6}$$

where e_1, e_1 are associated mean sqï»¿uared errors in ODE and boundary conditions, respectively.

$$e_1 = \frac{1}{N} \sum_{m=1}^{N} ((p_m \hat{y}'_m)' - q_m f(x_m, \hat{y}_m))^2, \tag{7}$$

where $x \epsilon (0, 1)$, step size $h = 0.05$, $N = \frac{1}{h}$, $p_m = p(x_m)$, $q_m = q(x_m)$, and $x_m = mh$.

Error due to boundary conditions is represented in Eq (8)

$$e_2 = \frac{1}{2}((\hat{y}_0 - a_1)^2 + (\hat{y}_N - c_1)^2). \tag{8}$$

If the boundary conditions are mixed as discussed in Eq (3), then e_2 can be approximated as,

$$e_2 = \frac{1}{2}((\hat{y}'_0)^2 + (a\hat{y}_N + b\hat{y}'_N - c_1)^2). \tag{9}$$

Search method of FO-DPSO

According to published findings, fractional calculus (FC) has received much interest to adapt it in the interpretation and solution of engineering challenges [43–45], applied mathematics, mechanical/dynamics [46, 47]. Grunwald-Letnikov defined a fractional derivative that con- tains fractional coefficients $\alpha \epsilon R$, a real number, by adjusting an unknown function $x(t)$ as in Eq (10),

$$D^{\alpha}[x(t)] = \lim_{h \to 0} \left[\frac{1}{h} \sum_{k=0}^{+\infty} \frac{(-1)^k \Gamma(\alpha + 1) x(t - kh)}{\Gamma(k+1)\Gamma(\alpha - k + 1)} \right], \tag{10}$$

where the symbol Γ represent gamma function.

It is further elaborated that the series is characterized by bounded terms in Eq (10), if the derivative is of integer order. If α is fractional, infinite terms represent the result. It is therefore important to note that ordinary derivatives are operators which are local / instantaneous, whereas fractional operators represent a memory of past variations. With time, the memory of past instances declines. The Eq (11) determines a derivative for discrete instances. [48–52],

$$D^{\alpha}[x(t)] = \frac{1}{T^{\alpha}} \left[\sum_{k=0}^{r} \frac{(-1)^k \Gamma(\alpha + 1) x(t - kh)}{\Gamma(k+1)\Gamma(\alpha - k + 1)} \right], \tag{11}$$

The term T refers to the time intervals of events and r is number of truncated terms. Because of their memory retention properties, methods found in fractional calculus are useful in irre- trievable and disorganised systems. Taking into account swarms 'chaotic behavior in the Dar- winian Particle swarms optimization algorithm, fractional calculus tools are appropriate to keep track of swarms' past movements.

Taking into account the inertial weight in FO-DPSO $w = 1$, T as 1 and the research per- formed in [35, 53, 54], we have the following expression:

$$D^{\alpha}[v_{t+1}^n] = \rho_1 r_1(\check{g}_t^n - x_t^n) + \rho_2 r_2(\check{x}_t^n - x_t^n) + \rho_3 r_3(\check{n}_t^n - x_t^n). \tag{12}$$

The empirical results of the algorithm are identical for $r \geq 4$. The computational complexity also increases almost linearly, and therefore takes up the memory of $O(r)$. Hence, it truncates the fifth term and onward for faster convergence. Thus r's value is kept as 4. The inclusion of these four differential derivative terms means that the velocity term in FO-DPSO is as in Eq (13),

$$\begin{aligned} v_{t+1}^n &= \alpha v_t^n + \frac{1}{2}\alpha v_{t-1}^n + \frac{1}{6}\alpha(1-\alpha)v_{t-2}^n + \frac{1}{24}\alpha(1-\alpha)(2-\alpha)v_{t-3}^n + \rho_1 r_1(\check{g}_t^n - x_t^n) \\ &+ \rho_2 r_2(\check{x}_t^n - x_t^n) + \rho_3 r_3(\check{n}_t^n - x_t^n). \end{aligned} \tag{13}$$

Test problems and empirical results

In this section, we present three doubly singular type of differential equations, and their nine case studies are considered here to test the efficiency of our new approach.

Problem 1

This problem is a linear ODE with a doubly singularity with a polynomial forcing term. It is a boundary value problem with a non-homogenous ODE. Mathematically, it can be represented as in Eq (14) [2],

$$\begin{cases} y''(x) + \frac{1}{x}y'(x) + \mu y = f(x), \\ y(0) = 1, \quad y(1) = 3, \end{cases} \quad (14)$$

where the exact solution is given in Eq (15)

$$y(x) = x^3 + x + 1. \quad (15)$$

Below we consider three cases of the problem by taking $\mu = -9, -1, 1$ and forcing term as $f_1(x) = -9 - 9x^3 + \frac{1}{x}$, $f_2(x) = -1 - x^3 + 8x + \frac{1}{x}$ and $f_3(x) = 1 + x^3 + 10x + \frac{1}{x}$ respectively.

Case 1: From Eq (14) with $\mu = -9$ and $f(x) = f_1(x)$ we get Eq (16),

$$y''(x) + \frac{1}{x}y'(x) - 9y = -9x^3 - 9 + \frac{1}{x}. \quad (16)$$

We give the error function which is used to measure the quality of the approximate solution in Eq(17)

$$E = \frac{1}{N}\sum_{m=1}^{N}(x_m\hat{y}''_m + \hat{y}'_m - 9x_m\hat{y}_m + 9x_m + 9x_m^4 - 1)^2 + \frac{1}{2}((\hat{y}_0 - 1)^2 + (\hat{y}_N - 3)^2) \quad (17)$$

Case 2: From Eq (14) with $\mu = -1$ and $f(x) = f_2(x)$ we get Eq (18),

$$y''(x) + \frac{1}{x}y'(x) - y = -x^3 + 8x - 1 + \frac{1}{x}. \quad (18)$$

We give the error function which is used to measure the quality of the approximate solution in Eq (19)

$$E = \frac{1}{N}\sum_{m=1}^{N}(x_m\hat{y}''_m + \hat{y}'_m - x_m\hat{y}_m + x_m + x_m^4 - 8x_m^2 - 1)^2 + \frac{1}{2}((\hat{y}_0 - 1)^2 + (\hat{y}_N - 3)^2) \quad (19)$$

Case 3: From Eq (14) with $\mu = 1$ and $f(x) = f_3(x)$ we get Eq (20),

$$y''(x) + \frac{1}{x}y'(x) + y = x^3 + 10x + 1 + \frac{1}{x}. \quad (20)$$

We give the error function which is used to measure the quality of the approximate solution in Eq (21)

$$E = \frac{1}{N}\sum_{m=1}^{N}(x_m\hat{y}''_m + \hat{y}'_m + x_m\hat{y}_m - x_m - x_m^4 - 10x_m^2 - 1)^2 + \frac{1}{2}((\hat{y}_0 - 1)^2 + (\hat{y}_N - 3)^2) \quad (21)$$

The unknown decision weights in error functions (17), (19) and (21) are determined by using the novel ANN based FO-DPSO approach. We briefly present approximate solutions for cases 1, 2, and 3 in Eqs (22), (23) and (24). We show the best values of weights along with convergence plots and step sizes in Figs 4, 5, 6 for Problems 1, 2, and 3. After getting the best weights, we used their values in approximate solutions given in Eq (4). For reproduction of our results, we have presented the complete solutions in the Appendix section without rounding off errors.

$$\hat{y}_1(x) = \frac{-0.8218}{1 + e^{-(0.6157x+1.57378)}} + \cdots + \frac{3.5514}{1 + e^{-(0.1380x+0.1581)}} \tag{22}$$

$$\hat{y}_2(x) = \frac{5.597}{1 + e^{-(2.3355x-3.0491)}} + \cdots + \frac{-0.3914}{1 + e^{-(4.1093x+2.4261)}} \tag{23}$$

$$\hat{y}_3(x) = \frac{5.7670}{1 + e^{-(1.9282x+4.0689)}} + \cdots + \frac{-2.8305}{1 + e^{-(0.1369x+0.3719)}} \tag{24}$$

Results got by FO-DPSO are compared with exact solutions, Genetic algorithm, its variant GA-SQP and are presented in Tables 1, 2, 3, 4, 5, 6, and 7 with step sizes h = 0.05 and 0.2, for problems 1, 2, 3 and 4. Input variable x is varied in interval [0 1]. The Absolute Error (AE) is calculated to highlight the better performance of our approach. Mathematically, it can be expressed as in Eq (25):

$$AE = |y(x) - \hat{y}(x)|. \tag{25}$$

Values of AEs show better results in terms of accuracy of our approach. We give all values of

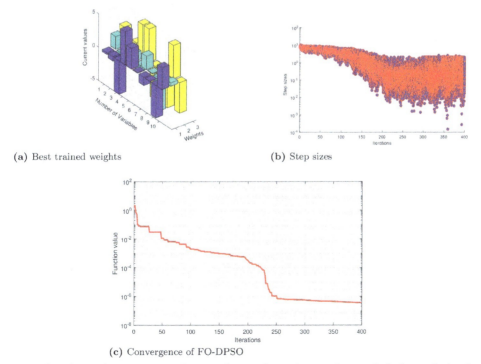

(a) Best trained weights   (b) Step sizes

(c) Convergence of FO-DPSO

Fig 4. Best weights obtained, convergence of error values and step sizes used to reach the best solution for problem 1 case 1, 2, 3 using feed-forward ANNs based on FO-DPSO algorithm.

AEs for step size h = 0.2 in Tables 2, 4 and 6. A graphical illustration of AEs with h = 0.05 is given in Fig 7.

FO-DPSO is better than GA-SQP and GA based approach as our approach is more accurate in solving problem 1. Results of AE show that ANN based FO-DPSO has produced values laying in ranges 10^{-7} to 10^{-12}, 10^{-8} to 10^{-12} and 10^{-8} to 10^{-11} for problem 1, case 1, 2, and 3 respectively. We establish that ANN based on FO-DPSO is a successful technique for solving the problem under consideration.

Problem 2

In this problem, we consider a doubly singular ODE with boundary values and variable coefficients. It is a homogenous differential equation of second order. Mathematically, this problem can be represented as in Eq (26) [2],

$$\begin{cases} (x^\mu y'(x))' = vx^{\mu+v-2}(vx^v + \mu + v - 1)y & 0 < x \leq 1 \quad \mu, v > 0 \\ y(0) = 1, \quad y(1) = e. \end{cases} \quad (26)$$

An exact solution for this problem is suggested in [2], Eq (27)

$$y = e^{x^v}. \quad (27)$$

We formulate three cases according to the variable coefficients μ, v. We have tested the proposed method by solving these three cases.

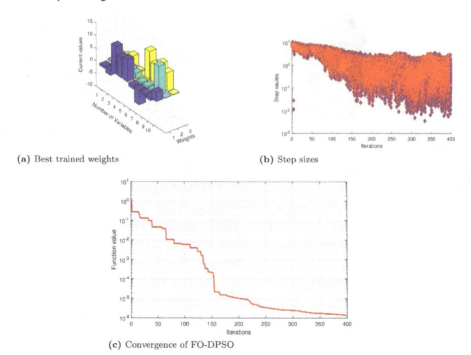

(a) Best trained weights

(b) Step sizes

(c) Convergence of FO-DPSO

Fig 5. Best weights obtained, convergence of error values and step sizes used to reach the best solution for problem 2 case 1, 2, 3 using feed-forward ANNs based on FO-DPSO algorithm.

Case 1: choosing $\mu = 0.5$ and $\nu = 1$ in problem (26), we get,

$$\sqrt{x}y''(x) + \frac{1}{2\sqrt{x}}y'(x) = \frac{1}{\sqrt{x}}\frac{2x+1}{2}y. \quad (28)$$

The error function to judge the quality of the solutions is formulated as

$$E = \frac{1}{N}\sum_{m=1}^{N}(2x_m\hat{y}''_m + \hat{y}'_m - 2x_m\hat{y}_m - \hat{y}_m)^2 + \frac{1}{2}((\hat{y}_0 - 1)^2 + (\hat{y}_N - e)^2). \quad (29)$$

Case 2: choosing $\mu = 0.75$ and $\nu = 1$ in problem (26), we get,

$$x^{3/4}y''(x) + \frac{3}{4x^{1/4}}y'(x) = \frac{1}{x^{1/4}}(x + \frac{3}{4})y. \quad (30)$$

The error function to judge the quality of the solutions is formulated as

$$E = \frac{1}{N}\sum_{m=1}^{N}(4x_m\hat{y}''_m + 3\hat{y}'_m - 4x_m\hat{y}_m - 3\hat{y}_m)^2 + \frac{1}{2}((\hat{y}_0 - 1)^2 + (\hat{y}_N - e)^2). \quad (31)$$

Case 3: choosing $\mu = 0.25$ and $\nu = 1$ in problem (26), we get,

$$x^{1/4}y''(x) + \frac{1}{4x^{3/4}}y'(x) = \frac{1}{x^{3/4}}(x + \frac{1}{4})y. \quad (32)$$

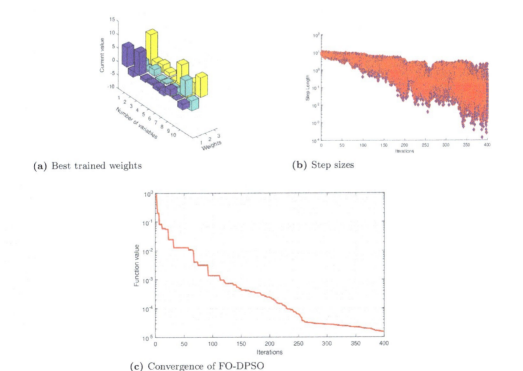

(a) Best trained weights (b) Step sizes

(c) Convergence of FO-DPSO

Fig 6. Best weights obtained, convergence of error values and step sizes used to reach the best solution for problem 3 case 1, 2, 3 using feed-forward ANNs based on FO-DPSO algorithm.

The error function to judge the quality of the solutions is formulated as

$$E = \frac{1}{N}\sum_{m=1}^{N}(4x_m\hat{y}''_m + \hat{y}'_m - 4x_m\hat{y}_m - \hat{y}_m)^2 + \frac{1}{2}((\hat{y}_0 - 1)^2 + (\hat{y}_N - e)^2). \tag{33}$$

The unknown decision weights in error functions (29), (31) and (33) are determined by using the novel ANN based FO-DPSO approach. Values of AE show better results in terms of accuracy of the corresponding approach. We give all values of solutions and AEs (for h = 0.2) in Tables 3 and 4 for problem 2. FO-DPSO is better than GA-SQP and GA based approach by solving problem 2 more accurately. Results of AE show that ANN-based FO-DPSO has produced, ranges in 10^{-8} to 10^{-11}, 10^{-8} to 10^{-11} and 10^{-9} to 10^{-11} respectively. We establish it that ANN-based on FO-DPSO is a successful technique for solving the prob- lem under consideration.

Problem 3

In this problem, we consider a nonlinear doubly singular ODE with boundary values and variable coefficients. It is a homogenous differential equation of second order. Mathematically, this problem can be represented as in Eq (34) [2],

$$\begin{cases} (x^\mu y'(x))' = vx^{\mu+v-2}e^y(e^y x^v - \mu - v + 1) & 0 < x \le 1 \quad \mu, v > 0 \\ y(0) = \ln(\frac{v}{4}), \quad y(1) = \ln(\frac{v}{5}). \end{cases} \tag{34}$$

Table 1. Empirical solutions for problem 1 (Case 1, 2, 3) achieved by FO-DPSO and GA. Which are compared with exact solutions for inputs x varying from 0 to 1 with a step size $h = 0.05$.

x		Case 1		Case 2		Case 3	
	Exact	FO-DPSO	GA	FO-DPSO	GA	FO-DPSO	GA
0	1	1.000031	1	1.000025	1	1	1
0.05	1.050125	1.050163	1.050124	1.050145	1.050125	1.050085	1.050122
0.1	1.101	1.101024	1.101	1.101008	1.101	1.100973	1.100999
0.15	1.153375	1.15338	1.153375	1.153373	1.153375	1.153365	1.153375
0.2	1.208	1.207991	1.208	1.207993	1.208	1.208002	1.208
0.25	1.265625	1.265613	1.265625	1.265617	1.265625	1.26563	1.265624
0.3	1.327	1.326993	1.327	1.326992	1.327	1.327002	1.326999
0.35	1.392875	1.392879	1.392875	1.392869	1.392875	1.39287	1.392875
0.4	1.464	1.464015	1.464	1.463995	1.464	1.463989	1.464
0.45	1.541125	1.541146	1.541125	1.54112	1.541125	1.541112	1.541125
0.5	1.625	1.625021	1.625	1.624993	1.625	1.62499	1.625001
0.55	1.716375	1.716391	1.716375	1.716365	1.716375	1.716369	1.716376
0.6	1.816	1.816008	1.816	1.815985	1.816	1.815998	1.816
0.65	1.924625	1.924625	1.924625	1.924607	1.924625	1.924626	1.924625
0.7	2.043	2.042996	2.043	2.042979	2.043	2.043	2.043
0.75	2.171875	2.171872	2.171875	2.171854	2.171875	2.171873	2.171876
0.8	2.312	2.312002	2.312	2.31198	2.312	2.311996	2.312001
0.85	2.464125	2.464133	2.464125	2.464105	2.464125	2.464121	2.464126
0.9	2.629	2.629013	2.629	2.62898	2.629	2.628997	2.629001
0.95	2.807375	2.807389	2.807375	2.807353	2.807375	2.807373	2.807376
1	3	3.000013	3	3	3	3	3.000001

An exact solution for this problem is suggested in [2], Eq (35)

$$y = \ln\frac{v}{4 + x^v}. \tag{35}$$

We formulate three cases according to the variable coefficients μ, v. We have tested the proposed method by solving these three cases,

Case 1: choosing $\mu = 0.25$ and $v = 1$ in problem (34), we get,

$$\begin{cases} x^{1/4}y''(x) + \dfrac{1}{4x^{3/4}}y'(x) = \dfrac{e^y}{x^{3/4}}(e^y x - \dfrac{1}{4}) & 0 < x \leq 1 \quad \mu, v > 0 \\ y(0) = \ln(\dfrac{1}{4}), \quad y(1) = \ln(\dfrac{1}{5}). \end{cases} \tag{36}$$

Table 2. Absolute errors in results for problem 1 (Case 1, 2, 3) achieved by FO-DPSO and GA-SQP. Which are matched with exact solutions for inputs x varying from 0 to 1 with a step size $h = 0.2$.

x	Case 1 (AE)		Case 2 (AE)		Case 3 (AE)	
	FO-DPSO	GA-SQP	FO-DPSO	GA-SQP	FO-DPSO	GA-SQP
0	9.98E-09	1.36E-07	4.19E-08	2.17E-07	4.69E-09	4.40E-07
0.2	6.25E-07	9.96E-09	8.94E-08	4.39E-08	3.71E-07	2.45E-07
0.4	1.43E-06	4.07E-08	1.17E-06	1.03E-07	1.04E-06	1.94E-08
0.6	1.19E-06	9.64E-10	1.50E-06	9.63E-08	9.23E-07	3.75E-07
0.8	3.30E-07	3.95E-08	5.96E-07	1.60E-07	2.53E-07	7.77E-07
1	1.34E-08	8.07E-08	2.26E-08	1.73E-07	1.85E-08	1.02E-06

Table 3. Empirical solutions for problem 2 (Case 1, 2, 3) achieved by FO-DPSO and GA. Which are compared with exact solutions for inputs x varying from 0 to 1 with a step size $h = 0.05$.

x		Case 1		Case 2		Case 3	
	Exact	FO-DPSO	GA	FO-DPSO	GA	FO-DPSO	GA
0	1	1.000001	1	1.000002	1.000014	1	1.000001
0.05	1.051271	1.051271	1.051245	1.051272	1.05129	1.051271	1.051273
0.1	1.105171	1.105171	1.10516	1.105172	1.105187	1.105171	1.105172
0.15	1.161834	1.161834	1.161832	1.161835	1.161844	1.161834	1.161835
0.2	1.221403	1.221403	1.221406	1.221403	1.221408	1.221403	1.221403
0.25	1.284025	1.284026	1.284031	1.284026	1.284027	1.284025	1.284026
0.3	1.349859	1.349859	1.349866	1.349859	1.349859	1.349859	1.349859
0.35	1.419068	1.419068	1.419074	1.419068	1.419068	1.419067	1.419068
0.4	1.491825	1.491825	1.49183	1.491825	1.491825	1.491825	1.491825
0.45	1.568312	1.568312	1.568317	1.568312	1.568314	1.568312	1.568312
0.5	1.648721	1.648721	1.648725	1.648721	1.648723	1.648721	1.648721
0.55	1.733253	1.733253	1.733257	1.733253	1.733254	1.733253	1.733253
0.6	1.822119	1.822119	1.822123	1.822118	1.822119	1.822119	1.822118
0.65	1.915541	1.915541	1.915547	1.91554	1.91554	1.915541	1.91554
0.7	2.013753	2.013753	2.013761	2.013752	2.01375	2.013753	2.013752
0.75	2.117	2.117	2.117011	2.117	2.116997	2.117	2.117
0.8	2.225541	2.225541	2.225555	2.22554	2.225537	2.225541	2.22554
0.85	2.339647	2.339647	2.339663	2.339646	2.339642	2.339647	2.339646
0.9	2.459603	2.459603	2.459621	2.459603	2.459598	2.459603	2.459602
0.95	2.58571	2.58571	2.585728	2.585709	2.585704	2.58571	2.585708
1	2.718282	2.718282	2.7183	2.718281	2.718276	2.718282	2.718281

The error function to judge the quality of the solutions is formulated as

$$E = \frac{1}{N}\sum_{m=1}^{N}(4x_m\hat{y}''_m + \hat{y}'_m - 4x_m e^{\hat{y}_m} + e^{\hat{y}_m})^2 + \frac{1}{2}((\hat{y}_0 - \ln\frac{1}{4})^2 + (\hat{y}_N - \ln\frac{1}{5})^2). \tag{37}$$

<u>Case 2</u>: choosing $\mu = 0.5$ and $\nu = 1$ in problem (34), we get,

$$\begin{cases} x^{1/2}y''(x) + \frac{1}{2x^{1/2}}y'(x) = \frac{e^y}{x^{1/2}}(e^y x - \frac{1}{2}) & 0 < x \le 1 \quad \mu, \nu > 0 \\ y(0) = \ln(\frac{1}{4}), \quad y(1) = \ln(\frac{1}{5}). \end{cases} \tag{38}$$

Table 4. Absolute errors in results for problem 2 (Case 1, 2, 3) achieved by FO-DPSO and GA. Which are matched with exact solutions for inputs x varying from 0 to 1 with a step size $h = 0.2$.

x	Case 1 (AE)		Case 2 (AE)		Case 3 (AE)	
	FO-DPSO	GA-SQP	FO-DPSO	GA-SQP	FO-DPSO	GA-SQP
0	6.45E-08	5.41E-07	2.48E-07	1.54E-06	8.66E-07	4.74E-07
0.2	9.19E-08	2.22E-07	2.67E-06	2.09E-07	3.23E-06	1.84E-07
0.4	1.05E-08	5.38E-08	8.41E-06	2.50E-08	1.88E-05	3.49E-08
0.6	2.22E-08	3.52E-08	5.33E-06	3.27E-07	2.65E-05	5.87E-08
0.8	5.91E-09	4.08E-08	9.07E-07	4.63E-07	1.17E-05	1.64E-07
1	2.77E-10	7.29E-08	2.94E-08	6.47E-07	1.05E-06	2.62E-07

Table 5. Empirical solutions for problem 3 (Case 1, 2, 3) achieved by FO-DPSO and GA. Which are compared with exact solutions for inputs x varying from 0 to 1 with a step size $h = 0.05$.

x		Case 1		Case 2		Case 3	
	Exact	FO-DPSO	GA	FO-DPSO	GA	FO-DPSO	GA
0	-1.38629	-1.386296	-1.38629	-1.386294	-1.38629	-1.386294	-1.38629
0.05	-1.39872	-1.398717	-1.39872	-1.398716	-1.39872	-1.398717	-1.39872
0.1	-1.41099	-1.410987	-1.41099	-1.410986	-1.41099	-1.410987	-1.41099
0.15	-1.42311	-1.423108	-1.42311	-1.423107	-1.42311	-1.423108	-1.42311
0.2	-1.43508	-1.435084	-1.43508	-1.435083	-1.43508	-1.435085	-1.43508
0.25	-1.44692	-1.446919	-1.44692	-1.446918	-1.44692	-1.446919	-1.44692
0.3	-1.45862	-1.458615	-1.45862	-1.458614	-1.45862	-1.458615	-1.45862
0.35	-1.47018	-1.470175	-1.47018	-1.470175	-1.47018	-1.470176	-1.47018
0.4	-1.4816	-1.481604	-1.4816	-1.481604	-1.4816	-1.481604	-1.4816
0.45	-1.4929	-1.492903	-1.4929	-1.492903	-1.4929	-1.492904	-1.4929
0.5	-1.50408	-1.504076	-1.50408	-1.504076	-1.50408	-1.504077	-1.50408
0.55	-1.51513	-1.515126	-1.51513	-1.515126	-1.51513	-1.515127	-1.51513
0.6	-1.52606	-1.526055	-1.52606	-1.526055	-1.52606	-1.526056	-1.52606
0.65	-1.53687	-1.536866	-1.53687	-1.536866	-1.53687	-1.536867	-1.53687
0.7	-1.54756	-1.547561	-1.54756	-1.547561	-1.54756	-1.547563	-1.54756
0.75	-1.55814	-1.558143	-1.55814	-1.558143	-1.55814	-1.558145	-1.55814
0.8	-1.56862	-1.568614	-1.56862	-1.568615	-1.56862	-1.568616	-1.56862
0.85	-1.57898	-1.578977	-1.57898	-1.578978	-1.57898	-1.578979	-1.57898
0.9	-1.58924	-1.589233	-1.58924	-1.589234	-1.58924	-1.589235	-1.58924
0.95	-1.59939	-1.599386	-1.59939	-1.599387	-1.59939	-1.599388	-1.59939
1	-1.60944	-1.609436	-1.60944	-1.609437	-1.60944	-1.609438	-1.60944

The error function to judge the quality of the solutions is formulated as

$$E = \frac{1}{N}\sum_{m=1}^{N}(2x_m\hat{y}_m'' + \hat{y}_m' - 2x_m e^{\hat{y}_m} + e^{\hat{y}_m})^2 + \frac{1}{2}((\hat{y}_0 - \ln\frac{1}{4})^2 + (\hat{y}_N - \ln\frac{1}{5})^2). \quad (39)$$

Case 3: choosing $\mu = 0.75$ and $v = 1$ in problem (34), we get,

$$\begin{cases} x^{3/4}y''(x) + \frac{1}{2x^{1/4}}y'(x) = \frac{e^y}{x^{1/4}}(e^y x - \frac{3}{4}) & 0 < x \leq 1 \quad \mu, v > 0 \\ y(0) = \ln(\frac{1}{4}), \quad y(1) = \ln(\frac{1}{5}). \end{cases} \quad (40)$$

Table 6. Absolute errors in results for problem 3 (Case 1, 2, 3) achieved by FO-DPSO and GA-SQP. Which are matched with exact solutions for inputs x varying from 0 to 1 with a step size $h = 0.2$.

x	Case 1 (AE)		Case 2 (AE)		Case 3 (AE)	
	FO-DPSO	GA	FO-DPSO	GA	FO-DPSO	GA
0	1.39E-08	1.30E-08	7.68E-12	2.06E-07	8.55E-10	1.22E-09
0.2	6.01E-09	1.14E-09	5.51E-12	1.70E-08	5.98E-09	1.16E-09
0.4	2.79E-10	6.65E-09	4.90E-11	6.57E-08	6.20E-09	3.97E-09
0.6	1.16E-08	8.70E-09	3.55E-13	1.57E-07	2.43E-09	2.83E-09
0.8	1.44E-08	1.23E-08	7.72E-11	1.90E-07	9.90E-09	3.27E-09
1	1.65E-09	1.60E-08	1.97E-11	2.31E-07	1.49E-09	3.65E-09

Table 7. Comparison of solution obtained for the problem 4 of porous fin designed model using FO-DPSO.

ξ	Analytical	FO-DPSO	GA-SQP	GA
0	0.700465898	0.701211994	0.701171286	0.701382837
0.1	0.703355803	0.704109143	0.704069922	0.704273632
0.2	0.712042101	0.712810621	0.712779543	0.71301169
0.3	0.72657335	0.727352198	0.727330423	0.727587757
0.4	0.747026504	0.747792755	0.747775149	0.74800451
0.5	0.773500952	0.774210653	0.774192242	0.774320572
0.6	0.806110215	0.806702499	0.806683296	0.806665864
0.7	0.844971293	0.845383557	0.845368094	0.845227794
0.8	0.890191724	0.890388584	0.890380987	0.890213813
0.9	0.941854371	0.941871225	0.941870115	0.941798879
1	1	0.99999949	0.999999331	1.000067362
MSE		6.50E-08	1.00E-07	1.00E-04

The error function to judge the quality of the solutions is formulated as

$$E = \frac{1}{N}\sum_{m=1}^{N}(4x_m\hat{y}_m'' + 3\hat{y}_m' - 4x_m e^{\hat{y}_m} + 3e^{\hat{y}_m})^2 + \frac{1}{2}((\hat{y}_0 - \ln\frac{1}{4})^2 + (\hat{y}_N - \ln\frac{1}{5})^2). \quad (41)$$

$$\hat{y}(x) = \frac{-1.7690}{1 + e^{-(-1.6186x - 4.3759)}} + \frac{0.6479}{1 + e^{-(-0.3783x - 0.1091)}} \quad (42)$$

$$\hat{y}(x) = \frac{-2.8314}{1+e^{-(0.8127x+2.7334)}} + \frac{0.1693}{1+e^{-(1.0845x+0.3734)}} \tag{43}$$

$$\hat{y}(x) = \frac{-0.2739}{1+e^{-(1.9006x+2.8037)}} + \frac{3.2499}{1+e^{-(0.4887x-4.5624)}} \tag{44}$$

The unknown decision weights in error functions (37), (39) and (41) are determined by using the novel ANN based FO-DPSO approach. We briefly present approximate solutions for all cases in Eqs (42), (43) and (44). Values of AEs show better performance in terms of accuracy of our approach. We give all solutions and AEs (for h = 0.2) in Tables 5, and 6 for problem 3. FO-DPSO is better than GA and its variant GA-SQP by solving problem 3 more accurately.

Results of AE show that ANN-based FO-DPSO has produced values laying in ranges 10^{-11} to 10^{-13}, 10^{-11} to 10^{-16} and 10^{-10} to 10^{-15} respectively. We establish it that ANN-based on FO-DPSO is a successful technique for solving the problem under consideration.

4 Mathematical model of the porous fin

Problem 4: The schematic diagram of straight fin problem possessing the arbitrary cross-sectional area A_c, perimeter P, and length b, is presented in Fig 8, [55]. The fin is joined with the base surface having the temperature T_b, and extends into fluid having temperature T_a, and its tip is insulated. The energy balance equation is written as:

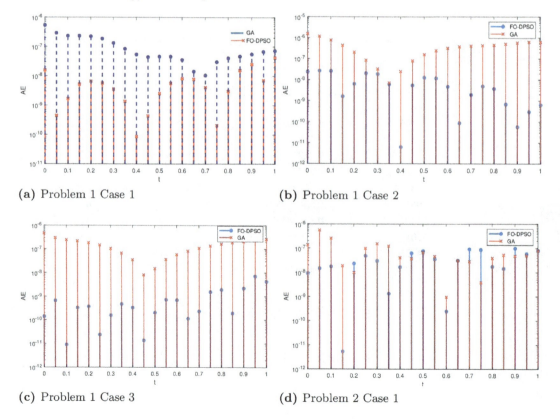

(a) Problem 1 Case 1

(b) Problem 1 Case 2

(c) Problem 1 Case 3

(d) Problem 2 Case 1

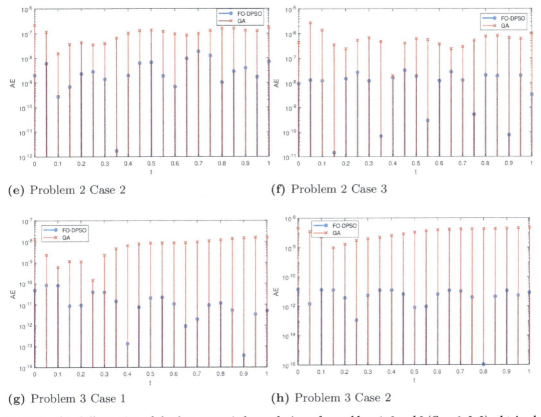

(e) Problem 2 Case 2

(f) Problem 2 Case 3

(g) Problem 3 Case 1

(h) Problem 3 Case 2

Fig 7. Graphical illustration of absolute errors in best solutions, for problem 1, 2 and 3 (Case 1, 2, 3), obtained by FO-DPSO and GA.

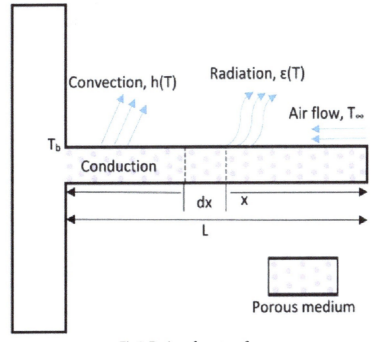

Fig 8. Design of a porous fin.

In the Eq (45), $k(T)$ indicates the temperature-dependent thermal conductivity and h represents the coefficient of heat transfer. It is considered that the thermal conductivity for the fin material is expressed:

$$k(T) = k_b[1 + \lambda(T - T_b)]. \tag{46}$$

In the expression (46), k_b represents the thermal conductivity at the ambient fluid temperature of the fin, and λ is standing for the variation of the thermal conductivity. Using the non- dimensional variables:

$$\theta = \frac{T - T_a}{T_b - T_a}, \quad \xi = \frac{x}{b}, \quad \mu = \lambda(T_b - T_a), \quad \text{and} \quad \psi = \left(\frac{Phb^2}{k_a A_c}\right)^{1/2}, \tag{47}$$

consequently the Eq (45) reduces into the following form:

$$\frac{d^2\theta}{d\xi^2} + \mu\theta\frac{d^2\theta}{d\xi^2} + \mu\left(\frac{d\theta}{d\xi}\right)^2 - \psi^2\theta, \quad 0 \leq \xi \leq 1, \tag{48}$$

with the boundary conditions

$$\left.\frac{d\theta}{d\xi}\right|_{\xi=0} = 0 \quad \text{and} \quad \theta|_{\xi=1} = 1. \tag{49}$$

We have solved a porous fin model using ANNs based FO-DPSO approach. The unknown weights are tuned by the FO-DPSO algorithm. Results obtained by FO-DPSO are compared with GA-SQP and GA and are given in Table 7. The graphical illustration of this model is given in Fig 8. Solutions obtained for this problem are plotted in Fig 9. Among the three algorithms, FO-DPSO performed well and gave us minimum error as compared to the other algorithms, see Fig 9.

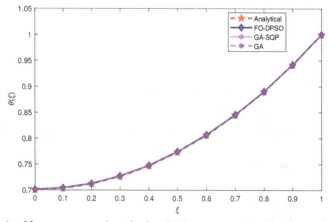

Fig 9. Solution obtained by our proposed method and other comparative algorithms for porous fin model.

Statistical analysis

Performance indicators like global mean absolute error (GMAE), mean absolute error (MAE), and mean value of fitness (M_{fit}) are used to asses the performance of ANNs based on FO-DPSO approach. We use these indicators on the data we have got through 100 indepen- dent simulations to determine the stability and robustness of our approach. We present MAE values in terms of sorted and unsorted form, see Figs 10, 11 and 12. Hence, the sorted results are presented in Figs 10b, 11b and 12b, while the unsorted errors in the solutions are given in Figs 10a, 11a and 12a, respectively. To further elaborate the difference between the errors obtained, and those reported in the literature, we have used the log scale plots for MAEs. From our graphical analysis, we get the minimum values of MAEs and better fitness for all problems. The performance of our approach is statistically analyzed in terms of the best minimum value, mean, and standard deviation (SD). This further validates our claim that our approach is better in convergence rate and has produced accurate results for all three BVPs with doubly singulari- ties, and porous fin model. We present statistical results in terms of GMAE, Mean-time, Max iterations in Tables 8, 9 and 10. Our experimental outcome dictates that the ANNs based FO-DPSO approach has consistently produced better solutions to the non-linear ODEs and a real application problem.

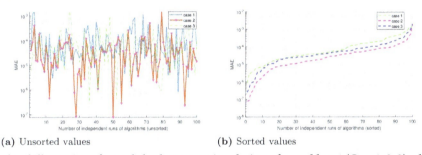

(a) Unsorted values (b) Sorted values

Fig 10. Graphical illustration of sorted absolute errors in solutions, for problem 1 (Case 1, 2, 3), obtained by FO-DPSO during 100 runs.

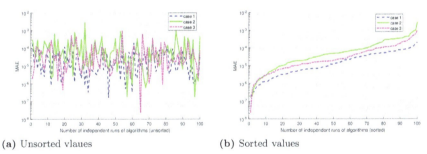

(a) Unsorted vlaues (b) Sorted values

Fig 11. Graphical illustration of sorted absolute errors in solutions, for problem 2 (Case 1, 2, 3), obtained by FO-DPSO during 100 runs.

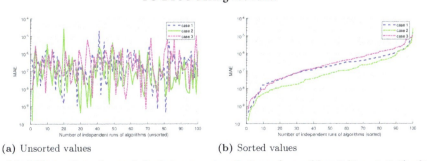

(a) Unsorted values (b) Sorted values

Fig 12. Graphical illustration of sorted absolute errors in solutions, for problem 3 (Case 1, 2, 3), obtained by FO-DPSO during 100 runs.

To verify the stability and robustness of the proposed technique, we got better values of global performance indicators; global mean absolute error (G_{MAE}) as in Eq (50) and mean of fitness values denoted as (M_{fit}) as in Eq (51). All results in terms of these global performance indicators, see Tables 8, 9, and 10 revealed the fact that our approach is better than state-of- the-art approaches reported in the literature [33].

$$G_{MAE} = \frac{1}{R}(\sum_{r=1}^{R}\frac{1}{P}(\sum_{i=1}^{P}|y_i - \hat{y}_{i,r}|)), \qquad (50)$$

Table 8. Performance indicators based on proposed results for Problem 1.

Type Parameters	Case 1	Case 2	Case 3
GMAE Values	9.40E-06	3.0273E-06	6.6613E-06
STD	5.6686E-06	2.7162E-06	5.3095E-06
Mfit VALUES	4.0931E-08	4.2931E-09	1.2824E-08
STD	3.2671E-08	4.6459E-09	9.6663E-09
Mean Time	27.8	30	28.4
Max Iteration	1000	1000	1000

Table 9. Performance indicators based on proposed results for Problem 2.

Type Parameters	Case 1	Case 2	Case 3
GMAE Values	1.51E-06	4.164E-06	3.207E-06
STD	9.542E-07	3.328E-06	2.047E-06
Mfit VALUES	7.747E-09	8.493E-09	1.037E-09
STD	1.005E-08	9.361E-09	1.729E-09
Mean Time	40.9	27.7	30.7
Max Iteration	1000	1000	1000

Table 10. Performance indicators based on proposed results for Problem 3.

Type Parameters	Case 1	Case 2	Case 3
GMAE Values	1.48E-08	5.4029E-09	1.2547E-08
STD	1.101E-08	3.7189E-09	1.0439E-08
Mfit VALUES	1.9996E-11	6.9589E-12	3.2591E-11
STD	2.4734E-11	4.8198E-12	6.6736E-11
Mean Time	29.5	29.9	28.9
Max Iteration	1000	1000	1000

$$M_{Fit} = \frac{1}{R}(\sum_{r=1}^{R}E_r), \qquad (51)$$

$$MeanTime = \frac{1}{R}(\sum_{r=1}^{R}T_r), \qquad (52)$$

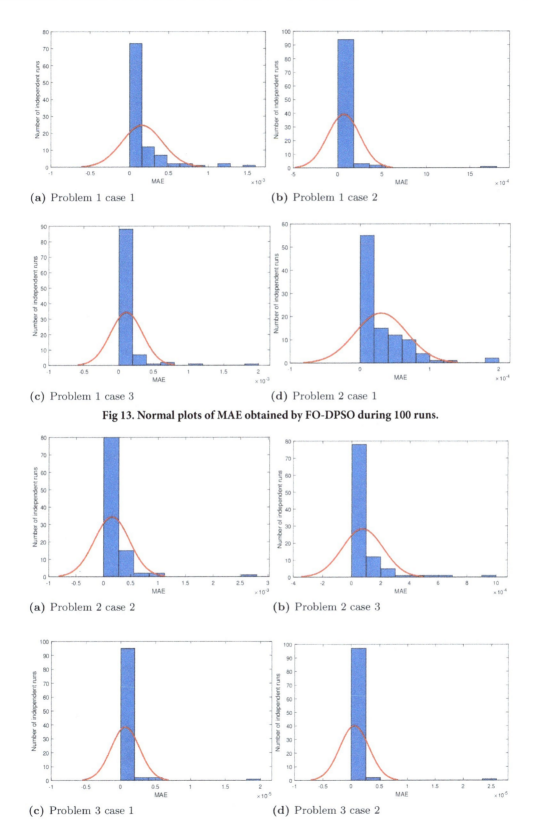

(a) Problem 1 case 1

(b) Problem 1 case 2

(c) Problem 1 case 3

(d) Problem 2 case 1

Fig 13. Normal plots of MAE obtained by FO-DPSO during 100 runs.

(a) Problem 2 case 2

(b) Problem 2 case 3

(c) Problem 3 case 1

(d) Problem 3 case 2

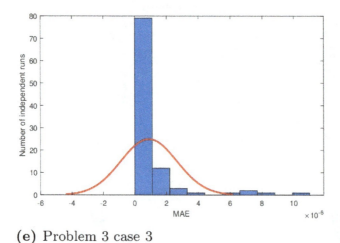

(e) Problem 3 case 3

Fig 14. Normal plots of MAE obtained by FO-DPSO during 100 runs.

where the number of inputs and the number of simulations is denoted by P, and R, respectively, and y_i denote the corresponding exact solutions and $\hat{y}_{i,r}$ the corresponding best approximate val- ues obtained with our approach. E_r represents the best objective value achieved during the rth sim- ulation. In our experiments, input variable varies from 0 to 1 with $h = 0.05$, and 0.2 as step size. Thus, the total grid points were 20, and 6. We repeated our simulations for 100 times. Moreover, mean absolute errors in all solutions are negligible, and histograms with normal distribution fit for mean absolute errors (MAEs) in solutions of problems 1, 2, 3 are given in Figs 13 and 14.

Conclusion

In this paper, we present a new soft computing approach that combines artificial neural networks with a fractional-order particle swarm optimization (FO-DPSO) algorithm. We conclude this research by following key findings.

- The proposed method to exploit ANNs based FO-DPSO accurately solves variants of non- linear, doubly singular ordinary differential equations according to the computational evaluations.

- We calculate absolute errors in our results using the exact reference solutions, and the errors. Experimental results show that our designed scheme is more accurate compared to the state- of-the-art algorithms GA, and its latest variant GA-SQP.

- We have considered three hard problems with nine cases. A real application is also consid- ered. We have analyzed a mathematical model of porous fins, and temperature profiles are studied for this model.

- We have analyzed several actual application problems. The Values of performance indica- tors, G_{MAE}, and M_{fit} dictate that our approach gives results with lower errors compared to other algorithms.

- Frequency graphs of 100 experiments in terms of MAD are presented with normal distribu- tion fittings. These graphs have proved that our approach is reliable and stable in terms of success rate.

- The solutions to the problems are in the appendix to help the reader reproduce the results presented in this paper.

- The proposed approach provides a more accurate solution to differential equations with multiple singularities and systems of such equations. Problems arising in thermodynamics, electromagnetic, and nanotechnology can be handled by the proposed method by changing the activation function of the artificial neural network.

Supporting information

S1 Appendix.

(PDF)

Author Contributions

Conceptualization: Waseem Waseem.

Formal analysis: Waseem Waseem.

Funding acquisition: Poom Kumam.

Methodology: Waseem Waseem.

Project administration: Muhammad Sulaiman.

Resources: Muhammad Sulaiman, Muhammad Asif Zahoor Raja, Saeed Islam.

Software: Muhammad Sulaiman.

Supervision: Muhammad Sulaiman.

Validation: Waseem Waseem, Muhammad Sulaiman, Muhammad Asif Zahoor Raja.

Visualization: Waseem Waseem.

Writing – original draft: Waseem Waseem.

Writing – review & editing: Muhammad Sulaiman, Poom Kumam, Muhammad Shoaib.

References

1. Singh R, Kumar J. An efficient numerical technique for the solution of nonlinear singular boundary value problems. Computer Physics Communications. 2014; 185(4):1282–1289. https://doi.org/10.1016/j.cpc. 2014.01.002

2. Singh R, Kumar J. The Adomian decomposition method with Green's function for solving nonlinear

sin- gular boundary value problems. Journal of Applied Mathematics and Computing. 2014; 44(1-2):397– 416. https://doi.org/10.1007/s12190-013-0699-4

3. Singh R, Kumar J, Nelakanti G. Approximate series solution of singular boundary value problems with derivative dependence using Green's function technique. Computational and Applied Mathematics. 2014; 33(2):451–467. https://doi.org/10.1007/s40314-013-0074-y

4. Singh R, Kumar J, Nelakanti G. New approach for solving a class of doubly singular two-point boundary value problems using Adomian decomposition method. Advances in Numerical Analysis. 2012; 2012. https://doi.org/10.1155/2012/541083

5. Verma AK, Singh M. Singular nonlinear three point BVPs arising in thermal explosion in a cylindrical reactor. Journal of Mathematical Chemistry. 2015; 53(2):670–684. https://doi.org/10.1007/s10910-014- 0447-5

6. Roul P. An improved iterative technique for solving nonlinear doubly singular two-point boundary value problems. The European Physical Journal Plus. 2016; 131(6):209. https://doi.org/10.1140/epjp/i2016- 16209-1

7. Roul P, Warbhe U. A novel numerical approach and its convergence for numerical solution of nonlinear doubly singular boundary value problems. Journal of Computational and Applied Mathematics. 2016; 296:661–676. https://doi.org/10.1016/j.cam.2015.10.020

8. Pandey RK, Gupta G. A note on fourth order method for doubly singular boundary value problems. Advances in Numerical Analysis. 2012; 2012. https://doi.org/10.1155/2012/349618

9. Pandey R, Verma AK. On solvability of derivative dependent doubly singular boundary value problems. Journal of Applied Mathematics and Computing. 2010; 33(1-2):489–511. https://doi.org/10.1007/ s12190-009-0299-5

10. Singh R, Wazwaz AM, Kumar J. An efficient semi-numerical technique for solving nonlinear singular boundary value problems arising in various physical models. International Journal of Computer Mathe- matics. 2016; 93(8):1330–1346. https://doi.org/10.1080/00207160.2015.1045888

11. Malek S. On the summability of formal solutions for doubly singular nonlinear partial differential equa- tions. Journal of dynamical and control systems. 2012; 18(1):45–82. https://doi.org/10.1007/s10883- 012-9134-7

12. Canalis-Durand M, Mozo-Ferna´ ndez J, Scha¨ fke R. Monomial summability and doubly singular differen- tial equations. Journal of Differential Equations. 2007; 233(2):485–511. https://doi.org/10.1016/j.jde. 2006.11.005

13. Mall S, Chakraverty S. Application of Legendre Neural Network for solving ordinary differential equa- tions. Applied Soft Computing. 2016; 43:347–356. https://doi.org/10.1016/j.asoc.2015.10.069

14. Mall S, Chakraverty S. Numerical solution of nonlinear singular initial value problems of Emden–Fowler type using Chebyshev Neural Network method. Neurocomputing. 2015; 149:975–982. https://doi.org/ 10.1016/j.neucom.2014.07.036

15. Mall S, Chakraverty S. Chebyshev neural network based model for solving Lane–Emden type equa- tions. Applied Mathematics and Computation. 2014; 247:100–114. https://doi.org/10.1016/j.amc.2014. 08.085

16. Chakraverty S, Mall S. Regression-based weight generation algorithm in neural network for solution of initial and boundary value problems. Neural Computing and Applications. 2014; 25(3-4):585–594. https://doi.org/10.1007/s00521-013-1526-4

17. Raja MAZ, Samar R, Alaidarous ES, Shivanian E. Bio-inspired computing platform for reliable solution of Bratu-type equations arising in the modeling of electrically conducting solids. Applied Mathematical Modelling. 2016; 40(11-12):5964–5977. https://doi.org/10.1016/j.apm.2016.01.034

18. Masood Z, Majeed K, Samar R, Raja MAZ. Design of Mexican Hat Wavelet neural networks for solving Bratu type nonlinear systems. Neurocomputing. 2017; 221:1–14. https://doi.org/10.1016/j.neucom. 2016.08.079

19. Raja MAZ. Solution of the one-dimensional Bratu equation arising in the fuel ignition model using ANN optimised with PSO and SQP. Connection Science. 2014; 26(3):195–214. https://doi.org/10.1080/ 09540091.2014.907555

20. Ahmad I, Raja MAZ, Bilal M, Ashraf F. Neural network methods to solve the Lane–Emden type equa- tions arising in thermodynamic studies of the spherical gas cloud model. Neural Computing and Applica- tions. 2017; 28(1):929–944. https://doi.org/10.1007/s00521-016-2400-y

21. Raja MAZ, Ahmad I, Khan I, Syam MI, Wazwaz AM. Neuro-heuristic computational intelligence for solv- ing nonlinear pantograph systems. Frontiers of Information Technology & Electronic Engineering. 2017; 18(4):464–484.   https://doi.org/10.1631/FITEE.1500393

22. Raja MAZ, Zameer A, Khan AU, Wazwaz AM. A new numerical approach to solve Thomas–Fermi model of an atom using bio-inspired heuristics integrated with sequential quadratic programming. SpringerPlus. 2016; 5(1):1400. https://doi.org/10.1186/s40064-016-3093-5 PMID: 27610319

23. Effati S, Pakdaman M. Artificial neural network approach for solving fuzzy differential equations. Infor- mation Sciences. 2010; 180(8):1434–1457. https://doi.org/10.1016/j.ins.2009.12.016

24. Effati S, Skandari MHN. Optimal control approach for solving linear Volterra integral equations. Interna- tional Journal of Intelligent Systems and Applications. 2012; 4(4):40. https://doi.org/10.5815/ijisa.2012. 04.06

25. Jafarian A, Measoomy S, Abbasbandy S. Artificial neural networks based modeling for solving Volterra integral equations system. Applied Soft Computing. 2015; 27:391–398. https://doi.org/10.1016/j.asoc. 2014.10.036

26. Raja MAZ, Farooq U, Chaudhary NI, Wazwaz AM. Stochastic numerical solver for nanofluidic problems containing multi-walled carbon nanotubes. Applied Soft Computing. 2016; 38:561–586. https://doi.org/ 10.1016/j.asoc.2015.10.015

27. Effati S, Buzhabadi R. A neural network approach for solving Fredholm integral equations of the second kind. Neural Computing and Applications. 2012; 21(5):843–852. https://doi.org/10.1007/s00521-010- 0489-y

28. Raja MAZ, Khan JA, Chaudhary NI, Shivanian E. Reliable numerical treatment of nonlinear singular Flierl–Petviashivili equations for unbounded domain using ANN, GAs, and SQP. Applied Soft Comput- ing. 2016; 38:617–636. https://doi.org/10.1016/j.asoc.2015.10.017

29. Sabouri J, Effati S, Pakdaman M. A neural network approach for solving a class of fractional optimal control problems. Neural Processing Letters. 2017; 45(1):59–74. https://doi.org/10.1007/s11063-016- 9510-5

30. Effati S, Mansoori A, Eshaghnezhad M. An efficient projection neural network for solving bilinear pro- gramming problems. Neurocomputing. 2015; 168:1188–1197. https://doi.org/10.1016/j.neucom.2015. 05.003

31. Kumar M, Yadav N. Numerical solution of Bratu's problem using multilayer perceptron neural network method. National Academy Science Letters. 2015; 38(5):425–428. https://doi.org/10.1007/s40009-015- 0359-3

32. Khan JA, Raja MAZ, Rashidi MM, Syam MI, Wazwaz AM. Nature-inspired computing approach for solv- ing non-linear singular Emden–Fowler problem arising in electromagnetic theory. Connection Science. 2015; 27(4):377–396. https://doi.org/10.1080/09540091.2015.1092499

33. Raja MAZ, Mehmood J, Sabir Z, Nasab AK, Manzar MA. Numerical solution of doubly singular nonlinear systems using neural networks-based integrated intelligent computing. Neural Computing and Applica- tions. 2019; 31(3):793–812. https://doi.org/10.1007/s00521-017-3110-9

34. Couceiro MS, Rocha RP, Ferreira NF, Machado JT. Introducing the fractional-order Darwinian PSO. Signal, Image and Video Processing. 2012; 6(3):343–350. https://doi.org/10.1007/s11760-012-0316-2

35. Couceiro M, Ghamisi P. Fractional-order Darwinian PSO. In: Fractional order darwinian particle swarm optimization. Springer; 2016. p. 11–20.

36. Yadav N, Yadav A, Kumar M, Kim JH. An efficient algorithm based on artificial neural networks and par- ticle swarm optimization for solution of nonlinear Troesch's problem. Neural Computing and Applica- tions. 2017; 28(1):171–178. https://doi.org/10.1007/s00521-015-2046-1

37. Raja MAZ. Stochastic numerical treatment for solving Troesch's problem. Information Sciences. 2014; 279:860–873. https://doi.org/10.1016/j.ins.2014.04.036

38. Raja MAZ, Khan MAR, Mahmood T, Farooq U, Chaudhary NI. Design of bio-inspired computing tech- nique for nanofluidics based on nonlinear Jeffery–Hamel flow equations. Canadian Journal of Physics. 2016; 94(5):474–489. https://doi.org/10.1139/cjp-2015-0440

39. Raja MAZ, Khan JA, Haroon T. Stochastic numerical treatment for thin film flow of third grade fluid using unsupervised neural networks. Journal of the Taiwan Institute of Chemical Engineers. 2015; 48:26–39. https://doi.org/10.1016/j.jtice.2014.10.018

40. Raja MAZ, Manzar MA, Samar R. An efficient computational intelligence approach for solving fractional order Riccati equations using ANN and SQP. Applied Mathematical Modelling. 2015; 39(10-11):3075– 3093. https://doi.org/10.1016/j.apm.2014.11.024

41. Raja MAZ, Samar R, Manzar MA, Shah SM. Design of unsupervised fractional neural network model optimized with interior point algorithm for solving Bagley–Torvik equation. Mathematics and Computers in Simulation. 2017; 132:139–158. https://doi.org/10.1016/j.matcom.2016.08.002

42. Lagaris IE, Likas A, Fotiadis DI. Artificial neural networks for solving ordinary and partial differential equations. IEEE transactions on neural networks. 1998; 9(5):987–1000. https://doi.org/10.1109/72. 712178 PMID: 18255782

43. Machado JT, Sabatier J. Advances in fractional calculus: theoretical developments and applications in physics and engineering. Springer Verlag; 2007.

44. Ortigueira M, Tenreiro Machado J. Special issue on fractional signal processing. Signal Process. 2003; 83(11):2285–2286.

45. Tenreiro Machado J, Silva MF, Barbosa RS, Jesus IS, Reis CM, Marcos MG, et al. Some applications of fractional calculus in engineering. Mathematical Problems in Engineering. 2010; 2010. https://doi. org/10.1155/2010/639801

46. Podlubny I. Fractional differential equations, vol. 198 of Mathematics in Science and Engineering; 1999.

47. Figueiredo Camargo R, Chiacchio AO, Capelas de Oliveira E. Differentiation to fractional orders and the fractional telegraph equation. Journal of Mathematical Physics. 2008; 49(3):033505. https://doi.org/ 10.1063/1.2890375

48. Waseem W, Sulaiman M, Alhindi A, Alhakamy H. A soft computing approach based on fractional order DPSO algorithm designed to solve the corneal model for eye surgery. IEEE Access. 2020;. https://doi. org/10.1109/ACCESS.2020.2983823

49. Bukhari AH, Raja MAZ, Sulaiman M, Islam S, Shoaib M, Kumam P. Fractional Neuro-Sequential ARFIMA-LSTM for Financial Market Forecasting. IEEE Access. 2020;. https://doi.org/10.1109/ACCESS.2020.2985763

50. Khan A, Sulaiman M, Alhakami H, Alhindi A. Analysis of oscillatory behaviour of heart by using a novel neuroevolutionary approach. IEEE Access. 2020; p. 1–1.

51. Bukhari AH, Raja MAZ, Sulaiman M, Islam S, Shoaib M, Kumam P. Fractional Neuro-Sequential ARFIMA-LSTM for Financial Market Forecasting. IEEE Access. 2020; 8:71326–71338. https://doi.org/ 10.1109/ACCESS.2020.2985763

52. Bukhari AH, Sulaiman M, Islam S, Shoaib M, Kumam P, Raja MAZ. Neuro-fuzzy modeling and predic- tion of summer precipitation with application to different meteorological stations. Alexandria Engineering Journal. 2020; 59(1):101–116. https://doi.org/10.1016/j.aej.2019.12.011

53. Pires ES, Machado JT, de Moura Oliveira P, Cunha JB, Mendes L. Particle swarm optimization with fractional-order velocity. Nonlinear Dynamics. 2010; 61(1-2):295–301. https://doi.org/10.1007/s11071- 009-9649-y

54. Couceiro M, Ghamisi P. Particle swarm optimization. In: Fractional Order Darwinian Particle Swarm Optimization. Springer; 2016. p. 1–10.

55. Waseem W, Sulaiman M, Islam S, Kumam P, Nawaz R, Raja MAZ, et al. A study of changes in temper- ature profile of porous fin model using cuckoo search algorithm. Alexandria Engineering Journal. 2020; 59(1):11–24.   https://doi.org/10.1016/j.aej.2019.12.001

8

Analytical solution to swing equations in power grids

HyungSeon Oh *

Department of Electrical and Computer Engineering, United States Naval Academy, Annapolis, Maryland, United States of America

Editor: Yang Li, Northeast Electric Power University, CHINA

Funding: The author received no specific funding for this work.

Competing interests: The author has declared that no competing interests exist.

Abbreviations: CCS, Cartesian coordinate system; COI, center of inertia; COM, coupled oscillation model; DAE, differential algebraic equation; DM, direct method; IEEE, Institute of Electrical and Electronics Engineers; MVA, Mega Volt Amp; PCS, polar coordinate system; TDS, time domain simulation; γ_{ik}, admittance angle of the line connecting Buses i and k; γ^K, cohesive angle for the coupled oscillation model; $\kappa_F()$, condition number of the matrix inside the parenthesis associated with the Frobenius norm, $\kappa_F(M) = \|M\|_F \|M^{-1}\|_F$; δ_i, rotor angle of a generator i; λ, eigenvalue; $\vartheta()$, order of the quantity inside the parenthesis; ς^K, coupling strength in the Kuramoto model; θ_k, terminal voltage angle at Bus k that a generator i is directly connected to; δi^K, phase of the ith oscillator in the Kuramoto model; ω_i, $d\delta_i/dt$ the speed of rotor angle of a generator i; ωi^K, natural frequency of the ith oscillator in the Kuramoto model; D_i, damping term associated with a generator i; E_i, voltage magnitude of the ith Ibus that remains unchanged over the transient; E_k, terminal voltage magnitude; Ibus, bus representing a generator; Kbus, bus that is directly connected to a generator or generators, i.e., slack bus or PV bus; Mbus, bus that is not directly connected to a generator, i.e., PQ bus; M_{ref}, reference inertia assigned for a frequency- or time-dependent load; M_i, inertia of a rotor in a generator i; λ, number of terminal buses in the system including Kbus and Mbus; ND, number of frequency- or time-dependent loads; NI, number of generators in the system including Ibus; Y, admittance matrix; bldiag, block-diagonal matrix; diag, diagonal matrix; e_i, i^{th} column vector in an identity matrix; g_{ii}, real component of the line admittance connecting a generator i and Bus k; h, scaling factor to adjust the update of $d^2 w_I/dt^2$; i_j, injection current at Bus j; $i°$, injection currents at Bus j when no voltages are applied; p_i, real power injection from a generator i, $p_{i \to k}^{elec}$, electrical power output from a generator i to Bus k; p_i^{mech}, mechanical power input to a generator i; p_{max}, maximum power injection, $p_{max}|\bar{y}_{ik}|E_iE_k$; q_k, reactive power injection from a generator I to Bus k; rad, radian; $s(\lambda)$, condition of λ, $s(\lambda) = |u_L^H u_R|$; time when a disturbance (a physical anomaly and/or a control action) occurs; u_L, left eigenvector of Mat, $Mat = \lambda u_L^H$; u_R, right eigenvector of Mat, $Matu_R = \lambda u_R$; v_j, voltages at Bus j; v_x, real part of the voltages v; v_y, imaginary part of the voltages v; x_I, real part of the loss-reflected voltage at Ibuses; y_I, imaginary part of the loss-reflected voltage at Ibuses; $|\bar{y}_{ik}|$, magnitude of line admittance connecting a generator i and Bus k.

* hoh@usna.edu

Abstract

Objective

To derive a closed-form analytical solution to the swing equation describing the power sys- tem dynamics, which is a nonlinear second order differential equation.

Existing challenges

No analytical solution to the swing equation has been identified, due to the complex nature of power systems. Two major approaches are pursued for stability assessments on sys- tems: (1) computationally simple models based on physically unacceptable assumptions, and (2) digital simulations with high computational costs.

Motivation

The motion of the rotor angle that the swing equation describes is a vector function. Often, a simple form of the physical laws is revealed by coordinate transformation.

Methods

The study included the formulation of the swing equation in the Cartesian coordinate sys- tem, which is different from conventional approaches that describe the equation in the polar coordi- nate system. Based on the properties and operational conditions of electric power grids referred to in the literature, we identified the swing equation in the Cartesian coordinate system and derived an analytical solution within a validity region.

Results

The estimated results from the analytical solution derived in this study agree with the results using conventional methods, which indicates the derived analytical solution is correct.

Conclusion

An analytical solution to the swing equation is derived without unphysical assumptions, and the closed-form solution correctly estimates the dynamics after a fault occurs.

Introduction

Electric power loads are expected to be fulfilled continuously in modern society, and when a load is not satisfied it is termed an "event". The infrastructure to generate and transport elec- tric- ity to end-consumers is called a power system. In the United States, this infrastructure comprises 19,023 individual, commercial generators (6,997 power plants) [1], 70,000 substations [2], and 360,000 miles of lines [3]. The number of power electronic devicess is more than a million, and non-anticipated losses of system components inevitably occur. The Federal Energy Regulatory

Commission in the United States regulates the interstate transmission and wholesale of electricity. According to a report submitted to them [4], the 1-in-10 standard is a widely used reliability standard across North America. To meet this standard in a large-scale network with many system components, the power system must be able to withstand sudden disturbances (such as electric short circuits or non-anticipated loss of system components). It should be noted that most disturbances (including the failure of components) do not lead to an event. When a disturbance occurs, a governor regulates the speed of a machine to adjust the output power of a generator according to the network conditions. In general, the timeframe of governor action is approximately 0.1–10 s [5]. Therefore, it is necessary to assess if the power system is stable approximately 10 s after a disturbance occurs, which is the subject of the tran- sient stability assessment.

Viewed overall, power systems consist of mechanical and electrical systems that obey energy conservation and Kirchhoff's laws, which are integrated as the so-called swing Eq [5],

$$M_i \frac{d^2 \delta_i}{dt^2} + D_i \frac{d\delta_i}{dt} = p_i^{mech} - p_{i \to k}^{elec}.$$

The swing equation is a heterogeneous nonlinear second-order differential equation with multi-variables. There is no known method to solve the differential equation in an analytical fashion. Instead, several approaches to analyze the problem are suggested: 1) simplify the problem by ignoring the difficult components; 2) solve the problem for a "simple" system and extend the knowledge to a complex system; and 3) adopt a numerical approach. While these three approaches provide practical assessment for some cases, their applicability is limited due to their assumptions.

This paper is structured as follows: the first section lists three approaches, assumptions, and limitations; the second section formulates the problem in a different coordinate system than the polar coordinate system (PCS), and discusses the differences between the two; the third section lists the solution process to solve the reconstructed problem; the fourth section pres- ents the analytical solution; the fifth section shows a set of examples; and the sixth section lists the conclusions and future studies.

Approaches for solving the swing equation

Coupled oscillator model

If the complexity associated with power systems is ignored, there might be a problem such as the swing equation. The first approach is to ignore the complexity of power systems, and to apply the knowledge from a different domain of science. It was found that in an oscillatory motion, if the frequency is spread more than the coupling between the oscillators, each oscilla- tor runs at its own frequency. Otherwise, the system spontaneously maintains synchronization [6]. The field of synchronization in networks is reviewed in [7]. The Kuramoto model [8] provides an analytic function as follows: $\frac{d\delta_i^K}{dt} = \omega_i^K + \frac{c_i^K}{N} \sum_{j=1}^{N} \sin(\delta_i^K - \delta_j^K)$.

For the coupled oscillator model (COM), a system of particles aims to minimize the energy function $E(\delta^K) = \sum_{\{i,j\}} a_{ij}[1 - \cos(\delta_i^K - \delta_j^K)] - \sum_k \omega_k^K \delta_k^K \cong \frac{1}{2} \sum_{\{i,j\}} a_{ij}(\delta_i^K - \delta_j^K)^2 - \sum_k \omega_k^K \delta_k^K$.

Term $E(\delta^K)$ features a phase-cohesive minimum with interacting particles no further than a certain angle γ^K. A solution has cohesive phases if there is an angle γ^K between 0 and $\pi/2$ that is the maximum phase distance among all the pairs of connected oscillators, i.e., $|\delta_i - \delta_j| \leq \gamma^K$ for a line connecting two nodes i and j. In [9], it is demonstrated that the mechanical analogy applies to the electric power grid to yield the phase cohesiveness. This approach involves a very light computation cost, but the applicability may be limited due to the difference between a generic network and the electric power grids.

Lyapunov stability

The second approach is to solve the equation for a simple problem without ignoring the nonlinear feature of the equation. To construct a simple problem, we consider a system with a single machine and infinite bus where the loss is ignored ($\gamma_{ik} = 0$). In physics, we often define zero potential at an infinitely remote location, so that the existing field is not affected. Similarly, we define an infinite bus so that the terminal voltages do not vary with the internal voltage angle (and therefore with time). However, the terminal voltages may change with the network condition. Then, the nonlinear component $|\tilde{y}_{ik}|E_i E_k \sin(-\theta_k + \delta_i - \gamma_{ik})$ is rewritten as $p_{max} \sin\delta_i$, where $p_{max}(= |\tilde{y}_{ik}|E_i E_k)$ is a constant dependent on the network condition. There are three different conditions regarding the condition for a disturbance: pre-fault, post-fault, and on-fault (See Fig 1).

Fig 1 shows that the corresponding p_{max} decreases in the order of pre-fault, post-fault, and on-fault conditions ($p_{max}^{pre-fault} > p_{max}^{post-fault} > p_{max}^{on-fault}$). In the pre-fault condition, the operation point is determined where the sine curve intersects with a horizontal line $p_i^{mech} - g_{ii} E_i^2$ Point a in Fig 1(B), note that losses are ignored, i.e., $g_{ii} = 0$). At a, the generator is synchronized with the system frequency.

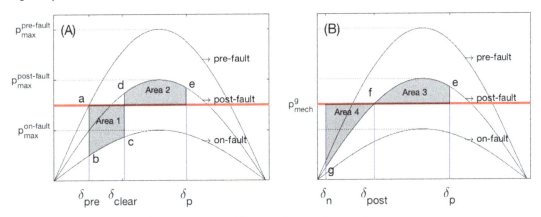

Fig 1. p-δ curve to illustrate the equal-area criterion.

During a fault, the sine function drops down the on-fault sine curve, but the internal voltage angle remains unchanged (δ_{pre}), because the angle must be continuous over time ($a \rightarrow b$). Therefore, the electric power output is less than the mechanical power input, and the difference must be stored in the rotor in terms of mechan- ical energy to meet the energy balance. The internal voltage angle increases accordingly ($b \to c$), which makes the rate of the internal voltage angle deviate from the system fre- quency. The fault will shortly be monitored and cleared, and the sine curve follows the post- fault curve that the internal voltage angle follows ($c \to d$). Similarly, the internal voltage moves to a new curve at δ_{clear} in which the mechanical power input is less than the electric power output. The angle will increase further until the stored energy is exhausted at δ_p ($d \to e$). The integration of the left-hand side between two synchronized points of the swing equation (after ignoring the damping term) yields zero (Area 1 = Area 2). This is called the equal-area criterion. At δ_p, the generator recovers the synchronization with the system frequency, but it cannot maintain the operation point because the electric output is greater than the mechanical power input and the stored energy is already exhausted. The rotor angle decreases until δ_n, which is set by the equal-area curve (Area 3 = Area 4) and the rotor, swings between the two angles $[\delta_n, \delta_p]$ if no damping exists. With the damping, the swing range decreases and settles down to δ_{post}, where the mechanical power input and elec- tric output are balanced.

Unfortunately, this equal-area criterion cannot generally be extended to a large-scale net- work with multiple machines. According to Lyapunov stability in nonlinear dynamic theory, one can tell if a system will be stable in the future if a Lyapunov function is identified to meet some conditions [10]. While there is no rule to construct a Lyapunov function, DM is devel- oped under various assumptions [10–17] that often includes a lossless network (zero damp- ing), no consideration of reactive power, and constant voltage magnitudes that are difficult to justify physically. Under these physically unjustifiable assumptions, the system is stable if one can find a Lyapunov function. On the other hand, the failure to find such a function (or the nonexistence of the function) does not necessarily mean the system will be unstable. Rather, the stability would be undetermined. Even though the advantage of this approach involves rel- atively inexpensive computation costs, it yields a stability region subset in addition to the phys- ically infeasible assumptions.

Both the COM and the DM ignore the losses and variations of the voltage magnitudes. Fur- ther, because both assumptions are not physically justifiable, the application of the approaches is either highly limited or a conservative application is performed.

Numerical approach

The third approach is to accommodate all the details of the swing equation and Kirchhoff's laws, and conduct a numerical computation. Often, a numerical integration or differentiation approach is applied to solve nonlinear differential equations and recent advancements in fast computation makes the numerical approach an attractive option [18], [19]. For the swing equation, a numerical method (such as the Euler or Runge–Kutta method) is used to find the temporal changes of internal and terminal voltages [20], [21]. The advantage of this approach is the capability to solve any complex function. However, in addition to the high computation costs, the numerical method involves a truncation error. For example, the errors associated with the Euler and modified Euler methods are in the order of $\vartheta(\|\Delta x\|^2)$ [20], and that of the Runge–Kutta method is $\vartheta(\|\Delta x\|^4)$ [22]. If there is a subset of modes that diverge exponentially, but in which the magnitudes are initially very small, the modes may be underestimated. The numerical approach also requires precise estimation of the model parameters. It should be noted that numerical simulation did not capture the impact of the 1996 WSCC (Western Sys- tem Coordinating Council) system outage [23].

Write down the problem

Feynman's algorithm

According to Dr. Thomas (an Emeritus Professor at Cornell University), Feynman's algorithm for solving hard problems is a process of thinking about problems in a different way. The swing equation, $M_i \frac{d^2 \delta_i}{dt^2} + D_i \frac{d\delta_i}{dt} = P_i^{mech} - g_{ii} E_i^2 - |\tilde{y}_{ik}| E_i E_k \sin(-\theta_k + \delta_i - \gamma_{ik})$, is formulated in the PCS because the equation describes the angular motion of a rotor inside a generator. The PCS provides a concise form of the equation of the motion; the differential terms are multi- plied with scalar constants that result in a linear form for the left-hand side; and on the right- hand side, the magnitudes of internal voltage and terminal voltage are linear. However, the nonlinearity involved in voltage angles makes the swing equation difficult to solve. In this paper, instead of the conventional approach using the PCS, we construct the problem in the Cartesian coordinate system (CCS) and aim to solve it without physically unjustifiable assumptions.

Observations in transient studies

We begin with a list of facts that are observed in transient studies. O1. E (internal voltage of a generator) at *Ibus* (see Section "Network flows" for its defini- tion) is constant [24], which is a general assumption applied to the transient studies in power system,

$$O1 = \left\| \left[1 - \sqrt{x_i^2 + y_i^2}/E_i \cdots \frac{1}{E_i} \sqrt{2x_i \left(\frac{dx_i}{dt}\right) + 2y_i \left(\frac{dy_i}{dt}\right)} \cdots \right]^T \right\|_F.$$

O2. $O2 = |O_i(t) - O_+^{0+}|$ where $O_i(t) = (-q_i + b_{ii} E_i^2)/M_i + (d\delta_i/dt)^2$; b_{ii} is the imaginary part of admittance; the superscript $0+$ represents "immediately after contingency"; and $t > t_0$, i.e., the sum of the reactive power injection and the rotor speed changes slowly over time strictly after a disturbance occurs before losing synchronization. The square of the rotor speed $(d\delta_i/dt)2$, the square of the rate to deviate from the synchronization, is in general smaller than $(-q_i + b_{ii} E_i^2)/M_i$ while maintaining synchronization. According to O2, while maintaining synchronization, $|O_i(t)|$ is a slowly varying function that follows the Karamata representation theorem [25]:

$$|O_i(t)| = \left| \frac{-q_i + b_{ii} E_i^2}{M_i} + \left(\frac{d\delta_i}{dt}\right)^2 \right| = \exp\left[\eta(t) + \int_{t_1}^{t} \frac{\mu(y)}{y} dy\right] \text{where}$$

$$\begin{cases} \lim_{t \to \infty} \eta(t) = \eta_\infty \\ \lim_{t \to \infty} \mu(t) = 0 \end{cases}$$ for $t \geq t_1 (> t_0)$. The variation of O (t) over a short time period is negligible so that $O_i(t) = (-q_i + b_{ii} E_i^2)/M_i + (d\delta_i/dt)^2 \approx (-q_i^0 + b_{ii} E_i^2)/M_i + (d\delta_i/dt)^2|_{t=0} = O_i(t_0)$. O2 is defined as $|O_i(t) - O_i(t_0)|$.

When a disturbance occurs it is possible for some generators to stray slightly off perfect synchronization. However, the rate of deviation from synchronization is always kept small before losing synchronization. The impacts of these approximations are discussed in a later section.

Write down the equation in CCS

Definition of variables. A natural choice of variables in CCS would be the real and the imaginary components of voltages. The choice leads to $E_i E_k \sin(-\theta_k + \delta_i - \gamma_{ik}) = -\cos\gamma_{ik}(E_i\cos\delta_i)(E_k\sin\theta_k) + \cos\gamma_{ik}(E_i\sin\delta_i)(E_k\cos\theta_k) - \sin\gamma_{ik}(E_i\cos\delta_i)(E_k\cos\theta_k) + \sin\gamma_{ik}(E_i\sin\delta_i)(E_k\sin\theta_k)$. A better choice is to rotate the voltage angles by the phase angle of the line between the internal and terminal buses ($\delta_i - \gamma_{ik}$), and the choice yields only two terms by $E_i E_k \sin(-\theta_k + \delta_i - \gamma_{ik}) = -x_i v_y^k + y_i v_x^k$ where $x_i = E_i\cos(\delta_i - \gamma_{ik})$, $y_i = E_i\sin(\delta_i - \gamma_{ik})$, $v_x^k = E_k\cos\theta_k$, and $v_y^k = E_k\sin\theta_k$. For CCS, two variables are necessary (x and y) instead of one (δ) to describe the angular motion of internal voltages, and an additional equation is imposed to preserve the constant internal voltage magnitudes $x_i^2 + y_i^2 = E_i^2$.

Load modeling. The development of a new load model is beyond the scope of this study. Instead, we attempt to integrate the existing load models for our proposed formulation. A widely used load model is outlined in [26]. Loads located at a same bus are integrated into a single load, and this load is separated and modeled into four subgroups: (Category I) an induction load, (Category II) a frequency-dependent or a time-dependent load, (Category III) a load with constant impedance or a load with constant current, and (Category IV) the remaining loads. While Categories I and II are integrated in the swing equation, the admittance matrix Y is modified to incorporate Categories III and IV in terms of the voltage–current relationship.

Category I: An induction load. This has nonzero inertia, meaning it appears in the swing equation as a synchronous machine that remains in the swing equation. The load is modeled similarly to that of a generator, or more specifically, to that of a negative generator.

Category II: A frequency dependent or a time dependent load. A frequency dependent load is $d_j = d_j^0 + m_j(d\delta_j/dt)$ [8], where the first component d_j^0 represents a load with a fixed impedance, and the second term $m_j(d\delta_j/dt)$ is the frequency-dependent load. If a synchronous machine is located at the same bus as the load, the coefficient of the first time-derivative term is the sum of both the damping and frequency-dependent terms. Similarly, a time dependent load is $d_j = m_j(dd_j/dt) + d_j^0$ [27]. For this type of load, the constant term is integrated into Category IV.

Category III: A load with constant impedance or with constant current. These are integrated in $i_l = i_l^{cc} + y_l^{ci} v_l$, where i_l^{cc} and y_l^{ci} model the constant current and the constant impedance at Bus l, respectively. This expression makes it possible to convert these loads into the diagonal shunt element in the admittance matrix $Y_{bus}(y_l)$ or a constant in $i_L = diag(y_L^{ci})v_L + i_L^{cc}$.

Category IV: A remaining load. The remaining load introduces nonlinear characteristics, and it cannot characterize load characteristics on a constant voltage node in the system due to non-dependency on the voltage angle at the node. The BIG model in [28] and [29] is an attempt to integrate the load as a linear model (similar to loads with constant impedance or current), and show a good match for static loads. Interested readers may wish to read a literature survey on modeling loads in [30]. The type of load may be interpreted with a Taylor series expansion in terms of voltages near the operation point: $I_l = \frac{s_l^*}{v_l^*} = \frac{(i_l^0 v_l + \bar{y}_l |v_l|^2)^*}{v_l^*} = i_l^0 + \bar{y}_l^* v_l$. combined with the load modeling constant current and constant impendence, the current is expressed in terms of voltage $i_L = i_L^0 + diag(y_L^0)v_L$.

Network flows. To establish a set of equations, an electric power grid is redefined, and a set of buses is introduced to model the internal voltage of a synchronous machine, *Ibus*; in other words, a *Kbus* is directly connected to an *Ibus*. Note that synchronous machines include induction motors, frequency- and time-dependent loads, and synchronous generators (but not asynchronous renewable generators). No *Ibus* is connected to multiple *Kbuses*, while a *Kbus* can be connected to multiple *Ibuses*. If two *Kbuses* are directly connected, an *Mbus* is inserted between them. Fig 2 illustrates the definitions of the *Ibus*, *Kbus*, and *Mbus*. Fig 2(A) shows the one-line diagram of a modified IEEE 3-bus system with three generators, and Fig 2(B) is the one-line diagram of the system modeled in this study. The generators are modeled as *Ibuses* (in red); the top two buses where the generators are directly connected are modeled as *Kbuses* (in green); and the bottom bus that is not connected to a generator is modeled as an *Mbus* (in blue), respectively. To prevent two *Kbuses* from being directly connected, another *Mbus* (verti- cal bus in blue) is inserted. Because all the buses in the original network remain in the model, and because *NI* of *Ibuses* and additional *Mbus* are introduced, the number of buses for this net- work model is always greater than that for the original network. It should be emphasized that unlike DM, in which the network reduction is applied to make the problem simple, this study preserves the topology of the network.

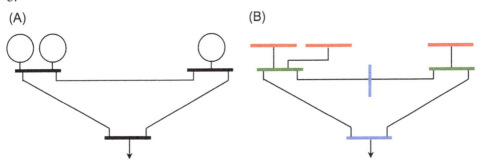

Fig 2. Schematic one-line diagrams of (A) the IEEE 3-bus system and of (B) the system with three *Ibuses* (in red), two *Kbuses* (in green), and two *Mbuses* (in blue).

Suppose a disturbance occurs between $t = 0$ and $t = 0^+$ (immediately after the disturbance). During the disturbance, the angular motion of the physical rotor must be continuous; that is, instantaneous change in the angular motion is zero. Unlike rotor angles, voltages do change abruptly, while Kirchhoff's laws are obeyed both before and after the disturbance. Using Kirchhoff's laws, the voltages are computed to meet the real and the reactive power balances at *Kbus* and *Mbus* immediately after the disturbance. After the disturbance, the angular motion of the rotors and voltages at all the buses correspondingly adjusts to meet both the swing equa- tions and Kirchhoff's laws. The admittance matrix of the redefined grid that accommodates Categories III and IV loads is in a block structure associated with both *Ibuses* and *Mbuses* because there is no direct connection between them. For the *Kbuses* and the *Mbuses* in the net- work model shown in Fig 2 (B), Ohm's law finds $i = Yv$ where the currents i and the voltages v at the buses are also expressed in terms of loads, i.e., $i_{KM} = i_{KM}^0 + diag(y_{KM})v_{KM}$. Note that induction, frequency-dependent, and time-dependent loads are not included in the current–voltage relationship. With no direct connection between *Ibus* and *Mbus*, Ohm's law over the network modeled in this work (Fig 2 (B)) is $\begin{pmatrix} i_{Ibus} \\ i_{Kbus} \\ i_{Mbus} \end{pmatrix} = \begin{bmatrix} Y_I^I & Y_I^K & 0 \\ Y_K^I & Y_K^K & Y_K^M \\ 0 & Y_M^K & Y_M^M \end{bmatrix} \begin{pmatrix} v_I \\ v_K \\ v_M \end{pmatrix}$. Because an *Ibus* is only connected to one *Kbus*, the inclusion of the buses is a radial extension of the original network. Therefore, the currents on the *Kbus* and the *Mbus* at a given voltage must be invariant to the choice of network model:

$$\begin{pmatrix} i_{Kbus} \\ i_{Mbus} \end{pmatrix} = \begin{bmatrix} Y_K^I & Y_K^K & Y_K^M \\ 0 & Y_M^K & Y_M^M \end{bmatrix} \begin{pmatrix} v_I \\ v_K \\ v_M \end{pmatrix} = Y_{KM} \begin{pmatrix} v_K \\ v_M \end{pmatrix} + \begin{pmatrix} i_{Kbus}^0 \\ i_{Mbus}^0 \end{pmatrix} \quad (1)$$

The parameters i_{Kbus}^0 and i_{Mbus}^0 represent the constant current and constant impedance components, respectively. Note that at *Kbuses* and *Mbuses*, only loads (including zero loads) exists because the loads in this study excludes the induction, frequency-dependent, and time-dependent loads. Eq (1) yields the following relationship in terms of modified voltages at *Ibus* (w_I):

$$\begin{pmatrix} v_K \\ v_M \end{pmatrix} = \left\{ Y_{KM} - \begin{bmatrix} Y_K^K & Y_K^M \\ Y_M^K & Y_M^M \end{bmatrix} \right\}^{-1} \left[\begin{pmatrix} Y_K^I \\ 0 \end{pmatrix} Rw_I - \begin{pmatrix} i_{Kbus}^0 \\ i_{Mbus}^0 \end{pmatrix} \right] = \begin{bmatrix} H_{KI} \\ H_{MI} \end{bmatrix} w_I + \begin{pmatrix} v_K^I \\ v_M^I \end{pmatrix} \quad (2)$$

where $R = \begin{bmatrix} \cos\gamma_{ik} & \sin\gamma_{ik} \\ -\sin\gamma_{ik} & \cos\gamma_{ik} \end{bmatrix}$; $v_I = Rw_I$; $w_I = \begin{pmatrix} x_I \\ y_I \end{pmatrix}$. Using (2), the real and the imaginary components of the bus voltages are

$$v_x^k = e_k^T v_K = e_k^T H_{KI} w_I + e_k^T v_K^I, \quad v_y^k = e_{k+NK}^T v_K = e_{k+NK}^T H_{KI} w_I + e_{k+NK}^T v_K^I \quad (3)$$

Revisiting swing equation. Typically, swing equations are written in the PCS, as this is convenient for describing angular motion. The PCS inevitably introduces the exponential function (i.e., sinusoidal function) to express the real power injection, which makes it difficult to solve analytically. In a DC power flow model [26], we often linearize the sinusoidal function by taking $\sin\theta \approx \theta$ and $\cos\theta \approx 1$ as $\theta \approx 0$. However, in the swing equation, δ to describe the angular motion is not small enough to apply the approximation. The power injection from *Ibus i* to *Kbus k* is as follows:

$$\begin{aligned} p_{i \to k}^{elec} &= g_{ii} E_i^2 + |\tilde{y}_{ik}| E_i E_k \sin(-\theta_k + \delta_i - \gamma_{ik}) = g_{ii} E_i^2 - |\tilde{y}_{ik}| x_i v_y^k + |\tilde{y}_{ik}| y_i v_x^k \\ q_{i \to k}^{elec} &= b_{ii} E_i^2 - |\tilde{y}_{ik}|(x_i v_x^k + y_i v_y^k) \to |\tilde{y}_{ik}|(x_i v_x^k + y_i v_y^k) = -q_{i \to k}^{elec} + b_{ii} E_i^2 \end{aligned} \quad (4)$$

The definitions of x_i and y_i lead to

$$\frac{d\delta_i}{dt} = -\frac{1}{y_i}\frac{dx_i}{dt} = \frac{1}{x_i}\frac{dy_i}{dt} = \frac{1}{E_i^2}\left(x_i\frac{dy_i}{dt} - y_i\frac{dx_i}{dt}\right) \text{ and } \frac{d^2\delta_i}{dt^2} = \frac{1}{E_i^2}\left(x_i\frac{d^2y_i}{dt^2} - y_i\frac{d^2x_i}{dt^2}\right) \quad (5)$$

Analytical solution to swing equations in power grids

With (4) and (5), the swing equation in CCS becomes

$$y_i\left(M_i\frac{d^2x_i}{dt^2}+D_i\frac{dx_i}{dt}\right)-x_i\left(M_i\frac{d^2y_i}{dt^2}+D_i\frac{dy_i}{dt}\right)+E_i^2(p_{i\to k}^{mech}-g_{ii}E_i^2-|\tilde{y}_{ik}|v_x^k y_i+|\tilde{y}_{ik}|v_y^k x_i)=0 \quad (6)$$

In comparison to the swing equation in the PCS, (6) is a heterogeneous nonlinear second-order differential equation with multi-variables; the time derivatives are multiplied with the variables; and the last term is nonlinear. Additionally, the conditions related to the constant internal voltage magnitude and to its derivatives are also nonlinear. If one combines the conditions with (6), the resulting equation becomes highly complex. Instead, we derive

$$x_i\left(M_i\frac{d^2x_i}{dt^2}+D_i\frac{dx_i}{dt}\right)+y_i\left(M_i\frac{d^2y_i}{dt^2}+D_i\frac{dy_i}{dt}\right)+E_i^2\left[M_i\left(\frac{d\delta_i}{dt}\right)^2\right]=0 \quad (7)$$

Eqs (6) and (7) lead to

$$M_i\frac{d^2x_i}{dt^2}+D_i\frac{dx_i}{dt}+\left[-q_i+b_{ii}E_i^2+M_i\left(\frac{d\delta_i}{dt}\right)^2\right]x_i+(p_i^{mech}-g_{ii}E_i^2)y_i-|\tilde{y}_{ik}|E_i^2 v_x^k=0$$

$$M_i\frac{d^2y_i}{dt^2}+D_i\frac{dy_i}{dt}-(p_i^{mech}-g_{ii}E_i^2)x_i+\left[-q_i+b_{ii}E_i^2+M_i\left(\frac{d\delta_i}{dt}\right)^2\right]y_i-|\tilde{y}_{ik}|E_i^2 v_y^k=0 \quad (8)$$

$$s.t. x_i^2+y_i^2=E_i^2; x_i\frac{dx_i}{dt}+y_i\frac{dy_i}{dt}=0$$

Note that (8) involves no approximation. It is interesting that the reactive power injection appears in (8), while the swing equation considers only the real power balance. The constraints are conditions that the differential equations hold under. From O2, $[(-q_i+b_{ii}E_i^2)/M_i+(d\delta_i/dt)^2]$ $-[(-q_i^{0+}+b_{ii}E_i^2)/M_i+(d\delta_i/dt)^2|_{t=0^+}]$ is approximately zero, or simply $O_i(t)=O_i(0+)=O_i^0$:

$$M_i\frac{d^2x_i}{dt^2}+D_i\frac{dx_i}{dt}+M_iO_i^{0+}x_i+(p_i^{mech}-g_{ii}E_i^2)y_i-|\tilde{y}_{ik}|E_i^2 v_x^k=0$$

$$M_i\frac{d^2y_i}{dt^2}+D_i\frac{dy_i}{dt}-(p_i^{mech}-g_{ii}E_i^2)x_i+M_iO_i^{0+}y_i-|\tilde{y}_{ik}|E_i^2 v_y^k=0 \quad (9)$$

Furthermore, the voltages at *Kbus* are expressed in terms of w_I as shown in (2)

$$e_i^T\frac{d^2w_I}{dt^2}+\frac{D_i}{M_i}e_i^T\frac{dw_I}{dt}+\left(O_i^{0+}e_i^T+\frac{p_i^{mech}-g_{ii}E_i^2}{M_i}e_{NI+i}^T-\frac{|\tilde{y}_{ik}|E_i^2}{M_i}e_k^T H_{KI}\right)w_I-\frac{|\tilde{y}_{ik}|E_i^2}{M_i}(e_k^T v_K^I)=0$$

$$e_{NI+i}^T\frac{d^2w_I}{dt^2}+\frac{D_i}{M_i}e_{NI+i}^T\frac{dw_I}{dt}+\left(-\frac{p_i^{mech}-g_{ii}E_i^2}{M_i}e_i^T+O_i^{0+}e_{NI+i}^T-\frac{|\tilde{y}_{ik}|E_i^2}{M_i}e_{NK+k}^T H_{KI}\right)w_I-\frac{|\tilde{y}_{ik}|E_i^2}{M_i}(e_{NK+k}^T v_K^I)=0 \quad (10)$$

$$s.t. O1^2=\left[\frac{1}{E_i^2}vec(e_ie_i^T+e_{i+NI}e_{i+NI}^T)\right]^T\left[w_I\otimes w_I \quad 2w_I\otimes\frac{dw_I}{dt}\right]-(1\ 0)=0; O2=O_i(t)-O_i^{0+}=0$$

$O1^2$ represents the element-wise square. If Bus *i* is directly connected to a frequency- and/or a time-dependent load, the damping coefficient is modified to accommodate the frequency and/or time-dependent constant m_i; that is, $D_i^{new} = D_i + m_i$. The nodal swing equation with only frequency- and/or time-dependent load without synchronized machine with O2 becomes [8]

$$e_i^T \frac{d\tilde{w}_I}{dt} + \left(\frac{M_{ref}}{D_i} O_i^{0+} e_i^T + \frac{p_i^{mech} - g_{ii} E_i^2}{D_i} e_{NI+i}^T - \frac{|\tilde{y}_{ik}| E_i^2}{D_i} e_k^T H_{KI} \right) \tilde{w}_I - \frac{|\tilde{y}_{ik}| E_i^2}{D_i} (e_k^T v_K^I) = 0$$

$$e_{NI+i}^T \frac{d\tilde{w}_I}{dt} + \left(-\frac{p_i^{mech} - g_{ii} E_i^2}{D_i} e_i^T + \frac{M_{ref}}{D_i} O_i^{0+} e_{NI+i}^T - \frac{|\tilde{y}_{ik}| E_i^2}{D_i} e_{NK+k}^T H_{KI} \right) \tilde{w}_I - \frac{|\tilde{y}_{ik}| E_i^2}{D_i} (e_{NK+k}^T v_K^I) = 0$$

(11)

Note that $\tilde{R}_I$ associated with $\tilde{w}_I$ is an identity matrix. To distinguish the modified voltages derived by the load from a physical internal voltage, $\tilde{w}_I$ is introduced in (11). The buses where the frequency- or time-dependent loads are located belong to the *Ibus*. For a system with *NI* of *Ibuses* and *NK* of *Kbuses*, (10) is further simplified with the constraints (O1 and O2 are small) as follows:

$$\frac{d^2 w_I}{dt^2} + diag\left(\frac{D_I}{M_I}\right) \frac{dw_I}{dt} + L w_I + l = 0 \qquad (12)$$

where $e_i^T L = O_i^{0+} I + \frac{p_i^{mech} - g_{ii} E_i^2}{M_i} e_i e_{i+NI}^T - \frac{|\tilde{y}_{ik}| E_i^2}{M_i} e_i e_k^T H_{KI}$, $e_{i+NI}^T L = -\frac{p_i^{mech} - g_{ii} E_i^2}{M_i} e_{i+NI} e_i^T + O_i^{0+} I$
$-\frac{|\tilde{y}_{ik}| E_i^2}{M_i} e_{i+NI} e_{k+NK}^T H_{KI}$, $e_i^T l = -\frac{|\tilde{y}_{ik}| E_i^2}{M_i} (e_k^T v_K^I)$, $e_{i+NI}^T l = -\frac{|\tilde{y}_{ik}| E_i^2}{M_i} (e_{k+NK}^T v_K^I)$, $e_i^T w_I = x_i$, $e_{i+NI}^T w_I = y_i$,

and the superscript T refers to transpose. Similarly, (11) becomes

$$\frac{d\tilde{w}_I}{dt} + \tilde{L}\tilde{w}_I + \tilde{l} = 0 \qquad (13)$$

where $e_i^T \tilde{L} = \frac{M_{ref}}{D_i} O_i^{0+} I + \frac{p_i^{mech} - g_{ii} E_i^2}{D_i} e_i e_{i+NI}^T - \frac{|\tilde{y}_{ik}| E_i^2}{D_i} e_i e_k^T H_{KI}$, $e_{i+NI}^T \tilde{L} = -\frac{p_i^{mech} - g_{ii} E_i^2}{D_i} e_{i+NI} e_i^T + \frac{M_{ref}}{D_i} O_i^{0+} I$
$-\frac{|\tilde{y}_{ik}| E_i^2}{D_i} e_{i+NI} e_{k+NK}^T H_{KI}$, $e_i^T \tilde{l} = -\frac{|\tilde{y}_{ik}| E_i^2}{D_i} (e_k^T v_K^I)$, $e_{i+NI}^T \tilde{l} = -\frac{|\tilde{y}_{ik}| E_i^2}{D_i} (e_{k+NK}^T v_K^I)$, $e_i^T \tilde{w}_I = \tilde{x}_i$, $e_{i+NI}^T \tilde{w}_I = \tilde{y}_i$,
$\tilde{\omega}_i^0 = \omega_i^0$.

and $\tilde{x}_i, \tilde{y}_i$ are the modified voltages derived from the frequency- or time-dependent load at Bus i. Let z be $[w_I; dw_I/dt; \tilde{w}_I]$, and then with the constraints (O1 and O2 are small) (12) and (13) becomes (14) is a constrained homogeneous linear first-order differential equation with multi-vari- ables. There is no known solution to the constrained differential equation.

$$\frac{dz}{dt} = Tz + b \text{ where } T = \begin{bmatrix} 0 & I & 0 \\ -L & -diag(D_I/M_I) & 0 \\ 0 & 0 & -\tilde{L} \end{bmatrix} \text{ and } b = \begin{pmatrix} 0 \\ -l \\ -\tilde{l} \end{pmatrix} \qquad (14)$$

Fig 3 illustrates our approach to solving the swing equation. The arrows with solid lines refer equivalent, and those with dashed lines represent similar in the proximity proportional to the relaxation applied. S1 is the solution to the conventional swing equation formulated in PCS, which is identical to S2 in CCS and S3 because (8) is equivalent to (10) with the constraint of

O2 = 0. Equality con- straints are identical to the inequality constraints if the upper bounds are zeros, i.e., S3 = S4. S5 is the solution to the problem relaxed by $\vartheta(\Delta_E)$. Note that $t_E + \delta t$ is the first time when any of the relaxed constraints is binding, and that δt is strictly positive. In the range of $[0, t_E]$, no con- straints are binding, which makes S6 involve nonbinding constraints. In the optimization the- ory, the solution to a constrained problem equals that of the same problem without the constraints that are not binding. Therefore, S7 is the same as that to S6 even though no con- straints are taken into consideration in solving S7. The process finds S7 is in the prox- imity of S1 by $\vartheta(\Delta_E)$ if $\delta t t_E$ and Δ_E is kept small enough to ensure the validity of the solution. The jus- tification is outlined in the section describing the validity region with projections so that the error remains small. This concludes the first step of Feynman's algorithm, *write down the problem*.

Think hard: Solve the swing equation

Initial condition. To solve a first-order differential equation, a set of initial conditions are re- quired. δ represents a physical angular motion that is continuous in time; i.e., $\delta_i^{0+} = \delta_i^0$. When a disturbance occurs, the mechanical power input does not change instantaneously, but the elec- tric power output $p_{i \to k}^{elec}$ exhibits a sudden but finite change. The change is represented by a step function $f(x \in R^1) = \sum_i m_i \chi_{A_i}(x)$ where m_i is a finite real number and χ_A is an indicator function over an interval A, $\chi_A(x) = \begin{cases} 1 \text{ if } x \in A \\ 0 \text{ otherwise} \end{cases}$. Because the time interval between 0 and 0+ has a fi- nite length, the Lebesgue integral of $p_{i \to k}^{elec}$ is $\int p_{i \to k}^{elec} dt = m\ell(A)$, where $\ell(A)$ is the length of the time interval between 0 and 0+, $\int_{t=0}^{t=0^+} p_{i \to k}^{elec} dt \simeq 0$. The value of $\omega_i (= d\delta_i/dt)$ immediately after the dis- turbance occurs is computed from the integration of the swing equation over the time interval $\omega_i^{0+} = \omega_i^0$. Similarly, for the frequency- or time-dependent loads, $\tilde{\omega}_i^{0+} = \tilde{\omega}_i^0$. The rotor angle δ (in terms of time) is differentiable, and its first derivative ω is bounded; hence, the rotor angle δ is Lipschitz continuous over time.

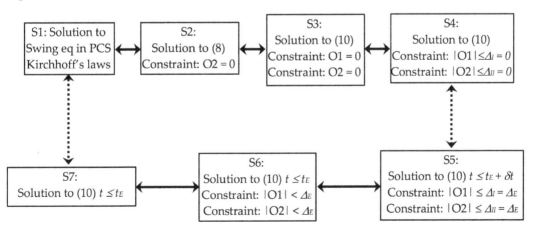

Fig 3. Solutions to various forms of swing equation and its relaxations.

General solution to the swing equation

The general solution to (14) that satisfies $\frac{dz_G}{dt} = Tz_G$ is as follows:

$$z_G = \Phi u \text{ where } u = \begin{pmatrix} e^{\tilde{\lambda}_1 t} \\ \vdots \\ e^{\tilde{\lambda}_{2N} t} \end{pmatrix} \text{ and } \begin{pmatrix} \tilde{\lambda}_1 \\ \vdots \\ \tilde{\lambda}_{2N} \end{pmatrix} = eig(T) \qquad (15)$$

where Φ is the coefficient matrix to be determined. The base solution u may contain terms associated with complex eigenvalues, because T is generally an asymmetric real-valued matrix. For any complex eigenvalue, its complex conjugate is also a complex eigenvalue. Let such a pair be is $\lambda_l + \lambda_{l+1}\sqrt{-1}$ and $\lambda_l - \lambda_{l+1}\sqrt{-1}$. Then, the lth corresponding pair in the base solution is

$$\begin{pmatrix} u_l \\ u_{l+1} \end{pmatrix} = \begin{bmatrix} e^{\lambda_l t}\cos(\lambda_{l+1} t) \\ e^{\lambda_l t}\sin(\lambda_{l+1} t) \end{bmatrix} \qquad (16)$$

It is possible to represent u in terms of a real-valued function as follows:

$$u = \begin{pmatrix} u_{Re} \\ u_{Co} \end{pmatrix} \text{ where } u_{Re} = \begin{pmatrix} u_{Re}^1 \\ \vdots \\ u_{Re}^{N_{Re}} \end{pmatrix}, u_{Co} = \begin{bmatrix} u_{Co}^{[1,2]} \\ \vdots \\ u_{Co}^{\{N_{Co}-1, N_{Co}\}} \end{bmatrix}, u_{Re}^n = e^{\lambda_{Re}^n t}, \text{ and } u_{Co}^{\{l,l+1\}}$$

$$= \begin{pmatrix} e^{\lambda_{Co}^{\{l\}} t}\cos(\lambda_{Co}^{\{l+1\}} t) \\ e^{\lambda_{Co}^{\{l\}} t}\sin(\lambda_{Co}^{\{l+1\}} t) \end{pmatrix} \qquad (17)$$

where n and l are the indices of real and complex eigenvalues, respectively. Differentiating u with respect to time introduces a block-diagonal matrix D_l such that $du/dt = D_l u$, where

$$D_l = \begin{bmatrix} diag(\lambda_{Re}) & 0 \\ 0 & bldiag\left(\begin{bmatrix} \lambda_{Co}^{Re} & -\lambda_{Co}^{Im} \\ \lambda_{Co}^{Im} & \lambda_{Co}^{Re} \end{bmatrix}\right) \end{bmatrix}, \lambda_{Re} \text{ is the real eigenvalues, } \lambda_{Co}^{Re} \text{ and } \lambda_{Co}^{Im} \text{ are}$$

real and imaginary components of the complex eigenvalues of T, respectively. It is noteworthy that D_l is obtained from the block Schur decomposition [31] $T = U_T^{-1} D_l U_T$, so that D_l has a block-diagonal structure. Therefore, T and D_l are related through the similarity transformation. In (15), Φ is the time-invariant coefficient matrix, and the differential equation yields

$$\frac{dz_G}{dt} = \frac{d}{dt}(\Phi u) = \Phi \frac{du}{dt} = \Phi D_l u = T\Phi u \qquad (18)$$

Because u is not a null vector, (18) introduces a Sylvester equation; that is, $\Phi D_l - T\Phi = 0$. The Bartels–Stewart algorithm is an efficient way to solve a Sylvester equation [31], and its computational cost is $\vartheta(N3)$. It involves the orthogonal reduction of D_l and T matrices into triangular form using the QR factorization, and then solving the resulting triangular system via back--

sub-stitution. Because D_λ is a block-diagonal matrix comprised of the eigenvalues of T, their eigen- spaces overlap. Therefore, the Bartels–Stewart algorithm is not applicable to solve (18). Con- sideration of the null space yields $vec(\Phi) \in null[(I \otimes T - D_\lambda^T \otimes I)^T]$, where represents the Kronecker product. Therefore, the QR-factorization of $I \otimes T - D_\lambda^T \otimes I$ reveals the space of Φ. Let ψ_l and Ψ_l be the lth vector spanning the null space of $I \otimes T - D_\lambda^T \otimes I$ and the matrix form of the vector, respectively. Then, Φ is a linear combination of all Ψs.

$$z_G = \left(\sum_l \beta_l \Psi_l\right) u, \text{ where } \Psi_l D_l - T\Psi_l = 0 \tag{19}$$

Solution to meet initial conditions. We claim that the following function is the solution to (14):

$$\hat{z}(t) = z_G + \begin{pmatrix} -L^{-1}l \\ 0 \\ -\tilde{L}^{-1}\tilde{l} \end{pmatrix} = \sum_{l=1}^{4NI+2ND} \beta_l \Psi_l u + \begin{pmatrix} -L^{-1}l \\ 0 \\ -\tilde{L}^{-1}\tilde{l} \end{pmatrix} \tag{20}$$

It is straightforward to prove this claim as follows:

$$\frac{d\hat{z}(t)}{dt} = \sum_{l=1}^{4NI+2ND} \beta_l \Psi_l \frac{du}{dt} = \sum_{l=1}^{4NI+2ND} \beta_l \Psi_l D_\lambda u = T \sum_{j=1}^{4NI+2ND} \beta_l \Psi_l u = T\left[\hat{z}(t) - \begin{pmatrix} -L^{-1}l \\ 0 \\ -\tilde{L}^{-1}\tilde{l} \end{pmatrix}\right]$$

$$= T\hat{z}(t) + \begin{pmatrix} 0 \\ -l \\ -\tilde{l} \end{pmatrix} = T\hat{z}(t) + b \tag{21}$$

Using $\omega_i^{0+} = \omega_i^0$ and $\tilde{\omega}_i^{0+} = \tilde{\omega}_i^0$, the initial condition of $\hat{z}(t = 0^+)$ is

$$w_I|_{t=0^+} = w_I^0, \frac{dw_I}{dt}\bigg|_{t=0} = -Jw_I^0, \text{ and } \frac{d\tilde{w}_I}{dt}\bigg|_{t=0} = -\tilde{L}\tilde{w}_I^0 - \tilde{l} \text{ where } J = \begin{bmatrix} 0 & diag(\omega_i^{0+}) \\ -diag(\omega_i^{0+}) & 0 \end{bmatrix} \tag{22}$$

With a set of vectors ξ, $\xi_l = \Psi_l u(t=0)$, β is determined from the following least square process:

$$\beta^* = \arg\min_\beta \left\| [\xi_1 \cdots \xi_{4NI} \cdots \xi_{4NI+2ND}]\beta - \begin{pmatrix} w_I^0 + L^{-1}l \\ -Jw_I^0 \\ -\tilde{L}\tilde{w}_I^0 - \tilde{l} \end{pmatrix} \right\|_2 \tag{23}$$

The analytical solution to the swing equation that satisfies Kirchhoff's laws is

$$w_I = \Theta u - L^{-1}l; v_K = H_{KI}\Theta u - H_{KI}L^{-1}l + v_K^I; v_M = H_{MI}\Theta u - H_{MI}L^{-1}l + v_M^I \text{ where } \Theta$$

$$= \sum_{r=1}^{4NI} \beta_r^*([I00]\Psi_r) \tag{24}$$

If all the real parts of the eigenvalues are negative, u converges to zero as time increases. Therefore, under the circumstance and when O1 and O2 remain sufficiently small,

$$w_I \to -L^{-1}l, \; \tilde{w}_I \to -\tilde{L}^{-1}\tilde{l}, \; v_K \to -H_{KI}L^{-1}l + v_K^I, \; v_M \to -H_{MI}L^{-1}l + v_M^I.$$

Write down the solution

Computational complexity

There are several steps to determining the complexity: finding voltages immediately after the disturbance, decomposing eigenvalues of T, and computing β. The first process has the same complexity as solving power flow equations $\vartheta(Nb^{1.5})$ [32]; the second and last processes involve $\vartheta[(NI+ND)^3]$ [31], because the dimension of T is $(4NI+2ND) \times (4NI+2ND)$. Therefore, the first process is predominant if $Nb > (NI+ND)^2$. Even though additional computations are necessary at τ, the first process does not need to be solved because the voltages are readily available from (2). Therefore, the additional computation may not necessarily increase the computational cost significantly.

Stability assessment

The conventional stability assessment based on COM or DM certifies when a state is eventually stable. We propose another stability assessment—that the eigenvalues of T play a key role.

Type I: stable if the real parts of all the eigenvalues are non-positive or $\Theta = 0$ (no disturbance).

Type II: operationally stable if the largest positive real eigenvalue λ_m is small enough such that $1/\lambda_m \leq T_{op}$, where T_{op} is a time scale where transient stability is concerned. Typical time scales of transient stability studies are between sub-seconds to tens of seconds [33].

Type III: operationally stable if the coefficients corresponding to the terms in the base solution with positive real eigenvalues are small enough for the system to remain stable within T_{op}.

Type IV: unstable if the system divulges rapidly.

In this assessment scheme, the positive real parts of eigenvalues and the corresponding coefficients play a key role. Apart from *Type III* (in the presence of a disturbance), assessment is possible with the eigenvalues of the T matrix. If no positive eigenvalues exist, the state is eventually stable.

Validity region

The general solution of (20) to (10) is obtained without considering the constraints. It would be ideal to solve the ordinary differential Eq (10) with the constraints–DAE. DAE integrates the differential variables and the algebraic variables. While DAE is complete, to the best of our knowledge, no analytical solution has currently been identified. Therefore, the approach in this study is to identify the validity region where the constraints hold. To discuss the change of the constraints (drift-off phenomena), if we do not explicitly consider them while solving the differential equation (such as P7 in Fig 3), suppose we want to solve the following pendulum problem, $dx/dt = u; du/dt = -\lambda x; dy/dt = v; dv/dt = -\lambda y - 1; x^2 + y^2 - 1 = 0$. One may eliminate λ, and find a differential equation as follows: $\frac{u'}{x} + \frac{u^2+v^2-y}{x^2+y^2} = 0; \frac{v'+1}{y} + \frac{u^2+v^2-y}{x^2+y^2} = 0$ with the following constraints: $x^2 + y^2 = 1$ and $xu + yv = 0$. The numerical solution to the differential equation is evaluated without considering the constraints explicitly. Fig 4(A) shows the drift-off phenomena of the constraints, and the errors associated with $x^2 + y^2 = 1$ and $xu + yv = 0$ grow quadratically and linearly, respectively [34].

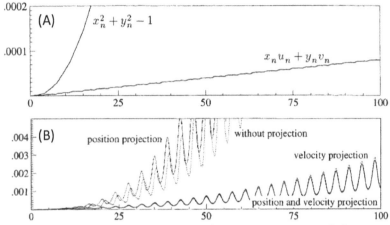

Fig 4. (A) Errors in the constraints and (B) global error with various projections. Both figures are from Ref. [34].

Eq (10) is analogous to constrained mechanical systems that can incorporate the repeated projection of a numerical solution onto the solution manifold; projections on position constraints and on velocity constraints for improving the stability (see Fig 4(B)). The error remains negligible with well-developed projections; hence, the solution (20) to (10) is valid without consideration of the constraints. However, the projections are not a linear process, which makes it difficult to integrate them into (10). Instead, we define a validity region where solution (20) is valid, and the projections are made at the boundary of the validity region to compensate for the drift-off. The induced errors and the approach to compensate the errors are discussed in the following section.

Internal voltages, E_I. Suppose, in a time range of $[0, \Delta t]$, the internal voltages deviate from the nominal values significantly, which affects z in (20), meaning the constraints do not hold. Because (20) yields the analytical expression of x_I and y_I in terms of t, one can evaluate a projection vector Δ_{con} that includes the error in E_I (position projection, $\tilde{p}_E i = 1 - \sqrt{x_i^2 + y_i^2}/E_i$) and the error in the first derivative (velcity projection, $\tilde{v}_E i = \sqrt{2x_i(dx_i/dt) + 2y_i(dy_i/dt)}/E_i$), i.e., Δ_{con} = The constraints hold if the 2-norm of the projection vector does not exceed a threshold Δ_E.

Error in T. The matrix T in (10) is exact except for one element each row by an approximation that $O_i(t) = (-q_i + b_{ii}E_i^2)/M_i + (d\delta_i/dt)^2 = (-q_i^0 + b_{ii}E_i^2)/M_i + (d\delta_i/dt)^2|_{t=0} = O_i^{0+}$. There might be a discrepancy between the true z and the estimated z due to the nonnegligible O2. The impact will affect two components: (1) the inaccuracy in estimating T and the eigen-values of T, and (2) the error in the coefficients according to Φ, $\delta\Phi$.

The impact of O2 in the inaccuracy is as follows: When the change in $O_i(t)$ is in the order of ε_T, the impact propagates to T, D_l, and correspondingly Φ and Ψ. The change in T due to $O_i(t)$ affects the eigenvalues of T that are in the block-diagonal in D_l. Suppose λ is a simple eigen-value of T and u_L and u_R are the corresponding left- and right- eigenvectors. $\vartheta(\varepsilon_T)$ changes in T can induce $\varepsilon_T/s(\lambda)$ in the eigenvalues [31],

$$|\lambda - \hat{\lambda}| \approx \frac{\varepsilon_T}{s(\lambda)} \text{ where } U_L^H T U_R = diag(\lambda_1, \cdots, \lambda_N),\ s(\lambda) = |u_L^H u_R|,\ \text{and}\ \varepsilon_T$$
$$= \|\delta T\|_F / \|T\|_F \tag{25}$$

Eq (25) indicates that the impact of the change in $O_i(t)$ is a linear change in the eigenvalues of T, the block-diagonal elements in D_l. Let $\hat{T}$ and $\hat{D}_l$ be the true matrices; δT and δD_l be the difference matrices between the true and the approximated matrices (i.e., $\hat{T} = T + \delta T$ and $\hat{D}_l = D_l + \delta D_l$); and $\hat{\Phi}$ and $\delta\Phi$ be the true solution matrix corresponding to $\hat{T}$ and $\hat{D}_l$, and the difference matrix, respectively. Further, define ρ and σ: $\rho = 2[\max(\|T\|_F, \|D_l\|_F)] \|(T \otimes I - I \otimes D_l^T)^{-1}\|_2$ and $\|\delta T\|_F \leq \sigma \|T\|_F$, $\|\delta D_l\|_F \leq \sigma \|D_l\|_F$. D_l is a similarity transformed matrix of T (i.e., $D_l = P^{-1}TP$); hence, $\|D_l\|_F = \|P^{-1}TP\|_F \leq \kappa_F(P)\|T\|_F$. From $\Phi D_l - \Phi T = 0$ and $\hat{\Phi}\hat{D}_l - \hat{T}\hat{\Phi} = 0$, one finds,

$$\hat{\Phi}\hat{D}_l - \hat{T}\hat{\Phi} \approx \Phi\delta D_l + \delta\Phi D_l - T\delta\Phi - \delta T\Phi = 0 \rightarrow \delta T\Phi - \Phi\delta D_l = T\delta\Phi - \delta\Phi D_l \tag{26}$$

Eq (26) can be rewritten in terms of the Kronecker product,

$$\delta T\Phi - \Phi\delta D_l = (\delta T \otimes I - I \otimes \delta D_l^T)vec(\Phi) \text{ and } T\delta\Phi - \delta\Phi D_l$$
$$= (T \otimes I - I \otimes D_l^T)vec(\delta\Phi) \tag{27}$$

Using the Kronecker product, one finds the upper/lower bounds of the Frobenius norm of a product between two matrices A and B as follows:

$$\|B\|_F = \|A^{-1}AB\|_F = \|vec(A^{-1}AB)\|_2 \leq \|I \otimes A^{-1}\|_2 \|vec(AB)\|_2 = \|A^{-1}\|_2 \|AB\|_F \rightarrow \|AB\|_F$$
$$\geq \|A^{-1}\|_2^{-1} \|B\|_F \tag{28}$$

Eqs (27) and (28), and the triangle inequality theorem yield

$$\|\delta T\Phi - \Phi\delta D_l\|_F = \|(\delta T \otimes I - I \otimes \delta D_l^T)vec(\Phi)\|_F \leq \|\delta T \otimes I - I \otimes \delta D_l^T\|_2 \|\Phi\|_F$$
$$\|T\delta\Phi - \delta\Phi D_l\|_F = \|(T \otimes I - I \otimes D_l^T)vec(\delta\Phi)\|_F \geq \|(T \otimes I - I \otimes D_l^T)^{-1}\|_2^{-1} \|\delta\Phi\|_F \tag{29}$$

and using the triangular inequality and the fact that the Frobenius norm is no less than 2-norm [31]:

$$\frac{\|\delta\Phi\|_F}{\|\Phi\|_F} \leq \|\delta T \otimes I - I \otimes \delta D_l^T\|_2 \left\|(T \otimes I - I \otimes D_l^T)^{-1}\right\|_2$$
$$\leq (\|\delta T\|_F + \|\delta D_l^T\|_F) \left\|(T \otimes I - I \otimes D_l^T)^{-1}\right\|_2 \tag{30}$$

With the definition of ρ and σ, (30) becomes:

$$\frac{\|\delta\Phi\|_F}{\|\Phi\|_F} \leq \rho\sigma \sim \vartheta(\varepsilon) \tag{31}$$

Because a time-varying $O_i(t)$ determines ε, one can compute the τ_ε that the relative error in T remain a threshold Δ_T, i.e., $\varepsilon = \|\delta T\|_F/\|T\|_F \leq \Delta_T$ for all $t \leq \tau_\varepsilon$. Let y_z be the first derivative of z with respect to time, then (14) yields: $y_z = Tz_0 + b$ where $y_z = dz/dt|_{t=0}$ and $z = z_0 + \int y dt$. In the time range, there might be an error in T involving the error that propagates in y_z and z as fol- lows:

$$\delta z \cong \delta z_0 + (T\delta z_0 + \delta T z_0)\Delta t = (I + T\Delta t)\delta z_0 + (\Delta t \delta T)z_0 \tag{32}$$

Beyond the validity region. The boundary of the validity region is defined either when the constraints do not hold, or when the error in T exceeds the threshold. At the boundary of the validity region, the values for z and T are updated as described. With the updated values, (14) still holds because the errors in (14) and the constraints remain less than the threshold. Therefore, the problem in (14) will be solved with the initial condition that is the updated values for z. It is nec- essary to ensure consistency among the initial values; hence, a consistent initialization problem. This problem has been studied widely; (1) a small artificial step with the backward Euler method [35], [36], (2) Taylor series expansion [37], [38], and (3) graph theoretic algorithm to obtain the minimal set of equations for differentiation to solve for consistent initial values [39], [40].

Eqs (12) and (13) and the constraints are rewritten as follows:

$$f(t, u_I, w_I, du_I/dt) = \begin{cases} \dfrac{du_1}{dt} + \text{diag}\left(\dfrac{D_I}{M_I}\right)u_1 + Lw_I + l = 0 \\ \\ \dfrac{du_2}{dt} + \tilde{L}u_2 + \tilde{l} = 0 \end{cases} \rightarrow \frac{du_I}{dt} + \hat{D}_I u_I + \hat{L}_I w_I + \hat{l}_I = 0$$

$$g(t, u_I, w_I) = \begin{cases} [\text{vec}(e_i e_i^T + e_{i+NI} e_{i+NI}^T)]^T(w_I \otimes w_I) - E_i^2 = 0 \\ \\ [\text{vec}(e_i e_i^T + e_{i+NI} e_{i+NI}^T)]^T(w_I \otimes u_I) = 0 \end{cases} \tag{33}$$

$$\rightarrow F(t, u_I, w_I, du_I/dt) = \begin{bmatrix} f(t, u_I, w_I, du_I/dt) \\ g(t, u_I, w_I) \end{bmatrix} = 0$$

where $u_1 = \dfrac{d\omega_I}{dt}; u_2 = \tilde{w}_I; u = \begin{pmatrix} u_1^{2NI} \\ u_2^{2ND} \end{pmatrix}; \hat{D}_I = \begin{bmatrix} \text{diag}\left(\dfrac{D_I}{M_I}\right) & 0 \\ 0 & \tilde{L} \end{bmatrix}; \hat{L}_I = \begin{bmatrix} L \\ 0 \end{bmatrix}; \hat{l}_I = \begin{pmatrix} l \\ \tilde{l} \end{pmatrix}$

The cardinalities of f and g are $2NI+2ND$ and $2NI$, respectively. By the definition of the variables in (33), u and s are classified as differential variables and algebraic variables. Accordingly (33) is called a DAE. At a given w_I^0 and u_I^0, the Newton-Raphson method finds:

$$\begin{bmatrix} \Delta f(t, u_I, w_I, du_I/dt) \\ \Delta g(t, u_I, w_I) \end{bmatrix} = \begin{bmatrix} hI & \hat{L}_I & \hat{D}_I \\ 0 & \hat{M}_w & \hat{M}_u \end{bmatrix} \begin{pmatrix} \Delta u_I'/h \\ \Delta w_I \\ \Delta u_I \end{pmatrix} \rightarrow \begin{pmatrix} \Delta u_I'/h \\ \Delta w_I \\ \Delta u_I \end{pmatrix} = \begin{bmatrix} hI & \hat{L}_I & \hat{D}_I \\ 0 & \hat{M}_w & \hat{M}_u \end{bmatrix}^{-1} \begin{pmatrix} \Delta f \\ \Delta g \end{pmatrix}$$

where $\hat{M}_w = \begin{bmatrix} \left(2w_I^{0T} M_1\right)^T \\ \vdots \\ \left(2w_I^{0T} M_N\right)^T \end{bmatrix}^T, \begin{pmatrix} 2u_I^{0T} M_1 \\ \vdots \\ 2u_I^{0T} M_N \end{pmatrix}^T \end{bmatrix}^T, \hat{M}_u = \begin{bmatrix} 0 & \begin{pmatrix} w_I^{0T} M_1 \\ \vdots \\ w_I^{0T} M_N \end{pmatrix}^T \end{bmatrix}^T$ (34)

The term in (34) is compensated to yield a consistent initial point immediately beyond the validity region. With the consistent initial point and updated T, the analytical solution in (20) is identified. Most components inside the curly bracket in (34) are constant. Therefore, only a few visits to the boundaries of the validity limits does not increase the computation time signif- icantly if efficient rank-update techniques are employed.

Insight on the dynamics

As in (12), T has a block structure, and it can be broken into two submatrices as follows:

$$T = \underbrace{\begin{bmatrix} 0 & I & 0 \\ -L_{sys} & -diag(D_I/M_I) & 0 \\ 0 & 0 & -\tilde{L}_{sys} \end{bmatrix}}_{T_{sys}} - \underbrace{\begin{bmatrix} 0 & 0 & 0 \\ L_{op} & 0 & 0 \\ 0 & 0 & \tilde{L}_{op} \end{bmatrix}}_{T_{op}=\sum(+)-\sum(-)}$$ (35)

The first term is invariant with the operation point of specific contingency, but the second term varies. We consider the eigenvalue decomposition of T as consecutive applications of the eigenvalue update and down-date of T_{op} from the eigenvalues of T_{sys}; that is, the sum of multiple rank-1 [41] or rank-2 [42] updates (+) and multiple down-dates (-). The impacts of T_{op} are to shift the eigenvalues of T_{sys}, which is bounded by the Bauer-Fike theorem [31]: $\min \min_{\lambda \in \lambda(T_{sys})} |\lambda - \hat{\lambda}| \leq (\|U_\Phi\|_p \|U_\Phi^{-1}\|_p) \|T_{op}\|_p$ where $U_\Phi^{-1} T_{sys} U_\Phi = diag(\hat{\lambda}_1, \cdots, \hat{\lambda}_N)$. If the diagonal elements in T_{op} are all zeros (or negligible), the Gershgorin circle theorem can be applied [31]. The Gershgorin disk is isolated from the other disks so that a disk contains precisely one eigen- value of T. Motivated readers may find the process in [43]. Note that T_{sys} only depends on the system, and T_{op} varies with the dynamics. The following section outlines the terms affecting the stability of the system.

Reactive power, q_i^c in $q_i^c + b_{ii} E_i^2$: This term negatively affects L_{con}, shifting the eigenvalues to the right. Reactive power supports voltages, which makes it difficult for a system to settle into new voltages. However, the reactive power injection is very small in comparison to the $b_{ii} E_i^2$ term; therefore, its impact on the transient stability is highly limited.

Unsettled mechanical power, p_i^{mech} in $p_i^{mech} - g_{ii} E_i^2$: These terms are associated with the mechanical power in L, and have the same magnitudes with opposite signs; hence, their impacts on shifting the real eigenvalues of T cancel out. However, their impacts are on shifting the imaginary components of the eigenvalues. It is intuitively correct that unsettled mechanical power enhances the oscillating motion. In many cases, the resistance of an internal generator model is zero, meaning g_{ii} is zero. Unlike the reactive power injection, p_i^{mech} plays a key role in stability assessment. A synchronization condition based purely upon the power injections is peoposed in [9].

Inter-voltage sensitivity matrix, H_{KI}: As the sensitivity of H_{KI} increases in a positive direc- tion, the voltages at *Kbus* swing tightly together with the generators. The negative sensitivities in H_{KI} imply that the voltages at *Kbus* swing against the angular motion of rotors. They act as a dragging force to stabilize the system. The impact of the inter-voltage sensitivities is to shift in the same direction as the sign of H_{KI}.

Damping coefficient. The damping term appears in the lower diagonal in T. Because the damping terms are non-negative, the impact is to shift the eigenvalues to the left. Therefore, the damping terms help to stabilize the system against disturbance.

Inertia. The terms reactive power, unsettled mechanical power, voltage sensitivity H_{KI}, and damping are all scaled by the inertia constant M_i. Regardless of their signs, the amplitudes are normalized in terms of inertia. The impact of inertia becomes clear in TDS in a way that agrees with this observation.

Loads. Loads are classified into three categories: synchronized induction load, frequency- or time-dependent load, and remaining loads. The remaining loads affect the sensitivity matrix H_{KI}, which is a part of the L, $\tilde{L}$, and T_{op} matrices. The frequency- and time-dependent loads, and the synchronized induction loads, are taken into consideration in the swing Eq (14). It is noteworthy that the T matrix has a block diagonal structure between the two loads; therefore, the Eigen-space of each block is independent so that their subspaces are orthogonal.

Illustrative examples and discussion

Simulations are performed for various IEEE model systems (IEEE 9, 14, 30, and 118-bus systems). To compare the results with both a coupled oscillator model [9] and DM, which assume a lossless system ($\gamma = 0$ for all the lines) and constant voltage magnitudes, the resistance components are all ignored. A phase cohesiveness for synchronization is also introduced: $|\theta_i - \theta_j| \leq m\epsilon[0,\pi/2]$. Electric power is generated at *Ibus* and injected into the grids; therefore, the angles at *Ibus* (δ_i) are greater than those at *Kbus*. ($\delta_i \geq \theta_k$). The power flow from *Ibus i* to *Kbus k* (power injection at *PV buses* of the original network) is proportional to $\sin(\delta_i \geq \theta_k)$ [26]. There- fore, the phase cohesiveness is equivalent to the constraints imposed on the injection between two directly connected oscillators. We found that as the system becomes unstable, the maximum angle differences after the disturbance are significantly higher than those in the stable case. Even though the values for the dynamics are not listed in [9], the phase cohesiveness for synchronization finds the trend correctly for the disturbances we tested. Similarly, the syn- chronization condition is also checked, and it was found that there is a γ^K appearing in [9] to satisfy $\|L^\dagger p\|_{\varepsilon,\infty} \leq \sin\gamma^K$ for the chosen equilibrium points. Because there can be many equilibrium points, it can be difficult to analyze the stability region of each point [13]. For the sake of visual presentation, the simulation results on the IEEE 9 bus system are discussed in this paper. Fig 5 shows the one-line diagram of the system, and Table 1 lists the data relevant to this study. In the proposed network modeling in this study, there are three *Ibuses*, three *Kbuses*, and six *Mbuses*. The pre-fault power flows are computed using the unified method based on the Kronecker product [44], and the threshold values for Δ_T and Δ_E are 1% and 10%, respectively. The scaling factor h in this study is 0.1. All the numerical computations are per- formed using a Mac pro with two 2.93 GHz 6-core Intel Xeon processors.

The results are summarized in Table 2. In the table, 10% d_8 in the first column refers to 10% loss of loads at Bus 8; Fig in the second column is the figures associated with the event at the first col- umn; $\Delta\delta_{max}$ under coupled oscillator column is the synchronization condition pro- posed in [9] where the threshold for the system is 0.129; ΔV_{st} under the DM column is the sta- bility margin that is defined in the Appendix. For some cases, the stability assessments are undetermined and shown as "-", because the certificate is not issued. Time under the time domain simulation (TDS) and proposed model columns represent the computation time for numerical computations.

Loss of loads

Three sets of cases are performed to simulate the loss of load at Bus 8; no loss, 10% loss, and entire loss. Prior to the loss of load, the system was in the steady state condition that was iden- ti- fied using the power flow study. When no disturbance occurs, the system should stay in the same steady state at $t = 0$. Immediately after the loss of load, a new operating point is found by solving (2) with the updated loads and w_I at the steady state, because the rotor angles do not change instantaneously.

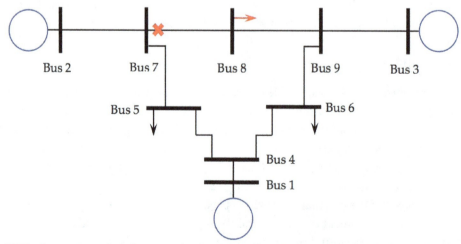

Fig 5. IEEE 9-bus system. The red arrow and red cross represent the loss of load and the line fault, respectively.

The stability of the post-fault operation point is estimated using DM and the synchroniza- tion condition proposed in [9]. The trajectory during the transient state is numerically calcu- lated in terms of TDS. The details of TDS and of DM based on an energy function are outlined in the Appendix. For the case with no disturbance, even though the eigenvalues of T are non-zeros, the coefficient Θ is zero (*Type I*) because the constant term in (20) is zero, meaning the system stays in the steady state at the pre-fault condition.

Slight change in the load at Bus 8. Fig 6 illustrates the trajectories of (A) rotor angles δ, of (B) voltage magnitudes E, of (C) $O_i(t)$, and of (D) computed magnitudes of the internal volt- ages when a 10% loss of the load at Bus 8 occurs at $t = 1$ s.

Stability of the system. In the post-fault state, the maximum angle differences in voltage angles immediately after the disturbance is 0.116 rad, while the synchronization condition reported from [9] is 0.129 rad. As shown Fig 6(B), the variations of the voltage magnitudes are not sig- nificant, which indicates that a condition of both COM and DM holds. Because the maximum angle difference is less than the threshold, the stability assessment predicts the con- vergence to a stable state. The energy functions $V(\delta, \omega)$ for DM are evaluated at all equilibrium points to check the stability of the system. Based on to the stability margin of 8.96 ($= V_{cr} - V_{cl} > 0$), a stability certificate is issued by DM. TDS is also performed with a time step Δt of 0.01 s. All three stability assessment approaches yield the same estimate—converging to a new equi- librium point.

The positive eigenvalues of T are small, and the corresponding Θ is numerically negligible (*Type II*). As shown in Fig 6(A), the rotor angles are all synchronous, and the voltage magni- tudes quickly settle down to a new equilibrium point. The analytical solution and the numeri- cal re- sults from TDS are visually indistinguishable, as shown in Fig 6(A). For clear presentation, the voltage magnitudes and the rotor speeds are generated by the proposed ana- lytical approach in Fig 6(B).

Validity region. In this study, the tolerance for $\varepsilon \, (= \|\delta T\|_F / \|T\|_F)$, Δ_T, is set to 1% for all the nu- merical evaluations. The value of $\|T\|_F$ in the case of 10% loss of the load at Bus 8 is 16.71; hence, the tolerance of the impact of O2 toward $\|\delta T\|_F$ is 1.67×10^{-2} ($= 1\% \times 16.71$). The variations of $O_i(t)$ are negligible, as shown in Fig 6(C)—the value of $\|\delta T\|_F$ is 1.21×10^{-4} ($\|\delta T\|_F / \|T\|_F = 7.24 \times 10^{-6}$) is $\vartheta(10^{-4})$.

Table 1. Machine data for the IEEE 9-bus system illustrated in Fig 2.

Generator location	M_i	D_i	g_{ii}	b_{ii}	$|E_i|$
Bus 1	2.364	0.0254	0.0	-16.45	1.057
Bus 2	0.640	0.0066	0.0	-8.35	1.050
Bus 3	0.301	0.0026	0.0	-5.52	1.017

Table 2. Summary of the simulation results.

Event	Fig	Coupled oscillator		DM		TDS		Proposed method	
		$\Delta\delta_{max}$	stability	V_{margin}	stability	time	stability	time	stability
10% d_8	6	0.116	stable	8.96	stable	3.80	stable	0.17	stable

entire d_8	7	0.113	stable		9.21	stable	2.62	stable	0.55	stable
on-fault	8	0.137	-		-34.87	-	3.45	unstable	0.11	unstable
postfault	10	0.155	-		-1.98	-	3.37	stable	1.91	stable

This means the impact of O2 is negligible, and the validity region extends to the entire time domain of the transient in this case. Fig 6(C) indicates that $\eta(t)$ for the generators are constant after assuming $\mu(t) = 0$. Therefore, $|O_i(t)|$ follows the Karamata representation theorem (i.e., a slowly varying function), which also leads to the extended validity region. Fig 6(D) shows that the estimated voltage magnitudes vary within the threshold ($\Delta_E = 10\%$) of the nominal voltages, and that O1 is less than the threshold most times. This implies only a few times of crossing the validity regions in 10 s. It is noteworthy to mention that in most simulations, O1 and O2 are small.

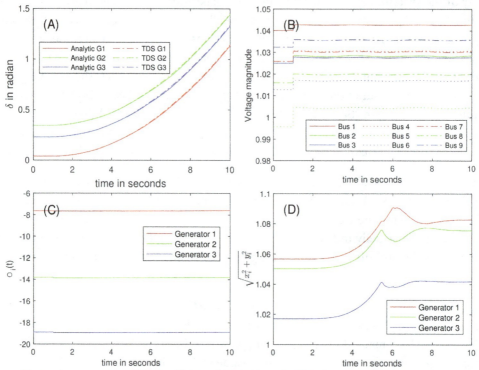

Fig 6. The trajectories of (A) δ, of (B) E, of (C) $O_i(t)$, and of (D) the magnitudes of the internal voltages after 10% of loss of the load at Bus 8 (red arrow in Fig 5).

Entire loss of the load at Bus 8. A significant change in the load at Bus 8 occurs at $t = 1$ s, and the generator dynamics are simulated. Fig 7 illustrates the trajectories of (A) δ, (B) the voltage magnitudes at the terminal voltages, (C) $O_i(t)$, and (D) the estimated magnitudes of the internal voltages.

Stability of the system. After the loss of load, the stability of the system is tested; (1) the maximum angle difference across the lines is 0.113 (<0.129), (2) the stability margin is 9.21 (>0), and (3) TDS with $\Delta t = 0.01$ s. All three methods find the stability of the system, and converge to a new equilibrium point after the disturbance. Due to the loss of loads at Bus 8, the reactive power injected to the grid exceeds the demands, and the uncompensated reactive power makes the voltages at the terminal buses increase (Fig 7(B)). A condition of COM and DM that the terminal voltage magnitudes are constant holds marginally.

Fig 7(A) shows the dynamics of rotor angle for the entire loss of the load at Bus 8, and exhibits no discrepancies between the analytic approach and TDS. The positive eigenvalues of T are small, but the corresponding Θ is numerically negligible (*Type II*). Therefore, the pro- posed analytical solution also estimates the stability of the system correctly.

Validity region. Because the value of $\|T\|_F$ in the case of the entire loss of the load at Bus 8 is 17.61, the tolerance to the impact of O2 toward $\|\delta T\|_F$ is 0.18 (= 1% × 17.61). The variation in $O_i(t)$ of three generators over 10 s is approximately 8, and the corresponding $\|\delta T\|_F$ is 9.88 × 10−2 ($\|\delta T\|_F/\|T\|_F$ = 5.61 × 10−3). Therefore, the validity region extends to the entire time domain of the transient if O1 is small. As shown in Fig 7(C), $\eta(t)$ (after assuming $\mu(t) = 0$) does not con- verge as t goes to infinity; therefore, the Karamata representation theorem may not be applicable. Con- sistent with this observation, $O_i(t)$ is not slowly varying, but the impacts on T are within the threshold. O1 often (5 peaks) reaches Δ_E (= 10%), as shown in Fig 7(D). The projection meth- od identifies the peaks (the boundaries of the validity regions), and the consistent initial points are evaluated using (34) for the application of (20), which increases the computation time.

Line fault: On-fault trajectory

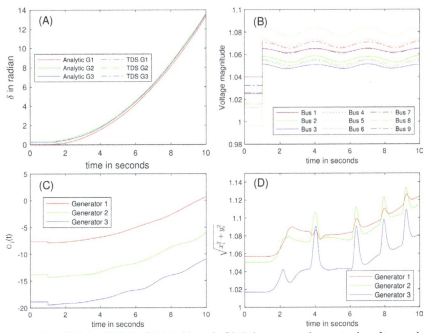

Fig 7. The trajectories of (A) δ, of (B) E, of (C) $O_i(t)$, and of (D) the estimated magnitudes of internal voltages due to the entire loss of the load at Bus 8 at $t = 1$ sec (red arrow in Fig 5).

At $t = 1$ s, a three-phase fault occurs on the line connecting Buses 7 and 8 near Bus 7 (red cross mark shown in Fig 4). Using the pre-fault power flow solution by applying the unified power flow analysis approach in [44], the parameters to formulate the swing equations are identified. Due to the continuation of the rotational movement, the rotor angle δ remain continuous in time regardless of the change in the network. The element in Y_{bus} corresponding to Bus 7 is increased to represent a high admittance to ground, and the voltage at Bus 7 collapses. With this modifica- tion, the on-fault voltages and the parameters for the swing equations in the on-fault trajectory are computed. Fig 8 illustrates the on-fault trajectories of (A) δ, of (B) E, of (C) $O_i(t)$, and of (D) the estimated magnitudes of the internal voltages when the fault is not cleared.

Stability of the system. After the disturbance, Bus 2 is isolated from the system that is directly connected to Generator 2, and its voltage magnitude reduces to zero (Fig 8(B)). Because no load is located at Bus 2, the electric power injection becomes zero, and the electrical power input from the generator is stored in the rotor in the form of mechanical power (increased rotor speed). For the rest of the system, the change in the power supply by the iso- lated Bus 2 (effectively the loss of Generator 2) is compensated by the other generators. This leads to a change in rotor angle, as shown in Fig 8(A). Clearly, the reactive power injection changes abruptly (see Fig 8(C)).

The maximum angle difference for COM is 0.137 rad, while the value for γ in [9] is 0.129, which does not meet the synchronization condition. If the method fails to issue a certificate, the stability of the system is undetermined. However, the short-circuit makes Bus 2 discon- nected from the rest of the network, and the remaining network different from the original network. Therefore, the prediction based on the synchronization condition may not be exact. The closest unstable equilibrium for DM is found, and the stability margin is found to be -34.87 (<0), which means the stability of the system is not certified. Similar to COM, the stabil- ity of the system is undetermined if a certificate is not issued.

Fig 8(A) shows large deviations in the results for Generator 2 between the proposed analytic solution and TDS after $t = 1.5$ s, but both indicate system instability. They both indicate that Generator 2 will lose synchronization quickly after the line fault, and be disconnected from the network (*Type IV, unstable*). In the on-fault condition, Generator 2 is isolated from the rest of the system. Therefore, as shown in Fig 8(A), the rotor angular motion becomes faster than motion of the other generators.

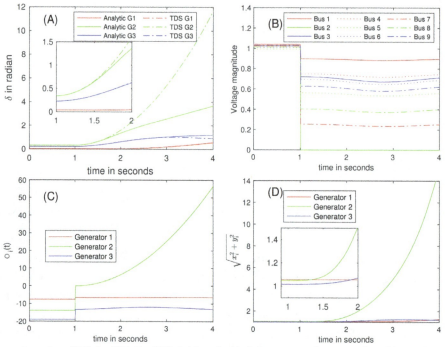

Fig 8. The trajectories of (A) δ, of (B) E, of (C) $O_i(t)$, and of (D) the estimated magnitudes of internal voltages due to the line fault at the line between Buses 7 and 8 near Bus 7 at $t = 1$ sec (red cross in Fig 5).

Validity region. Fig 8(C) shows the fault-on trajectories of $O_i(t)$ if the fault is not cleared. $O_i(t)$ for Generator 2 is not a slowly varying function, and it does not follow the Karamata representation theorem because $\eta(t)$ divulges for Generator 2, meaning $\eta(t)$ is not bound. $\|T\|_F$ is 7.88, and the threshold for $\|\delta T\|_F$ is 7.88×10^{-2} (1% × 7.88), but the impact of O2 is 1.00, which is greater than its threshold. Fig 8(D) lists the estimated magnitudes of the internal voltages. Shortly after the line fault, the internal voltages at Bus 2 increase too quickly, and O1 becomes large. Therefore, the validity region is limited quickly after 1 s, and neither O1 nor O2 are small during the on-fault trajectory.

Line fault: Post-fault trajectory

To prevent the loss of synchronization, the fault should be cleared. At $t = 1.1$ s, the fault is cleared by opening up the circuit breakers of the line between Buses 7 and 8. With the updated topology, the Y_{bus} is updated; accordingly, a new operation point is identified. However, the rotor angular motions are continuous. Fig 9 show the post-fault trajectories of (A) rotor angle, and (B) estimated magnitudes of the internal voltages, respectively. Fig 9(A) exhibits that the analytical approach does not predict the rotor dynamics properly after 2.5 s, and the discrep- ancy increases with time. The TDS has a time step of 0.01 s.

Stability of the system. The maximum angle difference in voltage angles immediately after clearing the fault is 0.155 rad, which COM concludes is the undetermined stability of the sys- tem. The closest unstable equilibrium point is observed, and the stability margin based on DM is -1.98 (<0), which also renders DM unable to estimate the stability of the system. There is a discrepancy in the results of the rotor angles between the proposed analytic method and TDS. This discrepancy increases after $t = 3$ s as the system evolves over time.

Validity region. The value of $\|T\|_F$ immediately after clearing the fault is 11.43; hence, the tolerance to the impact of O2 toward $\|\delta T\|_F$ is 0.114 (= 1% × 11.43). The impact of O2 is 1.79 ($\|\delta T\|_F / \|T\|_F = 0.156$), which is beyond the threshold of 1% after clearing the fault. Fig 9(D) indicates that the voltage magnitudes suddenly increase after 2 s, meaning O1 is not small after 2 s. The validity region boundaries are identified with the projections, and the consistent initial points are identified to correct the errors.

Post-fault trajectory: Beyond the validity region

At the boundary of the validity region, T and z are modified according to (32) and (34). Because they are still close to the true ones at the boundary, the consistent initial points are evaluated at the boundaries. With the updated values for T and z, one can compute the coeffi- cient to construct the analytical solution under the constraints that O1 and O2 are small with the updated values. Fig 10 show the post-fault trajectories beyond the validity region of (A) router, (B) the magnitudes of the bus voltages, (C) $O_i(t)$, and (D) the estimated magnitudes of the internal voltages, respectively.

Stability of the system. Fig 10(A) shows an improved similarity between two models. Fig 10(B) illustrates the post-fault trajectories of the terminal voltages that indicate non-negligible changes of voltage magnitudes. The left and the right eigenvectors are identified corresponding to each eigenvalue λ of T to compute the condition of the eigenvalue, $s(\lambda)$ from (25). The largest deviation of the eigenvalue corresponds to the smallest condition of the eigenvalue: $\max_j |\lambda_j - \hat{\lambda}_j| \approx \varepsilon [\min_j s(\lambda_j)]^{-1}$. It was found that the largest change in the eigenvalue was $\lambda_{max}^{Re} = 0.3879$ before the update, and the corresponding updated eigenvalue was $\lambda_{max}^{Re} = 0.3734$, and $s(\lambda)$ was approximately 1.8. It turns out that the eigenvalue has the largest real part in the positive, but the corresponding coefficients are not large enough to make the system divulge before $t = 10$ s (*Type III* stable). As shown Fig 10(A) and 10(B), both the analyti- cal approach and TDS expect the stability of the system.

Validity region. As discussed with Fig 10(D), the validity region boundaries are frequently revisited (15 times), and consistent initializations are performed by the projections using (34). Note that the codes are not currently optimized to utilize the sparse structure of the matrices and to explain the partial update of the Jacobian matrices $\hat{M}_w$ and $\hat{M}_u$ in (34), which makes the computation process inefficient. The voltage magnitudes of Generator 2 swings widely because the generator is closest to the location of the line fault among all the generators.

Future works

In this work, for the sake of simplicity, the dynamics of the generators are represented using the classical model. It is important to model the generators correctly to discuss the power sys- tem dynamics [23].

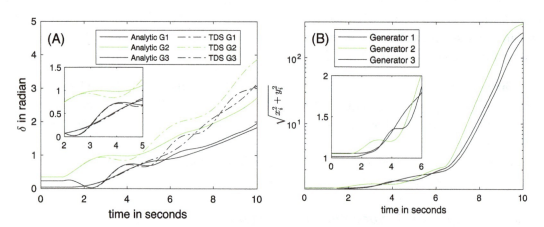

Fig 9. (A) Post-fault trajectories of the rotor angles and (B) the magnitudes of the estimated internal voltages.

We plan to accommodate generator models that have been widely used for decades in numerous commercial simulation programs for modeling round-rotor and salientpole synchronous generators [26] (the GENROU/GENSAL models) and for the detailed treat ment of saturation (the GENTPF/GENTPJ models) [45]. The immediate adjustment in modeling is on *Ibus* where the voltage magnitude E_i does not stay constant, which requires a modification in O1.

Conclusions

In this paper, we derived an analytical solution to the swing equations to assess the transient stability of power grids. To the best of our knowledge, our solution is a unique analytical solu- tion to the swing equations without physically unacceptable assumptions, while obeying Kirchhoff's laws. The solution indicates the factors affecting the stability after a disturbance occurs. Based on the solution, a new stability assessment approach is proposed. The assessment tool is different from the conventional assessment tools (COM, DM, and TDS approach) in that the derivation of our solution does not require the unphysical assumptions required for both COM and DM (such as lossless grids, constant voltages at all buses, and no consideration of reactive power). Moreover, its computational complexity is manageable. In addition to the low computational complexity, the approach proposed in this study explores the components affecting power sys- tem dynamics by examining the structure of the T matrix in terms of system dependent (T_{sys}) and operation point dependent (T_{op}) submatrices. The simulation results show that O1 and O2 are small in most cases. However, even in a case when O1 and O2 are large, it is possible to main- tain O1 and O2 small by introducing the validity region based on the projection methods. The consistent initializations make it possible to identify the trajecto- ries reliably.

Appendix

Time domain simulation, TDS

For comparison, TDS is performed using the Runge–Kutta method [46]. At first, the loads are converted to equivalent admittance to construct a linear model (i.e., the current is linear with the internal voltages). From the linear current and voltage relationship, the parameters in (2) are evaluated at the post-fault state.

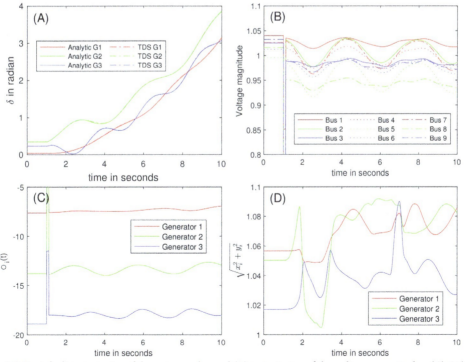

Fig 10. (A) Post-fault trajectories of the rotor angles and (B) trajectories of the voltage magnitudes, (C) $O_i(t)$, and (D) the estimated magnitudes of internal voltages after the fault is cleared at $t = 1.1$ seconds.

Because the magnitudes of internal voltages are assumed constant, the voltage at the *Ibuses* can be evaluated in terms of δ. Once the voltages at the *Ibuses* are computed, one can compute all the terminal voltages and real power injection at the *Ibuses*. Using the terminal voltages and the real power injection, it is possible to update δ.

We first define $z = [\delta^T \quad \omega^T \quad \tilde{w}^T]^T$, and rewrite the swing equation: $y = \frac{dz}{dt} = Az + b$

where $A = \begin{bmatrix} 0 & I & 0 \\ 0 & -diag(D/M) & 0 \\ 0 & 0 & -\tilde{L} \end{bmatrix}, b = \begin{pmatrix} 0 \\ f(z) \\ -\tilde{i} \end{pmatrix}$, and $f(z) = diag(1/M)[p_{mech} - p_{elec}(z)]$. At

the first sub step at iteration, with a step size $\Delta t > 0$, one finds

$$y_m^{\{1\}} = Az_m^{\{0\}} + b_{m-1}^{\{0\}}, z_m^{\{1\}} = z_{m-1} + \frac{1}{2}\Delta t y_m^{\{1\}} \text{ where } z_m^{\{0\}} = z_{m-1}, b_{m-1}^{\{0\}} = b_{m-1} \quad (A1)$$

The update of z allows finding a power flow solution corresponding to the value of z for updating $f(z_m^{\{1\}})$. (A1) is generalized at the kth sub step for the mth iteration as follows:

$$y_m^{\{k\}} = Az_m^{\{k-1\}} + b_{m-1}^{\{k-1\}}, z_m^{\{k\}} = z_{m-1} + r_k \Delta t y_m^{\{k\}} \text{ where } r_1 = r_2 = \frac{1}{2}, r_3 = 1 \quad (A2)$$

The update of $b_{m-1}^{\{k\}}$ follows (A2). Then the last update at the *mth* iteration is

$$z_{m+1} = z_m + \frac{\Delta t}{6}(y_m^{\{1\}} + 2y_m^{\{2\}} + 2y_m^{\{3\}} + y_m^{\{4\}}) \quad (A3)$$

In averaging the four increments, a larger weight is given to the middle increments. Multiple simulations are performed with various time steps Δt in a decreasing order until the simu- lation results are invariant.

Direct method based on an energy function

The Lyapunov stability theorem gives sufficient conditions to determine the stability of the system. Numerical computation of the underlying ordinary differential equations is not necessary to derive the stability properties. The Lyapunov stability theorem gives two conditions for stability for a Lyapunov function V on an open set U. (1) $V(x) = \begin{cases} = 0 \text{ if } x = 0 \\ > 0 \text{ otherwise} \end{cases}$ and (2) $\frac{dV(x)}{dt} = \sum_j \frac{\partial V}{\partial x_j} \frac{dx_j}{dt} = \nabla V \cdot \frac{dx}{dt} \leq 0 \forall x \neq 0$.

For a power system with multiple machines, the traditional DM utilizes the energy function under the assumptions of losses ($\gamma_{ik} = 0$) and of the fixed voltage magnitudes (E_k is fixed). The simplest energy function would be the mechanical interpretation [8] as follows:

$$V(\delta, \omega) = E_{KE} + E_{PE} = \frac{1}{2}\sum_k M_k \omega_k^2 + \left[-\sum_k p_k^{mech} \delta_k - \sum_{k,i} (|\tilde{y}_{ik}|E_i E_k)\cos\delta_{ik} \right] \quad (A4)$$

The corresponding swing equation is $\omega_i = \frac{d\delta_i}{dt}$, $M_i \frac{d\omega_i}{dt} = p_i^{mech} - (|\tilde{y}_{ik}|E_iE_k)\sin\delta_{ik} - D_i\omega_i$; there- therefore, V satisfies stability conditions as follows:

$$\frac{dV(\delta,\omega)}{dt} = \sum_i \left[\left(\frac{\partial V}{\partial \delta_i}\right)\left(\frac{d\delta_i}{dt}\right) + \left(\frac{\partial V}{\partial \omega_i}\right)\left(\frac{d\omega_i}{dt}\right)\right] = -\sum_i D_i\omega_i^2 \leq 0 \quad (A5)$$

The rotor angle δ^{ref} is the measure of the internal voltage angle with respect to the terminal voltage angle of the reference bus. For convenience, the reference frame for the rotor angle is redefined with respect to the rotating COI, i.e., $\delta_i = \delta_i^{ref} - \delta_{COI}$, $\omega_i = \omega_i^{ref} - \omega_{COI}$ where $\delta_{COI} = \frac{1}{M_T}\sum_i M_i\delta_i$, $\omega_{COI} = \frac{1}{M_T}\sum_i M_i\omega_i$, $M_T = \sum_i M_i$. Equilibrium points are found at a set of the rotor angles that satisfies $d\delta_i/dt = 0$ and $d\omega_i/dt = 0$, which leads to $\delta_{ik} = \sin^{-1}\left(\frac{p_i^{mech}}{|\tilde{y}_{ik}|E_iE_k}\right)$. Beside the stable equilibrium point, there will be the NI neighboring unstable equilibrium points to the stable equilibrium point. At the unstable equilibrium points, the "kinetic energy" (E_{KE}) is zero, and the energy equals the "potential energy", E_{PF}. Along the unstable equilibrium points, one can find the closest unstable equilibrium point. At the point, the critical energy V_{cr} is defined so that any trajectory starting from a point with a lower energy than V_{cr} is guaranteed to converge into the stable equilibrium point if no other equilibrium points are contained in the set [10]. As a result, one can certify the system stability based on the stability margin, $V_{margin} = V_{cr} - V_{cl}$ where Vcl is the current energy at the clearing time.

Author Contributions

Conceptualization: HyungSeon Oh.

Data curation: HyungSeon Oh.

Formal analysis: HyungSeon Oh.

Investigation: HyungSeon Oh.

Methodology: HyungSeon Oh.

Software: HyungSeon Oh.

Validation: HyungSeon Oh.

Writing – original draft: HyungSeon Oh.

Writing – review & editing: HyungSeon Oh.

References

1. U. S. Department of Energy. United States Electricity Industry Primer. July 2015; [Online] Available at https://www.energy.gov/sites/prod/files/2015/12/f28/united-states-electricity-industry-primer.pdf.

2. A Harris Williams & Co. Transmission & Distribution Infrastructure White Paper. Summer 2010, [Online] Available at https://www.harriswilliams.com/sites/default/files/industry_reports/final%20TD.pdf.

3. U.S. Department of Energy. Large Power Transformers and the U.S. Electric Grid. June 2012; [Online] Available at https://www.energy.gov/sites/prod/files/Large%20Power%20Transformer%20Study%20-%20June%202012_0.pdf

4. Pfeifenberger J., Spees K., and Carden K, Resource adequacy requirements: reliability and economic implications., September 2013; [Online] Available at https://www.ferc.gov/legal/staff-reports/2014/02- 07-14-consultant-report.pdf.

5. Kundur P. Power system stability and control. New York, NY, USA: McGraw-Hill, 1994.

6. Winfree A. Biological rhythms and the behavior of populations of coupled oscillators. Journal of Theoret- ical Biology. 1967; 16: 15–42. https://doi.org/10.1016/0022-5193(67)90051-3 PMID: 6035757

7. Rodrigues F., Peron T., Ji P., and Kurths J. The Kuramoto model in complex networks. Physics Reports. 2016; 610: 1–98.

8. Kuramoto Y, Araki H. Lecture notes in Physics: International Symposium on Mathematical Problems in Theoretical Physics. 1975. 39 Springer-Verlag, New York, 420.

9. Dorfler F., Chertkov M., and Bullo F. Synchronization in complex oscillator networks and smart grids., PNAS. 2013; 110: 2005–2010. https://doi.org/10.1073/pnas.1212134110 PMID: 23319658

10. Willems J, Willems J. The application of Lyapunov methods to the computation of transient stability regions for multimachine power systems. IEEE Pow. App. Syst. 1970; PAS-89: 795–801.

11. Pai M., Padiyar K., and Radihakrishna C. Transient stability analysis of multimachine ac-dc power sys- tems via energy function method. IEEE Pow. Eng. Rev. 1981; PAS-100: 49–50.

12. Chiang H. Direct methods for stability analysis of electric power systems. Hoboken, NJ, USA: Wiley, 2011.

13. Vu T, Turitsyn K. A framework for robust assessment of power grid stability and resiliency. IEEE T. Auto. Cont. 2017; 62

14. Chiang H. Study of the existence of energy functions for power systems with losses. IEEE T. Circ. Syst. 1989; 39: 1423–1429.

15. Vu T, Turitsyn K. Synchronization stability of lossy and uncertain power grids. 2015 Amer. Cont. Conf. Chicago, IL, USA, July, 2015.

16. Hill D, Chong C. Lyapunov functions of Lur'e-Postnikov form for structure preserving models of power systems. Automatica. 1989; 25: 453–460.

17. Filatrella G., Nielsen A., and Pedersen N. Analysis of a power grid using a Kuramoto-like model. Europ. Phys. J. 2008; 61: 485–491.

18. Huang Z. Jin S., amd Diao R. Predictive dynamic simulation for large-scale power systems through high-performance computing. High Perf. Comp. Net. Stor. Anal. (SCC). 2012; 347–354.

19. Nagel I., Fabre L, Pastre M., Krummenacher F., Cherkaoui R., and Kayal M. High-speed power system transient stability simulation using highly dedicated hardware. IEEE T. Pow. Syst. 2013; 28: 4218– 4227.

20. Butcher J. Numerical Methods for Ordinary Differential Equations. John Wiley & Sons, New York, USA, 2003.

21. Ascher U, Petzold L. Computer Methods for Ordinary Differential Equations and Differential-Algebraic Equations. Philadelphia, Society for Industrial and Applied Mathematics, 1998.

22. Runge C. Über die numerische Auflösung von Differentialgleichungen. Mathematische Annalen, 1895; 46: 167–178.

23. Kosterev D., Taylor C., and Mittelstadt W. Model validation for the August 10, 1996 WSCC system out- age. IEEE T. Pow. Syst. 1999; 14: 967–979.

24. Anderson P, Fouad A. Power system control and stability. Piscataway, NJ, USA: Wiley, 2003.

25. Karamata J. Sur un mode de croissance régulière. Théorèmes fondamentaux. Bull. Soc. Math. France. 1933; 61: 55–62.

26. Grainger J, Stevenson W. Power System Analysis, New York, NY USA: McGraw-Hill. 1994.

27. Karlsson D, Hill D. Modeling and identification of nonlinear dynamic loads in power systems. IEEE Transaction on Power Systems. 1994; 9: 157–166.

28. Jereminov M, Pandey A, Song HA, Hooi B, Faloutsos C, Pileggi L. Linear load model for robust power system analysis. IEEE PES Innovative Smart Grid Technologies. Torino Italy, September 2017.

29. Song HA, Hooi B, Jereminov M, Pandey A, Pileggi L, Faloutsos C. PowerCast: Mining and forecasting power grid sequences. Joint European Conference on Machine Learning and Knowledge Discovery in Databases, Springer 2017.

30. Arif A., Wang Z., Wang J., Mather B., Bashualdo H., and Zhao D. Load modeling–a review. IEEE Trans- action of Smart Grid. 2017; 9: 5986–5999.

31. Golub G, Van Loan C. Matrix computation. Baltimore, MD, USA: The Johns Hopkins Univ. Press, 2013.

32. Overbye T. Power system analysis lecture 14. [Online] Available at https://slideplayer.com/slide/9086637/

33. Sauer P, Pai M, Chow J. Power System Dynamics and Stability. Hoboken, NJ, USA: John Wiley and Sons, 2017.

34. Hairer E, Wanner G. Solving ordinary differential equations II–stiff and differential-algebraic problems. Springer series in computational mathematics, 2nd ed., Geneva, Switzerland, 2000.

35. Berzins M., Dew P., and Furzeland R. Developing software for time dependent problems using the method of lines and differential algebraic integrators. Applied Numerical Mathematics. 1988; 5: 375–397.

36. Kröner A., Marquardt W., and Gilles E. Computing consistent initial conditions for differential algebraic equations. Computers and Chemical Engineering. 1992; 16: 131–138 (suuplement).

37. Campbell S. Consistent initial conditions for linear time varying singular systems. Frequency Domain and State Space Methods for Linear Systems. 1986; 313–318.

38. Campbell S. A computational method for general higher index nonlinear singular systems of differential equations. IMACS Transactions on Scientific Computing. 1989; 12: 555–560.

39. Pantelides C. The consistent initialization of differential algebraic systems. SIAM Journal of Scientific and Statistical Computing. 1988; 9: 213–231.

40. Pantelides C. Speedup–recent advances in process simulation. Computers and Chemical Engineering. 1988; 12: 745–755.

41. Barlow J. Error analysis of update methods for the symmetric eigenvalue problem. SIAM J. Matrix Anal. Appl. 1993; 14: 598–618.

42. Oh H, Hu Z. Multiple-rank modification of symmetric eigenvalue problem. MethodsX. 2018; 5: 103– 117. https://doi.org/10.1016/j.mex.2018.01.001 PMID: 30619724

43. Wilkinson J. The algebraic eigenvalue problem. Oxford Univ. Press, London, 1965.

44. Oh H. A unified and efficient approach to power flow analysis. Energies, Multidisciplinary Digital Pub- lishing Institute. 2019; 12: 2425.

45. Weber J. Description of machine models GENROU, GENSAL, GENTPF and GENTPJ. December 3, 2015. [Online] Available at https://www.powerworld.com/files/GENROU-GENSAL-GENTPF-GENTPJ. pdf.

46. Runge C. Über die numerische Auflösung von Differentialgleichungen. Mathematische Annalen, Springer. 1895; 46: 167–178.

Neural minimization methods (NMM) for solving variable order fractional delay differential equations (FDDEs) with simulated annealing (SA)

Amber Shaikh 1*, M. Asif Jamal2, Fozia Hanif3, M. Sadiq Ali Khan4, Syed Inayatullah5

1 Department of Humanities and Sciences, National University of Computer and Emerging Sciences, Karachi, Pakistan, 2 Department of Basic Sciences Federal Urdu University of Art, Science and technology Karachi & Cadet College, Karachi, Pakistan, 3 Department of Mathematics, University of Karachi, Karachi, Pakistan, 4 Department of Computer Sciences, University of Karachi, Karachi, Pakistan, 5 Department of Mathematics, University of Karachi, Karachi, Pakistan

Editor: Bedreddine Ainseba, Universite de Bordeaux, FRANCE

Funding: The authors received no specific funding for this work.

Competing interests: The authors have declared that no competing interests exist.

* amb_shaikh@hotmail.com

Abstract

To enrich any model and its dynamics introduction of delay is useful, that models a precise description of real-life phenomena. Differential equations in which current time derivatives count on the solution and its derivatives at a prior time are known as delay differential equa- tions (DDEs). In this study, we are introducing new techniques for finding the numerical solu- tion of fractional delay differential equations (FDDEs) based on the application of neural minimization (NM) by utilizing Chebyshev simulated annealing neural network (ChSANN) and Legendre simulated annealing neural network (LSANN). The main purpose of using Chebyshev and Legendre polynomials, along with simulated annealing (SA), is to reduce mean square error (MSE) that leads to more accurate numerical approximations. This study provides the application of ChSANN and LSANN for solving DDEs and FDDEs. Proposed schemes can be effortlessly executed by using Mathematica or MATLAB software to get explicit solutions. Computational outcomes are depicted, for various numerical experiments, numerically and graphically with error analysis to demonstrate the accuracy and efficiency of the methods.

Introduction

In the old days, fractional calculus was only used by pure mathematicians due to its impercep- tible applications at that time. When mathematicians were trying to implement fractional cal- culus in modeling of physical phenomena then applicability of this marvelous tool was successfully revealed. Therefore, in recent years, fractional derivatives have been used in many phenomena in electromagnetic theory, fluid mechanics, viscoelasticity, circuit theory, control theory, biology, atmospheric physics, etc. Many real-world problems can be accurately mod- eled by fractional differential equations (FDEs) such as damping laws, fluid mechanics, rheol- ogy, physics, mathematical biology, diffusion processes, electrochemistry, and so on.

Nowadays, there has been a tremendous increase of interest in theory of FDDEs, due to the fact that it can express the system of many dynamics population with more accuracy as dem- onstrated by past research in science and engineering. In the systems of real world, delays can be recognized and implemented everywhere and due to this advancement in modeling it has captured a lot of attention of the scientific community towards it.

Literature review

Many researchers were inspired by the subject of fractional order and addressed these phe- nomena in different scenarios. For example, Lie et al. [1] used phase portraits, time domain waveform and bifurcation to examine the behavior of nonlinear system of fractional order. Modified Kalman filter was proposed by [2] to deal with fractional order system. To achieve the suitable generalized projective synchronization (GPS) of incommensurated fractional order system,[3] has introduced fuzzy approach. Coronel-Escamilla et al.[4] used Euler– Lagrange and Hamilton formalisms for describing the the fractional modeling and control of an industrial selective compliant assembly robot arm(SCARA).

Many researchers have made their efforts to investigate DEs and FDEs such as Zhang et al. [5] studied generalized Burgers equation and a generalized Kupershmidt equation through lie-group analysis method for similarity reductions and exact solutions, Zhang and Zhou [6] also carried out analysis of Drinfeld-Sokolov-Wilson system through symmetry analysis method, Yang et al.[7, 8] implemented travelling wave solution to local fractional two-dimensional Burgers-type equations and Boussinesq equation in fractal domain. Atangana and Go´mez-Aguilar [9] presented exact solution and semi group principle for evolutions equation by using three different definitions of fractional derivatives. Lie et al. [10] calculated the Haar wavelet operational matrix together with Block phase function to find the solution of FDEs. Li et al. [11] have also made a successful attempt to approximate the same problem by using Chebyshev wavelet operational matrix method. Adomian decomposition method and variational iteration method were implemented in [12–16] to solve a variety of FDEs. While differential transform method and power series method are also noteworthy for the solution of FDEs [17–22].

In recent years the approach of neural architecture has been used to solve Des and FDEs. In [23] Aarts and Veer implemented the multilayer neural algorithm for the solution of partial differential equations (PDEs) with evolutionary algorithm for training of weights. Feed for- ward NN technique, with the blend of piece splines of Lagrange polynomial, was proposed by [24]. Same approach was applied by [25] in which genetic algorithm was used as an evolution- ary algorithm for training of network of nonbed-catalytic gas reactor system. For solving PDEs [26] has put an effort by implementing NN together with Broden-Flecher-Goldfarb-Shanno algorithm. Nelder Meade optimization procedure with hybrid neural network (HNN) was adopted by [27] for the numerical simulation of higher order DEs. Levenberg-Marquardt algo- rithm with ANN and Mittag-Leffler kernel was implemented by [28] to solve FDEs.

Those systems that are governed by their past are modeled in the form of delay differential equations. DDEs are proved useful in control systems [29], lasers, traffic models [30], metal cutting, epidemiology, neuroscience, population dynamics [31], and chemical kinetics [32] condition. Due to the infinite dimensionality of delay systems it is very challenging to analyti- cally analyze DDEs, therefore numerical simulations of DDEs play a key role for study of such systems. A noteworthy study of FDDEs through neural network can be visualized in [33] Exis- tence and uniqueness theorems on FDDEs are discussed in [34–36]

This study will generate an approximate solution for solving the DDEs and FDDEs by using ChSANN and LSANN, which were first developed by Khan et al. [37,38] to solve Lane Emden equations and fractional differential equations on a discrete domain. In the following paper we have developed an approach to solve the higher order DDEs and FDDEs on a continuous domain. This paper is organized as follows: first section of paper describes introduction and literature review while second section concerns with details of the methodologies with well explained algorithm and implication procedure. Error analysis procedure is explained in third section where as fourth, fifth and sixth sections describe the numerical experiments along with results and their discussions.

Methodology

The proposed methodologies are based on functional link neural network with optimization through thermal minimization. In this study, the Caputo definition will be used for working out the fractional derivative in the subsequent procedure. These definitions of commonly used fractional differential operators are discussed in [39].

Initially introduced by Pao [40], ChSANN and LSANN are the revised version of functional link artificial NN coupled with optimization strategy for learning. Functional link architecture of NN was design to build the connection between the linearity in a single layer NN and the computationally challenging multilayer NN.

Inspired by the physical process of annealing, SA is basically a kind of combinatorial opti- mization process. This process is based on two steps: first to perturb and then to measure qual- ity of the solution. MSE is basically the fitness function denoted by E_r, that can be minimized by the use of SA.

Algorithm

Due to the structural similarity in ChSANN and LSANN, the steps of algorithm are described in combination for both methods. Accuracy of results depends on the selection of base polynomial.

Step 1: Initialize the network by applying number of Chebyshev polynomials or Legendre polynomials (to independent variable x) $k = 0$ to n.

Step 2: Provide in each polynomial a network adaptive coefficient (NAC).

Step 3: Calculate the summation of product of NAC and Chebyshev polynomial or Legen- dre polynomial and store the value in ϕ or ψ respectively.

Step 4: Activate ϕ and ψ, by first three terms of Taylor Series expansion of *tanh* function.

Step 5: As given by Lagaris and Fotiadias [41] trial solution will be generated by the help of initial conditions and activated ϕ (in case of LSANN ψ)

Step 6: Calculate the value of delay trail solution by repeating step 1 to 5 with delay inde- pendent variable.

Step 7: Calculate the MSE of DDEs or FDDEs by discretizing the domain in β number of points.

Step 8: Set Tolerance for accepting minimized value of MSE.

Step 9: Minimize the MSE by thermal minimizing methodology with the following **settings** from Mathematica 11.0.

- Level iterations$\rightarrow$50
- Perturbation Scale$\rightarrow$1.0
- Probability function$\rightarrow e^{\frac{-Log(i+1)\nabla MSE}{T}}$
- Random seed$\rightarrow$ 0
- Tolerance for accepting constraint violations$\rightarrow$ 0.001

Step 10: If the value of MSE falls in the range of pre-defined criteria, then substitute the value of NAC in trial solution to get the output otherwise go to step 1 change the value of n and repeat whole procedure till the acceptable MSE is obtained.

Pictorial presentation of above algorithm can be observed in Fig 1

Employment on delay differential equation

Now we apply ChSANN and LSANN on DDEs of the following type,

$$f^{(n)}(x) = F(f(x - \tau x), f'(x), f''(x), f'''(x), \ldots, f^{(n-1)}(x)) + g(x). \tag{1}$$

With initial conditions as follows,

$$f(0) = \alpha_0, f'(0) = \alpha_1, \ldots f^{(n-1)}(0) = \alpha_{n-1}. \tag{2}$$

For implementation of ChSANN and LSANN Eq (1) can be written as

$$f^{(n)}(x) - F(f(x - \tau x), f'(x), f''(x), f'''(x), \ldots, f^{(n-1)}(x)) - g(x) = 0. \tag{3}$$

For trial and delay trial solution of above differential equation consider,

$$\phi = \sum_{k=1}^{m} w_k T_{k-1}(x), \tag{4}$$

where T_k is Chebyshev polynomial with the following recursive formula defined as

$$T_{k+1}(x) = 2x T_k(x) - T_{k-1}, \quad k \geq 2$$

Here $T_0 = 1$ and $T_1 = x$ are the fundamental values of Chebyshev polynomials and

$$\psi = \sum_{j=1}^{n} w_k L_{j-1}(x), \tag{5}$$

where L_j is Legendre polynomial with the following recursive formula defined as:

$$L_{j+1}(x) = \frac{1}{(j+1)}(2j+1) x L_j(x) - \frac{1}{(j+1)} j L_{j-1}(x), \, j \geq 2, \tag{6}$$

where, $L_0(x) = 1$ and $L_1(x) = x$ are the fundamental values of Legendre polynomials. Here we are using first three terms of Taylor series expansion of *tanh* function to activate ϕ and ψ. As defined by Lagaris and Fotiadis [41], the trial solution of Eq (1) can be written as,

$$f_{trial}(x, w) = \alpha_0 + \alpha_1 x + \frac{x^2}{2!}\alpha_2 + \ldots + N(x, w)\frac{x^k}{k!}. \tag{7}$$

Where N is the activated ϕ or ψ, depends on the method, while the delay trial solution can be given by replacing the x by $x - \tau x$.

The MSE of the Eq (1) will be calculated from the following:

$$E_r(w) = \sum_{l=1}^{\beta} (f_{trial}^{(n)}(x_l, w) - F\begin{pmatrix} f_{trial}(x_l, w), f_{trial}(x_l - \tau x_l, w), f'_{trial}(x_l, w), f''_{trial}(x_l, w) \\ , \ldots, f^{(n-1)}_{trial}(x_l, w) \end{pmatrix} - g(x_l))^2 \tag{8}$$

here β represents the number of trial points while Eq (8) will be the fitness function for learn- ing of NAC. For implementation on FDDEs only the method of computing the fractional derivative of trial solution will vary, which will be taken in this study according to Caputo defi- nition as follows.

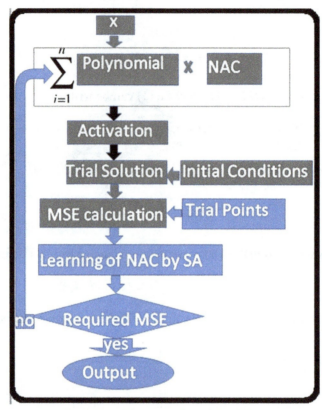

Fig 1. Pictorial presentation of algorithm.

Definition

According to [39] Caputo operator for $\lambda>0$ can be defined as:

$$D^\lambda g(\eta) = \frac{1}{\Gamma(m-\lambda)} \int_0^\eta (\eta - t)^{m-\eta-1} g^{(n)}(t) dt, \, m - 1 < \lambda < m, m \in N,$$

with

- $D^\lambda c = 0$, where c is a constant
- $D^\lambda(\eta^\beta) = \begin{cases} 0, & \lambda \in N_0, \beta < \lambda \\ \dfrac{\Gamma(\beta+1)}{\Gamma(\beta+1-\lambda)} \eta^{\beta-\lambda}, & \text{otherwise} \end{cases}$

Mathematica 11.0 is the minimization implementation tool in this study but details can be seen from [42].

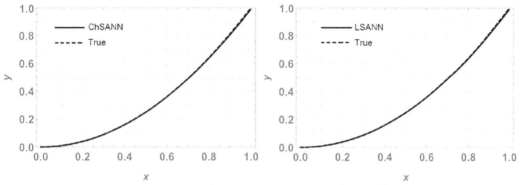

Fig 2. Comparison of ChSANN and LSANN with true solution.

Error analysis

The error analysis of numerical experiments for ChSANN and LSANN methodologies can be observed by following procedure. By substituting the values of NAC, after learning from SA algorithm, into the trial solution it will become ChSANN or LSANN solution that can be fur- ther substituted into Eq (9) for analyzing the accuracy of method on the domain of [0,1].

$$E(x) = |f''(x) - F(f(x - \tau x), f'(x), f''(x), f'''(x), \ldots, f^{(n-1)}(x)) - g(x)| \cong 0 \qquad (9)$$

While $f(x)$ is the obtained approximated continuous solution by ChSANN or LSANN. $E_r(x_i)$ tends to 0 as the value of MSE obtained by ChSANN and LSANN is in the predefined range. Convergence of solution is totally dependent on the learning methodology of respected NN architecture which is SA in the current case.

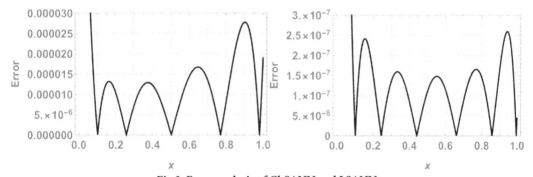

Fig 3. Error analysis of ChSANN and LSANN.

Table 1. ChSANN and LSANN results.

SNo	NAC	ChSANN	LSANN
1	w_1	1.70693523	1.710066834
2	w_2	-0.00101189	-0.00004054
3	w_3	-0.00220052	-0.00372515
4	w_4	-0.00028488	-0.00002618
5	w_5	0.000058430	-0.00003209
6	w_6	-0.00001175	-0.00002803

Numerical experiments

Experiment 1. Consider 2nd order DDE along with the initial conditions as:

$$y^{(\alpha)}(x) = \frac{3}{4}y(x) + y\left(\frac{x}{2}\right) - x^2 + 2, \ y(0) = 0; y'(0) = 0; \alpha = 2.$$

The exact solution when

$\alpha = 1$ is given as:

$$y(x) = x^2$$

In this experiment we employed proposed methodologies on above second order linear DDE on domain of [0,1].Both the methods were employed by dividing the domain with 10 equidistant training points and 6 NAC. For ChSANN and LSANN at $\alpha = 2$, the MSE at defined conditions is found to be $1.89855 \times 10-11$ and $1.32344 \times 10-14$ respectively. (Fig 2) depicts the comparison of both methods with true solution at continuous domain of [0,1], while (Fig 3) displays the error analysis for both methods at $\alpha = 2$. For the above experiment the trial and delay trail solutions are found to be following:

$$y_{trial} = \frac{x^2}{2}N$$

and

$$y_{dtrial} = \frac{(x/2)^2}{2}M.$$

Where, N and M are the structural outputs of both NNs. Table 1 displays the final values of NAC after training by SA algorithm and Figs 4 and 5 is displaying the data for 100 indepen- dent runs by altering the scale for random jumps.

Experiment 2. Consider 3rd order nonlinear DDE along with the initial condition as:

$$y^{(\alpha)}(x) = 2y^2\left(\frac{x}{2}\right) - 1, \ y(0) = 0, \ y'(0) = 1, y''(0) = 0, \ \alpha = 3$$

The exact solution when $\alpha = 3$ is given below:

$$y(x) = Sin(x)$$

3rd order nonlinear DDE is solved by ChSANN and LSANN on the continuous domain of [0,1]. ChSANN and LSANN were run for *10* and *6* NAC respectively while training points were *20* for both. With given predefined conditions MSE is found to be $2.5679 \times 10-5$ for ChSANN and $5.4843 \times 10-7$ for LSANN. Comparison of the methods with true solution can be visualized in

Fig 6 and error analysis can be observed in Fig 7. Table 2 represents the final values of weights after training by SA algorithm while Figs 8 and 9 are displaying the results for 100 independent runs for elapsed time in seconds, fitness and number of iterations.

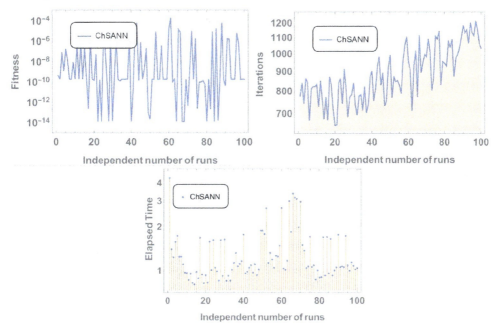

Fig 4. Results for 100 number of independent runs.

Following are trial and delay trail solution for the current experiment:

$$y_{trial} = x + \frac{x^3}{6} M$$

and

$$y_{dtrial} = \frac{x}{2} + \frac{(x/2)^3}{6} N.$$

While M and N are structural NN outputs of both the methods for in progress experiment.

Example 3. Consider FDDE along with the initial conditions as:

$$y^{(\alpha)}(x) = \frac{1}{2} e^{\frac{x}{2}} y(x) + \frac{1}{2} y(x), \ y(0) = 1, \ 0 \leq x \leq 1, \ 0 < \alpha \leq 1.$$

The exact solution at $\alpha = 1$ is given by:

$$y(x) = e^x$$

ChSANN and LSANN have been employed successfully on above FDDE with 10 training points and 6 NAC. (Fig 10) depicts the comparison of ChSANN and LSANN solutions with true values at $\alpha = 1$ while values of final NAC after learning by SA algorithm can be visualized in Table 3.

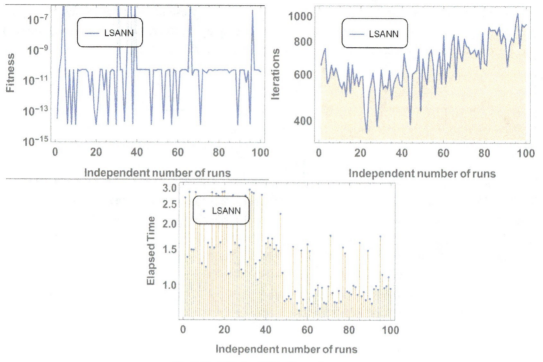

Fig 5. Results for 100 independent runs.

Both the methods were executed for different fractional values of α for which results can be visualized in Tables 4–6. Error analysis for all fractional values can be seen in (Fig 11) and Figs 12 and 13 is displaying the results for 100 independent runs for both the proposed methods Trial and delay trial solutions are taken to be:

$$y_{trial} = 1 + xN$$

and

$$y_{dtrial} = 1 + \frac{x}{2}M.$$

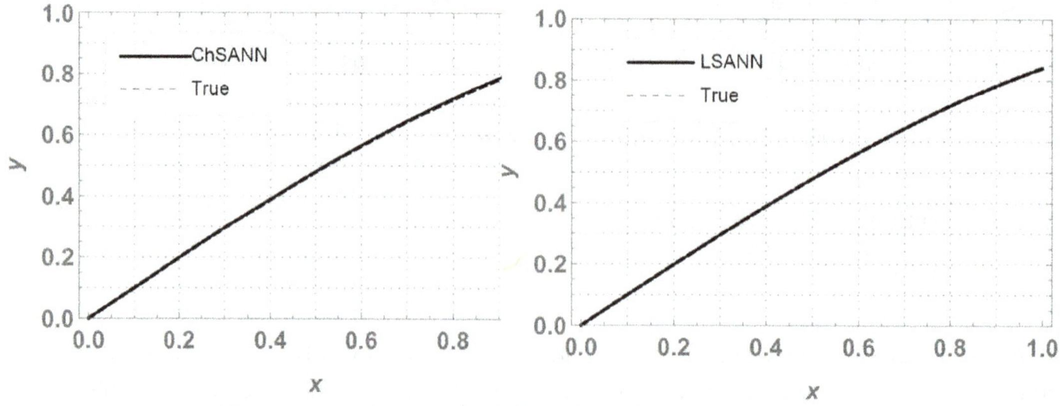

Fig 6. Comparison of ChSANN and LSANN with true solution.

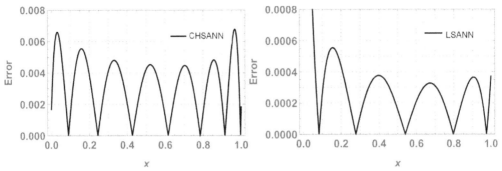

Fig 7. Error analysis of ChSANN and LSANN.

Example 4. Consider Non-Linear FDDE along with the initial condition as:

$$y^{(\alpha)}(x) = 1 - 2y^2\left(\frac{x}{2}\right),\ y(0) = 1,\ y'(0) = 0,\ 0 \le x \le 1,\ 1 < \alpha \le 2$$

The exact solution at $\alpha = 2$ is given by:

$$y(x) = Cos(x)$$

ChSANN and LSANN are implemented on the above experiment of nonlinear FDDE. (Fig shows the comparison of both methods with true values at $\alpha = 2$. While (Fig 15) represents the error analysis for different fractional values of α. Trial solution was taken in the same manner as in previous experiments. ChSANN and LSANN solutions with obtained MSE at different fractional values of α can be visualized in Tables 7–9 and Figs 16 and 17 are displaying the results for 100 independent runs.

Table 2. NAC values.

SNo	NAC	ChSANN	LSANN
1	w_1	-0.874541110	-1.2104577306
2	w_2	-0.478671752	-0.0384815109
3	w_3	0.0165454550	0.06652279388
4	w_4	0.4425599703	-0.0172612429
5	w_5	-0.594007418	0.00543949906
6	w_6	0.4767150261	-0.00064910894
7	w_7	-0.261651890	0.9618236111
8	w_8	0.0976657350	-1.0110550149
9	w_9	-0.022615861	2.3259303584
10	w_{10}	0.0024649552	-2.0381716440

Discussion

The above work is concerned with the successful implementation of ChSANN and LSANN on higher order DDEs and FDDEs. Some benchmark examples were considered for experimental cases and validity of implementation has been judged by standard error analysis procedure, data analysis of 100 number of runs of algorithm and comparison with other methods.

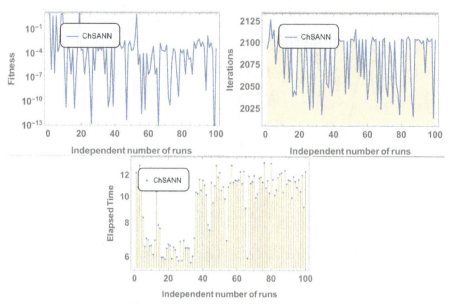

Fig 8. Results for 100 independent runs.

Error analysis

For test experiment 1, MSE for ChSANN and LSANN for $\alpha = 2$ were found to be 1.89855×10^{-11} and 1.32344×10^{-14} that gave the minimum error for ChSANN is 2.8×10^{-5} and for LSANN is 2.6×10^{-7} that can be easily visualized from (Fig 3). It shows that accuracy of both the methods is inversely proportional to the value of MSE and it can also be noticed that change of polynomials in both methods has strongly influenced the learning of NAC by SA algorithm that can be witnessed from Table 1. For experiment no 1, at similar conditions of training points and NAC, LSANN gave more promising results.

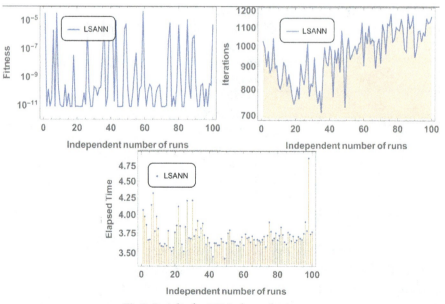

Fig 9. Results for 100 independent runs.

Similar trends can be depicted from the results of test experiment 2, in which MSE for ChSANN and LSANN for $\alpha = 3$ were found to be 2.5679×10^{-5} and 2.20049×10^{-7}. Error Analysis from (Fig 7) exhibiting the better performance of LSANN with less MSE as described above. Moreover it can also be visualized that this structure of neural network can be easily implemented on higher order nonlinear differential equations with ease.

In experiment no 3, both the proposed architectures were employed on linear delay fractional differential equation. MSE for ChSANN and LSANN at $\alpha = 1$ were found to be 4.63009×10^{-11} and 3.88356×10^{-11} respectively that gave excellent results that can be seen in (Fig 10).

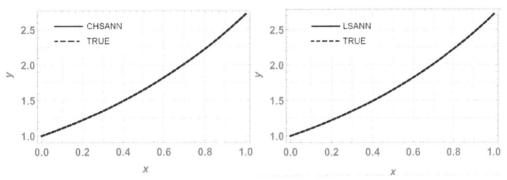

Fig 10. Comparison of ChSANN and LSANN with true solution at $\alpha = 1$.

Table 3. Values of NAC at $\alpha = 1$.

SNo	NAC	ChSANN	LSANN
1	w1	1.1669374428	1.1917375672
2	w2	0.5269817424	0.5257487109
3	w3	-0.079021346	-0.107447861
4	w4	0.0160510892	0.0265564411
5	w5	-0.0011808281	-0.003520220
5	w6	0.0000888300	0.000201124

Table 4. ChSANN and LSANN results at $\alpha = 0.5$.

X Values	ChSANN	LSANN
0.1	1.403383	1.403323
0.2	1.70248	1.702377
0.3	1.981465	1.981438
0.4	2.269594	2.269551
0.5	2.576318	2.576261
0.6	2.904964	2.904908
0.7	3.258834	3.258791
0.8	3.643645	3.643603
0.9	4.066385	4.066317
1.0	4.529678	4.529594
MSE For ChSANN	9.94644×10^{-6}	
MSE For LSANN	9.947919×10^{-6}	

Table 5. ChSANN and LSANN results at α = 0.7.

X Values	ChSANN	LSANN
0.1	1.22611657	1.22608151
0.2	1.42562311	1.42559953
0.3	1.62462243	1.62460268
0.4	1.83454966	1.83452219
0.5	2.06067256	2.06063551
0.6	2.30584961	2.30580978
0.7	2.57260285	2.57256722
0.8	2.86419774	2.86416399
0.9	3.18461364	3.18456933
1.0	3.5369036	3.53690470
MSE For ChNN	2.86781×10^{-6}	
MSE For LNN	2.86881×10^{-6}	

Table 6. ChSANN and LSANN at α = 0.8.

X Values	ChSANN	LSANN
0.1	1.17320141	1.1731957
0.2	1.33893581	1.3389313
0.3	1.50993727	1.5099335
0.4	1.69280737	1.6928027
0.5	1.89124814	1.8912421
0.6	2.10770277	2.1076959
0.7	2.34439824	2.3443915
0.8	2.60397962	2.6039730
0.9	2.88960028	2.8895924
1.0	3.20417343	3.2041639
MSE For ChSANN	1.2802637×10^{-6}	
MSE For LSANN	1.28041×10^{-6}	

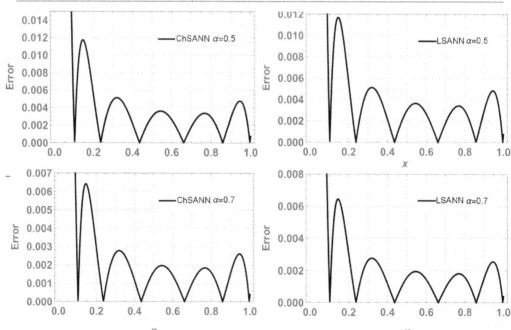

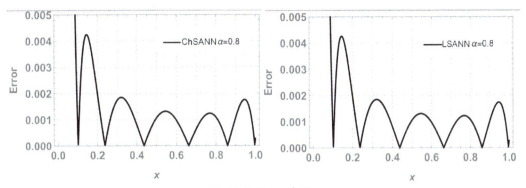

Fig 11. Error analysis.

Further both the methods were also executed for α = 0.7, 0.8 and 0.9. Obtained MSE for all the fractional values obtained by both the methods were in the same range so the accuracy of results for all the fractional values is approximately similar that can be visualized in (Fig 11).

Experiment no 4 is a case of nonlinear FDDE. For α = 2 values of MSE with 6 NAC were found to be *7.82099 ×10⁻⁶* and *1.050097 ×10⁻⁵*. For fractional values of α ChSANN is showing better results than LSANN at α = 1.5 and α = 1.9, with better values of MSE for ChSANN and for α = 1.7 both the methods are exhibiting the similar accuracy. Accuracy at fractional values can be visualized from (Fig 15).

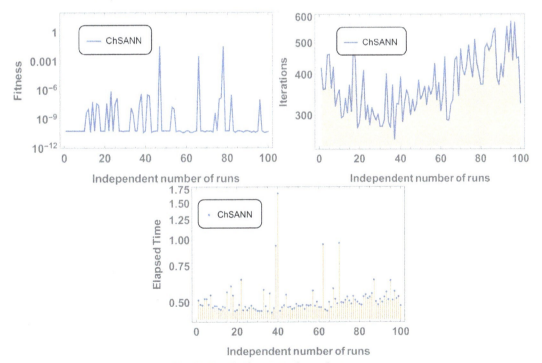

Fig 12. Results for 100 independent runs.

Data analysis for 100 numbers of independent runs. For each test experiment, algo- rithms of proposed techniques were executed 100 times by altering the scale of random jumps to assess the precision, performance and reliability. Results of obtained data can be visualized in form of figures which shows that for test experiment 1 fitness function is revolving between $10-4$ to 10^{-14} and 10^{-6} to 10^{-14} for CHSANN and LSANN respectively, Elapsed time in second is found to be within three seconds for both the methods while number of iterations were between *600–1200* and *400–1000* for CHSANN and LSANN respectively. Results of 100 inde- pendent runs for test experiments 2-4 can be visualized in Figs 7 and 8, Figs 11 and 12 and Figs 16 and 17 respectively, which demonstrate a similar trend except for the case of nonlinear models for which the maximum elapsed time is found to be 20 seconds due to computational complexity.

Comparison with other methods. We compared the proposed techniques in terms of accuracy, elapsed time, ease of calculation and error prediction with the methods presented in [43], [44] and [45]. Methods in [43–45] have been implemented on similar type of problems as in current study. Test example number 3 by Radial basis method presented in [43] and test example no 4 presented in [45] is similar to test experiment no 2 presented in following paper, Test example number 5 presented in [44] is similar to test experiment number 4 presented above and test experiment number 2 presented in [45] is similar to test experiment no 1 by proposed methods.

Following key points of comparison can be noticed.

- Radial basis method, method in [44] and method in [45] are providing results on colloca- tion points while proposed schemes are providing a continuous solution.

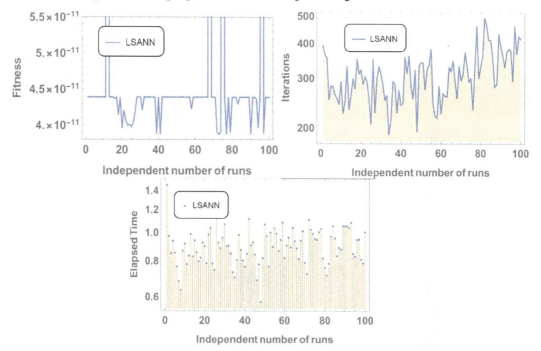

Fig 13. Results for 100 independent runs.

- On the other hand method in [43] is taking 10 to 85 seconds for solving a linear problem while proposed techniques are consuming 6 to 12 seconds and 3 to 5 seconds (Figs 8 and 9) for solving nonlinear problem by ChSANN and LSANN respectively.

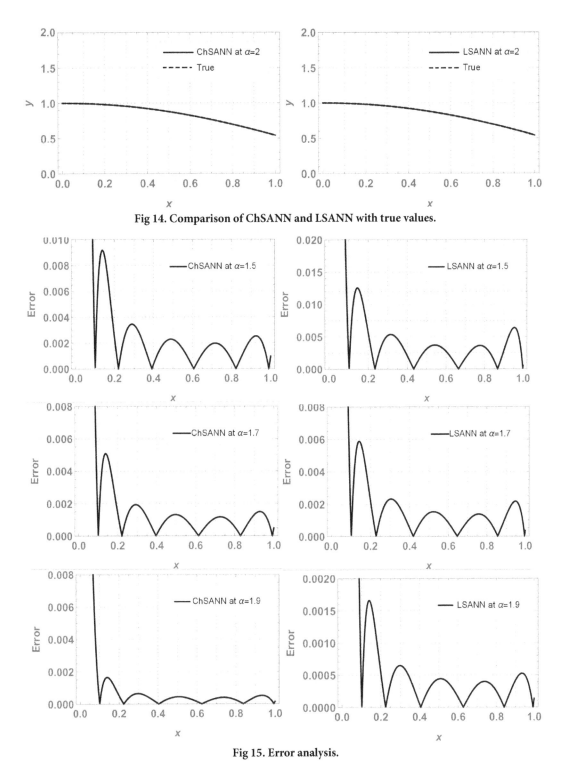

Fig 14. Comparison of ChSANN and LSANN with true values.

Fig 15. Error analysis.

- Computational complexity of method presented in [43–45] is very large due to solving a large system of nonlinear equations while proposed techniques are too simple in terms of implementation that can be observed through the computational time difference.

- There is no way to predict the accuracy in method proposed in [43–45] when there is no exact solution present while proposed schemes can predict accuracy of solution through fitness function. In terms of accuracy methods in [43–45] is giving more accurate results than proposed schemes but limitations of [43–45] is making proposed schemes more powerful.

Table 7. ChSANN and LSANN results for FDDE at $\alpha = 1.5$.

x	ChSANN	LSANN
0.1	0.9819344	0.9822488
0.2	0.9422867	0.9427229
0.3	0.8913654	0.8918057
0.4	0.8333956	0.8338889
0.5	0.7710079	0.7715448
0.6	0.7063767	0.7068606
0.7	0.6413646	0.6417259
0.8	0.5774801	0.5777550
0.9	0.5159194	0.5161458
1.0	0.4577416	0.4576439
MSE For ChSANN	3.32555×10^{-6}	
MSE For LSANN	1.05097×10^{-5}	

Table 8. ChSANN and LSANN results for FDDE at $\alpha = 1.7$.

x	ChSANN	LSANN
0.1	0.989296	0.9893501
0.2	0.962232	0.9623276
0.3	0.923181	0.9232965
0.4	0.874549	0.8746877
0.5	0.818103	0.8182648
0.6	0.755417	0.7555877
0.7	0.687973	0.6881368
0.8	0.617155	0.6173071
0.9	0.544234	0.5443786
1.0	0.470418	0.4705211
MSE For ChSANN	1.09296×10^{-6}	
MSE For LSANN	1.73111×10^{-6}	

Table 9. ChSANN and LSANN results for FDDE at $\alpha = 1.9$.

X Values	ChSANN	LSANN
0.1	0.9935924	0.99359236
0.2	0.9753719	0.97537194
0.3	0.9464374	0.94643741
0.4	0.9076268	0.90762677
0.5	0.8597139	0.87971386
0.6	0.8034944	0.80349434
0.7	0.7398116	0.73981153
0.8	0.6695533	0.66955325

0.9	0.5936416	0.59364162
1.0	0.5130313	0.51303120
MSE For ChSANN	1.26583×10^{-7}	
MSE For LNN	1.2960433×10^{-7}	

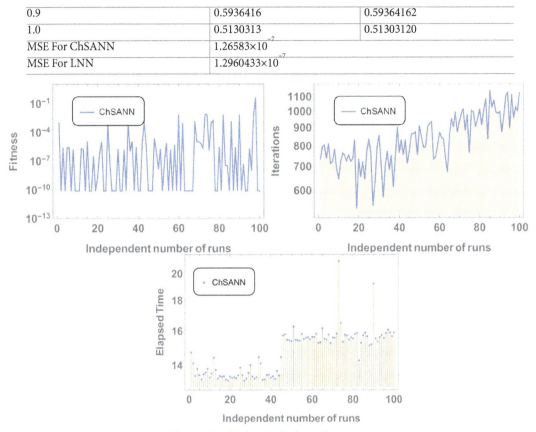

Fig 16. Results for 100 independent runs.

Conclusion

In above study we have developed two methods ChSANN and LSANN for simulation of fractional delay differential equation. After analyzing procedure and numerical experiments following points can be concluded.

- Proposed methods can be successfully implemented on linear and nonlinear FDDEs with ease of calculation.

- Accuracy of method can be increased by improving the learning methodology of NAC.

- Accuracy of both the methods is inversely proportional to MSE.

- Both the methods can easily handle the nonlinear terms.

- Accuracy prediction can be obtained for fractional values of derivatives by observing the MSE values.

In future the proposed schemes can be further developed for accuracy by refining the learn- ing methodology of NAC and by improving the neural architecture. However, it can also be successfully implemented on partial differential equations with some alterations in methodology.

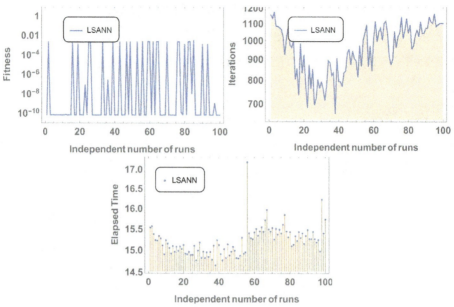

Fig 17. Results for 100 independent runs.

Author Contributions

Conceptualization: M. Asif Jamal.

Data curation: Amber Shaikh, M. Asif Jamal.

Formal analysis: Amber Shaikh, M. Sadiq Ali Khan, Syed Inayatullah.

Methodology: Amber Shaikh, M. Sadiq Ali Khan.

Project administration: Fozia Hanif, Syed Inayatullah.

Resources: M. Asif Jamal, Fozia Hanif, M. Sadiq Ali Khan, Syed Inayatullah.

Software: Amber Shaikh, M. Asif Jamal, M. Sadiq Ali Khan.

Supervision: Fozia Hanif, Syed Inayatullah.

Validation: Fozia Hanif.

Visualization: Amber Shaikh.

Writing – original draft: M. Asif Jamal.

Writing – review & editing: Amber Shaikh, Fozia Hanif, M. Sadiq Ali Khan, Syed Inayatullah.

References

1. Li Z, Chen D, Zhu J, Liu Y. Nonlinear dynamics of fractional order Duffing system. Chaos, Solitons & Fractals. 2015 Dec 1; 81:111–6.

2. Pourdehi S, Azami A, Shabaninia F. Fuzzy Kalman-type filter for interval fractional-order systems with finite-step auto-correlated process noises. Neurocomputing. 2015 Jul 2; 159:44–9.

3. Boulkroune A, Bouzeriba A, Bouden T. Fuzzy generalized projective synchronization of incommensu- rate fractional-order chaotic systems. Neurocomputing. 2016 Jan 15; 173:606–14.

4. Coronel-Escamilla A, Torres F, Gomez-Aguilar JF, Escobar-Jimenez RF, Guerrero-Ram´ırez GV. On the trajectory tracking control for an SCARA robot manipulator in a fractional model driven by induction motors with PSO tuning. Multibody System Dynamics. 2018 Jul 15; 43(3):257–77.

5. Zhang Y, Mei J, Zhang X. Symmetry properties and explicit solutions of some nonlinear differential and fractional equations. Applied Mathematics and Computation. 2018 Nov 15; 337:408–18.

6. Zhang Y, Zhao Z. Lie symmetry analysis, Lie-Ba¨cklund symmetries, explicit solutions, and conservation laws of Drinfeld-Sokolov-Wilson system. Boundary Value Problems. 2017 Dec 1; 2017(1):154.

7. Yang XJ, Gao F, Srivastava HM. Exact travelling wave solutions for the local fractional two-dimensional Burgers-type equations. Computers & Mathematics with Applications. 2017 Jan 15; 73(2):203–10.

8. Yang XJ, Machado JT, Baleanu D. Exact traveling-wave solution for local fractional Boussinesq equa- tion in fractal domain. Fractals. 2017 Aug; 25(04):1740006.

9. Atangana A, Go´mez-Aguilar JF. Decolonisation of fractional calculus rules: Breaking commutativity and associativity to capture more natural phenomena. The European Physical Journal Plus. 2018 Apr; 133:1–22.

10. Li Y, Zhao W. Haar wavelet operational matrix of fractional order integration and its applications in solv- ing the fractional order differential equations. Applied Mathematics and Computation. 2010 Jun 15; 216 (8):2276–85.

11. Yuanlu L. Solving a nonlinear fractional differential equation using Chebyshev wavelets. Communica- tions in Nonlinear Science and Numerical Simulation. 2010 Sep 1; 15(9):2284–92.

12. Odibat Z, Momani S. Numerical methods for nonlinear partial differential equations of fractional order. Applied Mathematical Modelling. 2008 Jan 1; 32(1):28–39.

13. Momani S, Odibat Z. Numerical approach to differential equations of fractional order. Journal of Compu- tational and Applied Mathematics. 2007 Oct 1; 207(1):96–110.

14. El-Wakil SA, Elhanbaly A, Abdou MA. Adomian decomposition method for solving fractional nonlinear differential equations. Applied Mathematics and Computation. 2006 Nov 1; 182(1):313–24.

15. Hosseinnia SH, Ranjbar A, Momani S. Using an enhanced homotopy perturbation method in fractional differential equations via deforming the linear part. Computers &Mathematics with Applications. 2008 Dec 1; 56(12):3138–49.

16. Dhaigude DB, Birajdar GA. Numerical solution of system of fractional partial differential equations by discrete Adomian decomposition method. J. Frac. Cal. Appl. 2012 Jul; 3(12):1–1.

17. Arikoglu A, Ozkol I. Solution of fractional differential equations by using differential transform method. Chaos, Solitons & Fractals. 2007 Dec 1; 34(5):1473–81.

18. Arikoglu A, Ozkol I. Solution of fractional integro-differential equations by using fractional differential transform method. Chaos, Solitons & Fractals. 2009 Apr 30; 40(2):521–9.

19. Darania P, Ebadian A. A method for the numerical solution of the integro-differential equations. Applied Mathematics and Computation. 2007 May 1; 188(1):657–68.

20. Ertürk VS, Momani S. Solving systems of fractional differential equations using differential transform method. Journal of Computational and Applied Mathematics. 2008 May 15; 215(1):142–51.

21. Erturk VS, Momani S, Odibat Z. Application of generalized differential transform method to multi-order fractional differential equations. Communications in Nonlinear Science and Numerical Simulation. 2008 Oct 1; 13(8):1642–54.

22. Odibat ZM, Shawagfeh NT. Generalized Taylor's formula. Applied Mathematics and Computation. 2007 Mar 1; 186(1):286–93.

23. Aarts LP, Van Der Veer P. Neural network method for solving partial differential equations. Neural Pro- cessing Letters. 2001 Dec 1; 14(3):261–71.

24. Meade AJ Jr, Fernandez AA. The numerical solution of linear ordinary differential equations by feedfor- ward neural networks. Mathematical and Computer Modelling. 1994 Jun 1; 19(12):1–25.

25. Parisi DR, Mariani MC, Laborde MA. Solving differential equations with unsupervised neural networks. Chemical Engineering and Processing: Process Intensification. 2003 Aug 1; 42(8–9):715–21.

26. Lagaris IE, Likas A, Fotiadis DI. Artificial neural networks for solving ordinary and partial differential equations. IEEE Transactions on Neural Networks. 1998 Sep; 9(5):987–1000. https://doi.org/10.1109/ 72.712178 PMID: 18255782

27. Malek A, Beidokhti RS. Numerical solution for high order differential equations using a hybrid neural net- work—optimization method. Applied Mathematics and Computation. 2006 Dec 1; 183(1):260–71.

28. Zúñiga-Aguilar CJ, Romero-Ugalde HM, Gómez-Aguilar JF, Escobar-Jiménez RF, Valtierra-Rodríguez M. Solving fractional differential equations of variable-order involving operators with Mittag-Leffler ker- nel using artificial neural networks. Chaos, Solitons & Fractals. 2017 Oct 1; 103:382–403.

29. Davis LC. Modifications of the optimal velocity traffic model to include delay due to driver reaction time. Physica A: Statistical Mechanics and its Applications. 2003 Mar 1; 319:557–67.

30. Epstein IR, Luo Y. Differential delay equations in chemical kinetics. Nonlinear models: The cross- shaped phase diagram and the Oregonator. The Journal of chemical physics. 1991 Jul 1; 95(1):244–54.

31. Kuang Y, editor. Delay differential equations: with applications in population dynamics. Academic Press; 1993 Mar 5.

32. Benchohra M, Henderson J, Ntouyas SK, Ouahab A. Existence results for fractional order functional dif- ferential equations with infinite delay. Journal of Mathematical Analysis and Applications. 2008 Feb 15; 338(2):1340–50.

33. Zúñiga-Aguilar CJ, Coronel-Escamilla A, Gómez-Aguilar JF, Alvarado-Martínez VM, Romero-Ugalde HM. New numerical approximation for solving fractional delay differential equations of variable order using artificial neural networks. The European Physical Journal Plus. 2018 Feb 1; 133(2):75.

34. Henderson J, Ouahab A. Fractional functional differential inclusions with finite delay. Nonlinear Analy- sis: Theory, Methods & Applications. 2009 Mar 1; 70(5):2091–105.

35. Maraaba TA, Jarad F, Baleanu D. On the existence and the uniqueness theorem for fractional differen- tial equations with bounded delay within Caputo derivatives. Science in China Series A: Mathematics. 2008 Oct 1; 51(10):1775–86.

36. Maraaba T, Baleanu D, Jarad F. Existence and uniqueness theorem for a class of delay differential equations with left and right Caputo fractional derivatives. Journal of Mathematical Physics. 2008 Aug; 49(8):083507.

37. Khan NA, Shaikh A. A smart amalgamation of spectral neural algorithm for nonlinear Lane-Emden equations with simulated annealing. Journal of Artificial Intelligence and Soft Computing Research. 2017 Jul 1; 7(3):215–24.

38. Khan NA, Shaikh A, Sultan F, Ara A. Numerical Simulation Using Artificial Neural Network on Fractional Differential Equations. In Numerical Simulation-From Brain Imaging to Turbulent Flows 2016. InTech.

39. Yang XJ, Baleanu D, Srivastava HM. Local fractional integral transforms and their applications. Aca- demic Press; 2015 Oct 22.

40. Pao YH, Takefuji Y. Functional-link net computing: theory, system architecture, and functionalities. Computer. 1992 May; 25(5):76–9.

41. Lagaris IE, Likas A, Fotiadis DI. Artificial neural networks for solving ordinary and partial differential equations. IEEE Transactions on Neural Networks. 1998 Sep; 9(5):987–1000. https://doi.org/10.1109/ 72.712178 PMID: 18255782

42. Ledesma S, Aviña G, Sanchez R. Practical considerations for simulated annealing implementation. InSimulated Annealing 2008. InTech.

43. Saeed U. Radial basis function networks for delay differential equation. Arabian Journal of Mathemat- ics. 2016 Sep 1; 5(3):139–44.

44. Saeed U. Hermite wavelet method for fractional delay differential equations. Journal of Difference Equa- tions. 2014; 2014.

45. Iqbal MA, Saeed U, Mohyud-Din ST. Modified Laguerre wavelets method for delay differential equations of fractional-order. Egypt. J. Basic Appl. Sci. 2015 Mar 1; 2:50.

Permissions

All chapters in this book were first published in PLOS ONE, by The Public Library of Science; hereby published with permission under the Creative Commons Attribution License or equivalent. Every chapter published in this book has been scrutinized by our experts. Their significance has been extensively debated. The topics covered herein carry significant findings which will fuel the growth of the discipline. They may even be implemented as practical applications or may be referred to as a beginning point for another development.

The contributors of this book come from diverse backgrounds, making this book a truly international effort. This book will bring forth new frontiers with its revolutionizing research information and detailed analysis of the nascent developments around the world.

We would like to thank all the contributing authors for lending their expertise to make the book truly unique. They have played a crucial role in the development of this book. Without their invaluable contributions this book wouldn't have been possible. They have made vital efforts to compile up to date information on the varied aspects of this subject to make this book a valuable addition to the collection of many professionals and students.

This book was conceptualized with the vision of imparting up-to-date information and advanced data in this field. To ensure the same, a matchless editorial board was set up. Every individual on the board went through rigorous rounds of assessment to prove their worth. After which they invested a large part of their time researching and compiling the most relevant data for our readers.

The editorial board has been involved in producing this book since its inception. They have spent rigorous hours researching and exploring the diverse topics which have resulted in the successful publishing of this book. They have passed on their knowledge of decades through this book. To expedite this challenging task, the publisher supported the team at every step. A small team of assistant editors was also appointed to further simplify the editing procedure and attain best results for the readers.

Apart from the editorial board, the designing team has also invested a significant amount of their time in understanding the subject and creating the most relevant covers. They scrutinized every image to scout for the most suitable representation of the subject and create an appropriate cover for the book.

The publishing team has been an ardent support to the editorial, designing and production team. Their endless efforts to recruit the best for this project, has resulted in the accomplishment of this book. They are a veteran in the field of academics and their pool of knowledge is as vast as their experience in printing. Their expertise and guidance has proved useful at every step. Their uncompromising quality standards have made this book an exceptional effort. Their encouragement from time to time has been an inspiration for everyone.

The publisher and the editorial board hope that this book will prove to be a valuable piece of knowledge for researchers, students, practitioners and scholars across the globe.

List of Contributors

Kevin L. McKee and Michael C. Neale
Virginia Commonwealth University, Virginia Institute of Psychiatric and Behavioral Genetics, Richmond, Virginia, United States of America

Kun She
School of Information and Software Engineering, University of Electronic Science and Technology of China, Chengdu, China

Peng Zhang
School of Information and Software Engineering, University of Electronic Science and Technology of China, Chengdu, China
School of Science, Southwest University of Science and Technology, Mianyang, China

Xin Ma
School of Science, Southwest University of Science and Technology, Mianyang, China
State Key Laboratory of Oil and Gas Reservoir Geology and Exploitation, Southwest Petroleum University, Chengdu, China

Annika Hoyer and Sophie Kaufmann
Institute for Biometrics and Epidemiology, German Diabetes Center, Leibniz Center for Diabetes Research at Heinrich Heine University Düsseldorf, Düsseldorf, Germany

Ralph Brinks
Institute for Biometrics and Epidemiology, German Diabetes Center, Leibniz Center for Diabetes Research at Heinrich Heine University Düsseldorf, Düsseldorf, Germany
Hiller ResearchUnit for Rheumatology, Heinrich Heine University Düsseldorf, Düsseldorf, Germany

Elvan Akın
Department of Mathematics and Statistics, Missouri University of Science and Technology, Rolla, Missouri, United States of America

Neslihan Nesliye Pelen
Department of Mathematics, Ondokuz Mayıs University, Arts and Science Faculty, Samsun, Turkey

Ismail Uğur Tiryaki
Department of Mathematics, Bolu Abant Izzet Baysal University, Faculty of Arts and Science, Bolu, Turkey

Fusun Yalcin
Department of Mathematics, Faculty of Science, Akdeniz University, Antalya, Turkey

Sufang Han and Guoxin Liu
School of Mathematics and Statistics, Central South University, Changsha, China

Tianwei Zhang
City College, Kunming University of Science and Technology, Kunming, China

Daniel Lill and Daniel Kaschek
Institute of Physics, University of Freiburg, Freiburg, Germany

Jens Timmer
Institute of Physics, University of Freiburg, Freiburg, Germany
BIOSS Centre For Biological Signalling Studies, University of Freiburg, Freiburg, Germany

Waseem Waseem and Muhammad Sulaiman
Department of Mathematics, Abdul Wali Khan University Mardan, KP, Pakistan

Poom Kumam
KMUTTFixed Point Research Laboratory, Department of Mathematics, Faculty of Science, King Mongkut's University of Technology Thonburi (KMUTT), Bangkok, Thailand
KMUTT-Fixed Point Theory and Applications Research Group, Theoretical and Computational Science Center (TaCS), Faculty of Science, King Mongkut's University of Technology Thonburi (KMUTT), Bangkok, Thailand
Department of Medical Research, China Medical University Hospital, China Medical University, Taichung, Taiwan

Muhamad Shoaib
Department of Mathematics, COMSATS University Islamabad, Attock, Pakistan

Muhammad Asif Zahoor Raja
Future Technology Research Center, National Yunlin University of Science and Technology, Yunlin, Taiwan, R.O.C.
Department of Electrical and Computer Engineering, COMSATS University Islamabad, Attock, Pakistan

Saeed Islam
Department of Mathematics, Abdul Wali Khan University Mardan, KP, Pakistan
Informetrics Research Group, Ton Duc Thang University, Ho Chi Minh City, Vietnam
Faculty of Mathematics & Statistics, Ton Duc Thang University, Ho Chi Minh City, Vietnam

HyungSeon Oh
Department of Electrical and Computer Engineering, United States Naval Academy, Annapolis, Maryland, United States of America

Amber Shaikh
Department of Humanities and Sciences, National University of Computer and Emerging Sciences, Karachi, Pakistan

Syed Inayatullah
Department of Mathematics, University of Karachi, Karachi, Pakistan

Fozia Hanif
Department of Mathematics, University of Karachi, Karachi, Pakistan

M. Sadiq Ali Khan
Department of Computer Sciences, University of Karachi, Karachi, Pakistan

M. Asif Jamal
Department of Basic Sciences Federal Urdu University of Art, Science and technology Karachi & Cadet College, Karachi, Pakistan

Index

B
Brownian Motion, 113, 125, 133

C
Carbon Emission Forecasting, 65
Christoffel Symbols, 137-138, 142, 145-149
Combinatorial Optimization, 107

D
Data-generating Configurations, 1
Descriptive Statistics, 3-4
Direct Estimation, 1, 3-4
Dynamic Structure, 4, 25

E
Empirical Application, 3, 5
Empirical Identification, 1, 25
Empirically Unidentified, 14, 24
Enzymatic Reaction, 145-146
Epidemiological Indices, 71-73, 78
Euler Scheme, 107, 128
Extrapolation, 6, 19

F
Finite Confidence Intervals, 138, 146
Fractal Property, 12
Fractional Order, 35-36, 40, 43, 46, 56-57, 63-64, 69, 155, 176-177, 212, 230-232
Fractional-order Particle Swarm Optimization Technique, 152
Fuzzy Operations, 107, 131-132

G
Gaussian Process, 4, 31
Geodesic Coordinates, 137
Geodesic Equation, 137, 142, 147, 149
Geometric Framework, 137
Global Exponential Stability, 107, 129-136
Gompertz Dynamic Equations, 83
Gompertz-liard Curve, 104

H
Harmonic Oscillation, 1, 26

Hemodynamic Perturbations, 3, 31
Hessian Matrix, 25, 142, 148
Human Postural Sway, 1, 3, 31

I
Insensitivity Radius, 6, 14-15, 24
Instantaneous Feedback, 2, 19, 24
Intermittent Activation, 1-3, 26, 30
Intermittent Control Model, 4, 31
Intermittent Feedback Control Model, 2
Intermittent Postural Control, 2, 4, 30
Interpolation, 6, 12-13

K
Kinematic Data, 29
Knee Joint Friction, 4
Krasnoselskii's Fixed Point Theorem, 131-132

L
Langevin Dynamics, 1, 26, 28
Levi-civita Connection, 138
Linear Dynamics, 6

M
Mechanical Resistance, 1
Mechanistic Theories, 29
Meta-heuristic Algorithm, 153
Model Manifold, 137-138, 142, 145, 147-148
Multi-variable Grey Model, 35, 67

N
Neural Networks, 34, 107, 128-129, 131-136, 151-153, 155, 172, 175-176, 232-233
Non-equilibrium Langevin Dynamics, 1, 26

O
Optimal Order, 34-36, 43, 46, 52-53, 61-62, 64-65

P
Parameterization, 142, 148

Parkinson's Disease, 28, 32
Periodic Coefficients, 107
Periodic Sequence Solution, 107, 131-132, 136
Post-hoc Analyses, 3
Postural Sway, 1-3, 24, 29-32

R
Robotics Associative Memory, 107

S
Semi-discrete Stochastic Models, 107
Semi-discretization Technique, 107, 131-132
Soft Computing, 66-68, 151-155, 172, 174-175, 177, 233

Square-mean Global Exponential Stability, 129-130
Stabilogram, 28-29
Stochastic Differential Equations, 133
Stochastic Disturbance, 131-132
Stochastic Perturbations, 107, 131-132
Structural Parameters, 3, 25

T
Thermodynamics, 152-153, 173
Toppling Torque, 4-5, 25
Trajectories, 2, 11, 24, 26, 30, 128, 200-205

U
Unstable Manifolds, 5, 26

CPSIA information can be obtained
at www.ICGtesting.com
Printed in the USA
BVHW061956260822
645617BV00004B/254